New Challenges in Mathematical Modelling and Control of COVID-19 Epidemics: Analysis of Non-Pharmaceutical Actions and Vaccination Strategies

About the Editors

Cristiano Maria Verrelli

Cristiano Maria Verrelli was born in Italy on 12 September 1977. He is an Associate Professor (dynamical systems, control theory and applications) at the Department of Electronic Engineering at the University of Rome "Tor Vergata" and Coordinator of the M.Sc. Degree Course in Mechatronics Engineering. He is co-author (with R. Marino, P. Tomei) of the book "Induction Motor Control Design" (Springer, 2010) and author of "Mathematical Control Design for Linear Systems" (Esculapio, 2020, 2023). He is a co-inventor of the international patent: "Individually tailored exercise training and rehabilitation technique: Medical Personal Trainer". His research interests are in dynamic system analysis, robust adaptive nonlinear control, and learning control theory with application in electrical machines, electrical vehicles, robots, physiological systems, epidemic models, and harmonic structures in human movement. He is a co-recipient of the IFAC grant for the activity: Kids in Control. He is a member of the Technical Committee 9.2. of IFAC "Systems and Control for Societal Impact" for the triennium of 2020–2023.

Fabio Della Rossa

Fabio Della Rossa is an Associate Professor at Politecnico di Milano, Italy, specializing in the analysis and regulation of complex and nonlinear systems. His research interests encompass bifurcation analysis, chaos theory, and emergent phenomena in complex networks. He serves as an associate editor for the International Journal of Bifurcation and Chaos. He has authored over 100 scientific publications, as well as the book "Modeling Love Dynamics," published in 2015 by World Scientific.

Yan Li, Samreen, Laique Zada, Emad A. A. Ismail, Fuad A. Awwad and Ahmed M Hassan
Assessing the Impact of Time-Varying Optimal Vaccination and Non-Pharmaceutical
Interventions on the Dynamics and Control of COVID-19: A Computational Epidemic
Modeling Approach
Reprinted from: *Mathematics* **2023**, *11*, 4253, doi:10.3390/math11204253 . . . . **187**

Cristiano Maria Verrelli and Fabio Della Rossa
Two-Age-Structured COVID-19 Epidemic Model: Estimation of Virulence Parameters through
New Data Incorporation
Reprinted from: *Mathematics* **2024**, *12*, 825, doi:10.3390/math12060825 . . . . **212**

Contents

About the Editors vii

Preface ix

Cristiano Maria Verrelli and Fabio Della Rossa
New Challenges in the Mathematical Modelling and Control of COVID-19 Epidemics: Analysis of Non-Pharmaceutical Actions and Vaccination Strategies
Reprinted from: *Mathematics* **2024**, *12*, 1353, doi:10.3390/math12091353 1

Cristiano Maria Verrelli and Fabio Della Rossa
Two-Age-Structured COVID-19 Epidemic Model: Estimation of Virulence Parameters to Interpret Effects of National and Regional Feedback Interventions and Vaccination
Reprinted from: *Mathematics* **2021**, *9*, 2414, doi:10.3390/math9192414 7

Zubair Ahmad, Zahra Almaspoor, Faridoon Khan and Mahmoud El-Morshedy
On Predictive Modeling Using a New Flexible Weibull Distribution and Machine Learning Approach: Analyzing the COVID-19 Data
Reprinted from: *Mathematics* **2022**, *10*, 1792, doi:10.3390/math10111792 19

Askar Akaev, Alexander I. Zvyagintsev, Askar Sarygulov, Tessaleno Devezas, Andrea Tick and Yuri Ichkitidze
Growth Recovery and COVID-19 Pandemic Model: Comparative Analysis for Selected Emerging Economies
Reprinted from: *Mathematics* **2022**, *10*, 3654, doi:10.3390/math10193654 45

Victor Zakharov, Yulia Balykina, Igor Ilin and Andrea Tick
Forecasting a New Type of Virus Spread: A Case Study of COVID-19 with Stochastic Parameters
Reprinted from: *Mathematics* **2022**, *10*, 3725, doi:10.3390/math10203725 63

Isra Al-Shbeil, Noureddine Djenina, Ali Jaradat, Abdallah Al-Husban, Adel Ouannas and Giuseppe Grassi
A New COVID-19 Pandemic Model Including the Compartment of Vaccinated Individuals: Global Stability of the Disease-Free Fixed Point
Reprinted from: *Mathematics* **2023**, *11*, 576, doi:10.3390/math11030576 81

Alexandru Topîrceanu
On the Impact of Quarantine Policies and Recurrence Rate in Epidemic Spreading Using a Spatial Agent-Based Model
Reprinted from: *Mathematics* **2023**, *11*, 1336, doi:10.3390/math11061336 96

Zehba Raizah and Rahat Zarin
Advancing COVID-19 Understanding: Simulating Omicron Variant Spread Using Fractional-Order Models and Haar Wavelet Collocation
Reprinted from: *Mathematics* **2023**, *11*, 1925, doi:10.3390/math11081925 115

Svetozar Margenov, Nedyu Popivanov, Iva Ugrinova and Tsvetan Hristov
Differential and Time-Discrete SEIRS Models with Vaccination: Local Stability, Validation and Sensitivity Analysis Using Bulgarian COVID-19 Data
Reprinted from: *Mathematics* **2023**, *11*, 2238, doi:10.3390/math11102238 145

Mohammad Anamul Haque and Giuliana Cortese
Cumulative Incidence Functions for Competing Risks Survival Data from Subjects with COVID-19
Reprinted from: *Mathematics* **2023**, *11*, 3772, doi:10.3390/math11173772 171

Editors
Cristiano Maria Verrelli
Electronic Engineering
Department
University of Rome
Tor Vergata
Rome
Italy

Fabio Della Rossa
Diparimtento di Elettronica
Politecnico di Milano
Milano
Italy

Editorial Office
MDPI AG
Grosspeteranlage 5
4052 Basel, Switzerland

This is a reprint of articles from the Special Issue published online in the open access journal *Mathematics* (ISSN 2227-7390) (available at: https://www.mdpi.com/journal/mathematics/special_issues/New_Challenges_Math_Model_Control_COVID_Epidemics).

For citation purposes, cite each article independently as indicated on the article page online and as indicated below:

Lastname, A.A.; Lastname, B.B. Article Title. *Journal Name* **Year**, *Volume Number*, Page Range.

ISBN 978-3-7258-1175-5 (Hbk)
ISBN 978-3-7258-1176-2 (PDF)
doi.org/10.3390/books978-3-7258-1176-2

© 2024 by the authors. Articles in this book are Open Access and distributed under the Creative Commons Attribution (CC BY) license. The book as a whole is distributed by MDPI under the terms and conditions of the Creative Commons Attribution-NonCommercial-NoDerivs (CC BY-NC-ND) license.

New Challenges in Mathematical Modelling and Control of COVID-19 Epidemics: Analysis of Non-Pharmaceutical Actions and Vaccination Strategies

Editors

Cristiano Maria Verrelli
Fabio Della Rossa

Basel • Beijing • Wuhan • Barcelona • Belgrade • Novi Sad • Cluj • Manchester

Preface

From its emergence in China in December 2019, COVID-19 escalated into a global pandemic spanning six continents and over 195 countries. This Special Issue delves into its worldwide ramifications, offering not only insights into the economic and health crises precipitated by the pandemic, but also providing the essential mathematical tools for its comprehension and management. Researchers from different fields have contributed to it. They have presented scientific inquiries that span a spectrum of crucial topics including clinical manifestations, variant dissemination, economic resurgence, mortality prognostication, and intervention tactics. This interdisciplinary approach facilitates the regulation of epidemics, thereby ensuring the continuous refinement of strategies by assimilating evolving data. The methodological depth of the papers encapsulated within the Special Issue amplifies its significance. These papers serve as an indispensable foundation for ongoing updates in this field, showcasing how this methodology governs the interpretation of data.

As Guest Editors of the Special Issue, we thank the authors for their quality contributions, the reviewers for their valuable comments for improving the submitted works, and the administrative staff of MDPI publications for their support in completing this project. Moreover, special thanks are due to the Managing Editor of the Special Issue, Dr. Nemo Guan, for her excellent collaboration and valuable assistance.

Cristiano Maria Verrelli and Fabio Della Rossa
Editors

Editorial

New Challenges in the Mathematical Modelling and Control of COVID-19 Epidemics: Analysis of Non-Pharmaceutical Actions and Vaccination Strategies

Cristiano Maria Verrelli [1,*] and Fabio Della Rossa [2]

[1] Electronic Engineering Department, University of Rome Tor Vergata, Via del Politecnico 1, 00133 Rome, Italy

[2] Department of Electronic, Information and Biomedical Engineering, Politecnico di Milano, 20133 Milan, Italy; fabio.dellarossa@polimi.it

* Correspondence: verrelli@ing.uniroma2.it; Tel.: +39-(0)6-72597410

Following its official appearance in China in December 2019, COVID-19 (SARS-CoV-2) infection immediately reached pandemic proportions on six continents and in over 195 countries. It was the cause of a worldwide economic and health crisis inflicting USD one trillion in global economic damage in 2020 (to name just one figure) and had a very uneven impact on different age groups of society. Long-term lockdowns were not feasible, and contact-tracing procedures dramatically lost effectiveness with high case numbers. The availability of approved COVID-19 vaccines then led to mass vaccinations around the world. At the end of 2020, the US Food and Drug Administration granted emergency use authorization for COVID-19 vaccines, and several countries, including the US, began a mass vaccination campaign.

Vaccine efficacy, the duration of immunity, the escape of vaccine-induced immunity, the mechanisms of virus transmission, the effect of variants, the effects of age stratification, spatial effects, social networks, contact patterns, and the specificity of individual behaviours have provoked researchers to implement new strategies by combining innovative approaches, interpreting current epidemic scenarios and forecasting new ones.

Since the beginning of the pandemic, the entire scientific community across various fields of expertise has striven to help alleviate the associated problems. To provide just one figure, the call for papers proposed by *Nature Communications* collected more than 700 papers on the issue from 2019 to 2022. In addition to the sciences that are closely related to diseases, such as medicine, biology, and epidemiology, the mathematical modeling and control community has also endeavoured to contribute. Many journals have devoted Special Issues to the topic, mainly to give a prompt answer to the problem (see, for example, the Special Issue of *Frontiers in Physics* titled *Mathematical Modelling of the Pandemic of 2019 Novel Coronavirus (COVID-19): Patterns, Dynamics, Prediction, and Control*, composed of 34 general scope papers, or the Special Issue of *Mathematical Modelling of Natural Phenomena* titled *Coronavirus: Scientific Insights and Societal Aspects*, with 26 manuscripts). More-specialized journals also tried to share their perspectives, proposing Special Issues that helped us to better understand the impact of the pandemic from a retrospective point of view. For example, note the following instances:

- The *Bulletin of Mathematical Biology* proposed a Special Issue titled *Mathematics and Covid 19* in which they provided a general analysis of the various aspects of COVID-19, ranging from the forecasting of its evolution [1–3] to vaccination strategies [4–7] and the effectiveness of different control strategies [8–12].
- The *Journal of Theoretical Biology* proposed a Special Issue titled *Modelling COVID-19 and Preparedness for Future Pandemics* specifically focused on studying different modeling techniques to better understand which is the best model to use for each question and how efficient a model can be for disease prediction [13–22]; better understand the

Citation: Verrelli, C.M.; Della Rossa, F. New Challenges in the Mathematical Modelling and Control of COVID-19 Epidemics: Analysis of Non-Pharmaceutical Actions and Vaccination Strategies. *Mathematics* **2024**, *12*, 1353. https://doi.org/10.3390/math12091353

Received: 7 November 2023
Revised: 16 March 2024
Accepted: 16 March 2024
Published: 29 April 2024

Copyright: © 2024 by the authors. Licensee MDPI, Basel, Switzerland. This article is an open access article distributed under the terms and conditions of the Creative Commons Attribution (CC BY) license (https://creativecommons.org/licenses/by/4.0/).

cyclic phenomena observed in this pandemic [23–29]; explore the roles of different virus strains and methods with which to rapidly detect them [30–34]; and relate social problems to the pandemic [35–38].

- Even less-specialized journals decided to pay attention to COVID-19, such as the *Journal of Mathematics in Industry*, which proposed the Special Issue *Mathematical Models of the Spread and Consequences of the SARS-CoV-2 Pandemics. Effects on Health, Society, Industry, Economics and Technology*, principally devoted to the effect of the pandemic on society [39–41] and the estimation of the cost of the pandemic [42–44] and therefore particularly devoted to understanding the effects of non-pharmaceutical interventions [45–49].

Our Special Issue (SI) is part of this vibrant and dynamic context. It consists of eleven papers, namely, ten articles and one feature paper [50] (i.e., a substantial original article that covers several techniques or approaches, gives an outlook on future research directions, and describes possible research applications). Researchers from fields such as applied mathematics, data science, engineering, statistical sciences, computer science, biology, communication and information technology, economics, and management have been successfully integrated.

Indeed, this SI collects innovative results, tools for mathematical modelling and epidemic control based on a multidisciplinary approach (which is typical of complex systems), and analyses related to the following aspects:

1. The clinical characteristics and risk factors of COVID-19 related to patient survival;
2. The spread of variants;
3. The recovery of economic systems;
4. The prediction of mortality rates;
5. The impact of non-pharmaceutical interventions and their relative weight in relation to vaccination strategies in terms of deaths and infections;
6. Coordinated actions that simultaneously include non-pharmaceutical interventions and vaccination strategies.

A novel mathematical model for evaluating the effects of administering multiple constant and time-varying vaccines (including first and second doses and boosters) on the incidence and persistence of infections is presented in [51]. The authors also developed a control model to suggest the optimal strategy and thus minimize disease by introducing time-varying controls: by effectively adjusting vaccination strategies, such as the timing and frequency of doses, this model showed that the spread of disease can be significantly controlled. These time-varying controls allow for a more adaptive and responsive approach, enabling health authorities to adjust vaccination campaigns based on evolving epidemiological conditions, the emergence of new variants, and overall vaccination coverage in a population.

Regression models for competing risks can generally provide a crucial basis for precise individualized predictions. This is presented in [50], which shows how patients can be prioritized accordingly for vaccination and/or how clinical decisions can be made either for close monitoring, ICU admission, or the approval of new interventions. In addition, ref. [50] illustrates how competing risk survival analyses can be used to estimate the Cumulative Incidence Function of dying from COVID-19 and the Cumulative Incidence Function of dying from other causes for Brazilian subjects with COVID-19. In particular, exposure to asthma, diabetes, obesity, older age, male gender, being black or indigenous, lack of influenza vaccination, admission to an intensive care unit, and the presence of other risk factors, such as immuno-suppression and chronic kidney, neurological, liver, and lung diseases, significantly increase the probability of dying from COVID-19. Finally, it is noted that the highest hazard ratio was observed for people aged over 70 years (in comparison to people aged 50–60 years).

A continuous-time category model SEIRS-VB (S, susceptible; E, exposed; I, infectious; R, recovered; S, susceptible, where V stands for susceptible vaccinated individuals and B denotes individuals with vaccine-induced immunity) was used in [52] in order to model the long-term behaviour of the COVID-19 pandemic, and the basic reproduction number associated with this model in the autonomous case is defined: the single disease-free equilibrium point is locally asymptotically stable when the basic reproduction number is less than one and unstable when it is greater than one. Furthermore, a family of discrete-time models with weights is proposed that preserves the biological properties of the differential model. This study aimed to better understand why Bulgaria has the lowest COVID-19 vaccination rate in the European Union and the second highest COVID-19 mortality rate in the world.

A novel approach to simulating the spread of the Omicron variant of SARS-CoV-2 using fractional-order COVID-19 models is presented in [53]. Through the use of the Haar wavelet collocation method, the authors aimed to account for the various factors influencing virus transmission, and, using data from Pakistan, they illustrate the effectiveness of the proposed approach.

Epidemics can be modeled using both macroscopic (compartmental) and microscopic models. An example of the latter is presented in [54], where a novel agent-based simulation framework based on unique mobility patterns for agents between their home location and a point of interest and the extended SICARQD epidemic model (S, susceptible; I, incubating; C, contagious; A, aware; R, recovered; Q, quarantined; D, dead) is applied. This paper provides a qualitative assessment of the impact of quarantine policies and patient relapse rates on society in relation to the proportion of the population infected. The role of three possible quarantine policies (proactive, reactive, and no quarantine) is investigated, along with variable quarantine restriction (0–100%) and three recurrence scenarios (short, long, and no recurrence). The results show that proactive quarantine combined with a higher quarantine rate (i.e., a stricter quarantine policy) triggers a phase transition that reduces the total infected population by over 90% compared to reactive quarantine. Non-pharmaceutical guidelines are also proposed that can be directly applied by global policy-makers.

A new COVID-19 discrete-time compartmental model is presented in [55], where the number of vaccinated people is considered a new state variable. This study shows the existence of two fixed points, a disease-free fixed point (with global asymptotic stability properties) and an endemic fixed point: if a certain inequality with respect to the vaccination rate is satisfied, then the pandemic disappears. These findings can help decision-makers to better understand the epidemiological behavior of this disease over time.

The use of emergent intelligence tools to model complex network systems is the key idea proposed in [56]. This paper focuses on the possibility of analyzing the dynamics of changes in the indicators of the spread of COVID-19 during the first wave of the epidemic (when there was a lack of sufficient statistics because vaccination had not yet started). Percentage growth is therefore used as the most important parameter of the model, which has a stochastic nature. The principle of dynamic balance of epidemiological processes is also innovatively used. This principle is based on the fact that the past values of the total number of cases are close enough to the values of the total number of recovered and deceased patients at the current time. The problem of predicting the future dynamics of the exactly random values of the model parameters is then addressed to determine the future values of the total number of cases and the recovered and deceased as well as active cases. Data from Russia and European countries (Germany and Italy) collected during the first wave of the epidemic are used.

Medical shocks, such as the COVID-19 pandemic, significantly affect countries' economic systems. To understand the recovery from such shocks, ref. [57] proposes a mathematical model that accounts for the interplay between pandemic dynamics and economic development, including management considerations. This model's effectiveness was validated by applying it to five emerging economies: India, Brazil, Indonesia, South Africa, and Kazakhstan. These countries were selected to provide broad geographical coverage while representing scenarios in which developing economies may face greater challenges in allocating sufficient funds to combat the pandemic compared to developed economies. The results obtained from these applications, especially in the early stages of the pandemic's spread, underscore the crucial importance of implementing proactive corrective measures promptly to overcome the pandemic and facilitate economic recovery.

A novel statistical model for capturing the mortality rates of COVID-19 patients is proposed in [58]. The model, the New Modified Flexible Weibull Extension (NMFWE) distribution, offers a fresh perspective in the realm of data modeling. The study includes the derivation of maximum likelihood estimators for the NMFWE model and their assessment through a simulation study using specific datasets from Mexico and Canada. A comparison with other statistical models reveals that the NMFWE model outperforms them in terms of seven key statistical metrics when applied to mortality rate data.

Using a deterministic two-age-structured COVID-19 epidemic compartmental model, ref. [59] illustrates how age plays a crucial role in disease transmission dynamics. Two age classes are considered, those under 60 years old and those aged 60 and older, to align with vaccination strategies and address identifiability issues. This research examines six distinct disease transmission scenarios alongside social distancing measures and feedback interventions, including an age-stratified vaccine prioritization strategy. These interventions impact age-dependent patterns of social contact and the spread of COVID-19. Notably, this study innovatively identifies virulence parameters within these age groups during different phases. The findings from [59] contribute to our understanding of how human contact networks and behavior influence the spread of infectious diseases, without aiming for predictive applications, while also assessing the implications for public health policy planning. In real epidemics, behavioral changes not only reduce contact and intensity but also alter the structure of contact networks.

Finally, the significance of this SI cannot be overstated owing to the methodological nature of the papers contained within it. Indeed, such papers can serve as a foundational resource for continuous updates as new data become available. This concept is successfully illustrated in [60], in which the same methodology used in [59] is applied to new data, showing how the significance of the principle governs the data itself.

Author Contributions: All authors have contributed equally. All authors have read and agreed to the published version of the manuscript.

Conflicts of Interest: The authors declare no conflicts of interest.

References

1. Musa, S.S.; Wang, X.; Zhao, S.; Li, S.; Hussaini, N.; Wang, W.; He, D. The heterogeneous severity of COVID-19 in African countries: A modeling approach. *Bull. Math. Biol.* **2022**, *84*, 32. [CrossRef] [PubMed]
2. Hu, L.; Wang, S.; Zheng, T.; Hu, Z.; Kang, Y.; Nie, L.F.; Teng, Z. The effects of migration and limited medical resources of the transmission of SARS-CoV-2 model with two patches. *Bull. Math. Biol.* **2022**, *84*, 55. [CrossRef]
3. Cui, J.; Wu, Y.; Guo, S. Effect of non-homogeneous mixing and asymptomatic individuals on final epidemic size and basic reproduction number in a meta-population model. *Bull. Math. Biol.* **2022**, *84*, 38. [CrossRef] [PubMed]
4. Betti, M.I.; Abouleish, A.H.; Spofford, V.; Peddigrew, C.; Diener, A.; Heffernan, J.M. COVID-19 vaccination and healthcare demand. *Bull. Math. Biol.* **2023**, *85*, 32. [CrossRef] [PubMed]
5. Xue, Y.; Chen, D.; Smith, S.R.; Ruan, X.; Tang, S. Coupling the Within-Host Process and Between-Host Transmission of COVID-19 Suggests Vaccination and School Closures are Critical. *Bull. Math. Biol.* **2023**, *85*, 6.
6. Wang, X.; Wu, H.; Tang, S. Assessing age-specific vaccination strategies and post-vaccination reopening policies for COVID-19 control using SEIR modeling approach. *Bull. Math. Biol.* **2022**, *84*, 108. [CrossRef] [PubMed]
7. Zou, Y.; Yang, W.; Lai, J.; Hou, J.; Lin, W. Vaccination and quarantine effect on COVID-19 transmission dynamics incorporating Chinese-spring-festival travel rush: Modeling and simulations. *Bull. Math. Biol.* **2022**, *84*, 30. [CrossRef]
8. Tang, B.; Zhou, W.; Wang, X.; Wu, H.; Xiao, Y. Controlling multiple COVID-19 epidemic waves: An insight from a multi-scale model linking the behaviour change dynamics to the disease transmission dynamics. *Bull. Math. Biol.* **2022**, *84*, 106. [CrossRef]
9. Gharouni, A.; Abdelmalek, F.M.; Earn, D.J.; Dushoff, J.; Bolker, B.M. Testing and isolation efficacy: Insights from a simple epidemic model. *Bull. Math. Biol.* **2022**, *84*, 66. [CrossRef] [PubMed]
10. Feng, S.; Zhang, J.; Li, J.; Luo, X.F.; Zhu, H.; Li, M.Y.; Jin, Z. The impact of quarantine and medical resources on the control of COVID-19 in Wuhan based on a household model. *Bull. Math. Biol.* **2022**, *84*, 47. [CrossRef] [PubMed]
11. Zhou, L.; Rong, X.; Fan, M.; Yang, L.; Chu, H.; Xue, L.; Hu, G.; Liu, S.; Zeng, Z.; Chen, M.; et al. Modeling and evaluation of the joint prevention and control mechanism for curbing COVID-19 in Wuhan. *Bull. Math. Biol.* **2022**, *84*, 28. [CrossRef] [PubMed]
12. Lou, Y.; Salako, R.B. Control strategies for a multi-strain epidemic model. *Bull. Math. Biol.* **2022**, *84*, 1–47. [CrossRef] [PubMed]

13. Dangerfield, C.E.; Abrahams, I.D.; Budd, C.; Butchers, M.; Cates, M.E.; Champneys, A.R.; Currie, C.S.; Enright, J.; Gog, J.R.; Goriely, A.; et al. Getting the most out of maths: How to coordinate mathematical modelling research to support a pandemic, lessons learnt from three initiatives that were part of the COVID-19 response in the UK. *J. Theor. Biol.* **2023**, *557*, 111332. [CrossRef] [PubMed]
14. Southall, E.; Ogi-Gittins, Z.; Kaye, A.; Hart, W.; Lovell-Read, F.; Thompson, R. A practical guide to mathematical methods for estimating infectious disease outbreak risks. *J. Theor. Biol.* **2023**, *562*, 111417. [CrossRef] [PubMed]
15. KhudaBukhsh, W.R.; Bastian, C.D.; Wascher, M.; Klaus, C.; Sahai, S.Y.; Weir, M.H.; Kenah, E.; Root, E.; Tien, J.H.; Rempała, G.A. Projecting COVID-19 cases and hospital burden in Ohio. *J. Theor. Biol.* **2023**, *561*, 111404. [CrossRef] [PubMed]
16. Duan, M.; Jin, Z. The heterogeneous mixing model of COVID-19 with interventions. *J. Theor. Biol.* **2022**, *553*, 111258. [CrossRef] [PubMed]
17. Whittaker, D.G.; Herrera-Reyes, A.D.; Hendrix, M.; Owen, M.R.; Band, L.R.; Mirams, G.R.; Bolton, K.J.; Preston, S.P. Uncertainty and error in SARS-CoV-2 epidemiological parameters inferred from population-level epidemic models. *J. Theor. Biol.* **2023**, *558*, 111337. [CrossRef] [PubMed]
18. Nugent, A.; Southall, E.; Dyson, L. Exploring the role of the potential surface in the behaviour of early warning signals. *J. Theor. Biol.* **2022**, *554*, 111269. [CrossRef] [PubMed]
19. Glasser, J.W.; Feng, Z.; Vo, M.; Jones, J.N.; Clarke, K.E. Analysis of serological surveys of antibodies to SARS-CoV-2 in the United States to estimate parameters needed for transmission modeling and to evaluate and improve the accuracy of predictions. *J. Theor. Biol.* **2023**, *556*, 111296. [CrossRef] [PubMed]
20. Korosec, C.S.; Betti, M.I.; Dick, D.W.; Ooi, H.K.; Moyles, I.R.; Wahl, L.M.; Heffernan, J.M. Multiple cohort study of hospitalized SARS-CoV-2 in-host infection dynamics: Parameter estimates, identifiability, sensitivity and the eclipse phase profile. *J. Theor. Biol.* **2023**, *564*, 111444. [CrossRef] [PubMed]
21. Aristotelous, A.C.; Chen, A.; Forest, M.G. A hybrid discrete-continuum model of immune responses to SARS-CoV-2 infection in the lung alveolar region, with a focus on interferon induced innate response. *J. Theor. Biol.* **2022**, *555*, 111293. [CrossRef] [PubMed]
22. Tatematsu, D.; Akao, M.; Park, H.; Iwami, S.; Ejima, K.; Iwanami, S. Relationship between the inclusion/exclusion criteria and sample size in randomized controlled trials for SARS-CoV-2 entry inhibitors. *J. Theor. Biol.* **2023**, *561*, 111403. [CrossRef] [PubMed]
23. Keeling, M.J. Patterns of reported infection and reinfection of SARS-CoV-2 in England. *J. Theor. Biol.* **2023**, *556*, 111299. [CrossRef] [PubMed]
24. Hill, E.M. Modelling the epidemiological implications for SARS-CoV-2 of Christmas household bubbles in England. *J. Theor. Biol.* **2023**, *557*, 111331. [CrossRef] [PubMed]
25. Whitfield, C.A.; Hall, I. Modelling the impact of repeat asymptomatic testing policies for staff on SARS-CoV-2 transmission potential. *J. Theor. Biol.* **2023**, *557*, 111335. [CrossRef] [PubMed]
26. Zhong, H.; Wang, K.; Wang, W. Spatiotemporal pattern recognition and dynamical analysis of COVID-19 in Shanghai, China. *J. Theor. Biol.* **2022**, *554*, 111279. [CrossRef] [PubMed]
27. Chen, A.; Wessler, T.; Forest, M.G. Antibody protection from SARS-CoV-2 respiratory tract exposure and infection. *J. Theor. Biol.* **2023**, *557*, 111334. [CrossRef] [PubMed]
28. Are, E.B.; Song, Y.; Stockdale, J.E.; Tupper, P.; Colijn, C. COVID-19 endgame: From pandemic to endemic? Vaccination, reopening and evolution in low-and high-vaccinated populations. *J. Theor. Biol.* **2023**, *559*, 111368. [CrossRef] [PubMed]
29. Hurford, A.; Martignoni, M.M.; Loredo-Osti, J.C.; Anokye, F.; Arino, J.; Husain, B.S.; Gaas, B.; Watmough, J. Pandemic modelling for regions implementing an elimination strategy. *J. Theor. Biol.* **2023**, *561*, 111378. [CrossRef] [PubMed]
30. Bramble, J.; Fulk, A.; Saenz, R.; Agusto, F.B. Exploring the role of superspreading events in SARS-CoV-2 outbreaks. *J. Theor. Biol.* **2023**, *558*, 111353. [CrossRef] [PubMed]
31. Tu, Y.; Wang, X.; Tang, S. Exploring COVID-19 transmission patterns and key factors during epidemics caused by three major strains in Asia. *J. Theor. Biol.* **2023**, *557*, 111336. [CrossRef] [PubMed]
32. Gao, S.; Shen, M.; Wang, X.; Wang, J.; Martcheva, M.; Rong, L. A multi-strain model with asymptomatic transmission: Application to COVID-19 in the US. *J. Theor. Biol.* **2023**, *565*, 111468. [CrossRef] [PubMed]
33. Creswell, R.; Robinson, M.; Gavaghan, D.; Parag, K.V.; Lei, C.L.; Lambert, B. A Bayesian nonparametric method for detecting rapid changes in disease transmission. *J. Theor. Biol.* **2023**, *558*, 111351. [CrossRef] [PubMed]
34. Cueno, M.E.; Wada, K.; Tsuji, A.; Ishikawa, K.; Imai, K. Structural patterns of SARS-CoV-2 variants of concern (alpha, beta, gamma, delta) spike protein are influenced by variant-specific amino acid mutations: A computational study with implications on viral evolution. *J. Theor. Biol.* **2023**, *558*, 111376. [CrossRef] [PubMed]
35. Colman, E.; Puspitarani, G.A.; Enright, J.; Kao, R.R. Ascertainment rate of SARS-CoV-2 infections from healthcare and community testing in the UK. *J. Theor. Biol.* **2023**, *558*, 111333. [CrossRef] [PubMed]
36. Li, Z.; Zhao, J.; Zhou, Y.; Tian, L.; Liu, Q.; Zhu, H.; Zhu, G. Adaptive behaviors and vaccination on curbing COVID-19 transmission: Modeling simulations in eight countries. *J. Theor. Biol.* **2023**, *559*, 111379. [CrossRef] [PubMed]
37. Avusuglo, W.; Mosleh, R.; Ramaj, T.; Li, A.; Sharbayta, S.S.; Fall, A.A.; Ghimire, S.; Shi, F.; Lee, J.K.; Thommes, E.; et al. Workplace absenteeism due to COVID-19 and influenza across Canada: A mathematical model. *J. Theor. Biol.* **2023**, *572*, 111559. [CrossRef] [PubMed]

38. Kadelka, C.; Islam, M.R.; McCombs, A.; Alston, J.; Morton, N. Ethnic homophily affects vaccine prioritization strategies. *J. Theor. Biol.* **2022**, *555*, 111295. [CrossRef] [PubMed]
39. Micheletti, A.; Araújo, A.; Budko, N.; Carpio, A.; Ehrhardt, M. Mathematical models of the spread and consequences of the SARS-CoV-2 pandemics: Effects on health, society, industry, economics and technology, 2021. *J. Math. Ind.* **2021**, *11*, 1–2.
40. McCarthy, Z.; Xiao, Y.; Scarabel, F.; Tang, B.; Bragazzi, N.L.; Nah, K.; Heffernan, J.M.; Asgary, A.; Murty, V.K.; Ogden, N.H.; et al. Quantifying the shift in social contact patterns in response to non-pharmaceutical interventions. *J. Math. Ind.* **2020**, *10*, 1–25.
41. Bambusi, D.; Ponno, A. Linear behavior in Covid19 epidemic as an effect of lockdown. *J. Math. Ind.* **2020**, *10*, 27. [CrossRef] [PubMed]
42. Thron, C.; Mbazumutima, V.; Tamayo, L.V.; Todjihounde, L. Cost effective reproduction number based strategies for reducing deaths from COVID-19. *J. Math. Ind.* **2021**, *11*, 11. [CrossRef] [PubMed]
43. Langfeld, K. Dynamics of epidemic diseases without guaranteed immunity. *J. Math. Ind.* **2021**, *11*, 1–8.
44. Wijaya, K.P.; Ganegoda, N.; Jayathunga, Y.; Götz, T.; Schäfer, M.; Heidrich, P. An epidemic model integrating direct and fomite transmission as well as household structure applied to COVID-19. *J. Math. Ind.* **2021**, *11*, 1–26. [CrossRef] [PubMed]
45. Tarrataca, L.; Dias, C.M.; Haddad, D.B.; De Arruda, E.F. Flattening the curves: On-off lock-down strategies for COVID-19 with an application to Brazil. *J. Math. Ind.* **2021**, *11*, 1–18.
46. Kantner, M.; Koprucki, T. Beyond just "flattening the curve": Optimal control of epidemics with purely non-pharmaceutical interventions. *J. Math. Ind.* **2020**, *10*, 23. [CrossRef] [PubMed]
47. Colombo, R.M.; Garavello, M.; Marcellini, F.; Rossi, E. An age and space structured SIR model describing the COVID-19 pandemic. *J. Math. Ind.* **2020**, *10*, 22. [CrossRef] [PubMed]
48. Götz, T.; Heidrich, P. Early stage COVID-19 disease dynamics in Germany: Models and parameter identification. *J. Math. Ind.* **2020**, *10*, 20. [CrossRef] [PubMed]
49. Wu, J.; Tang, B.; Bragazzi, N.L.; Nah, K.; McCarthy, Z. Quantifying the role of social distancing, personal protection and case detection in mitigating COVID-19 outbreak in Ontario, Canada. *J. Math. Ind.* **2020**, *10*, 1–12. [CrossRef] [PubMed]
50. Haque, M.A.; Cortese, G. Cumulative Incidence Functions for Competing Risks Survival Data from Subjects with COVID-19. *Mathematics* **2023**, *11*, 3772. [CrossRef]
51. Li, Y.; Samreen; Zada, L.; Ismail, E.A.; Awwad, F.A.; Hassan, A.M. Assessing the Impact of Time-Varying Optimal Vaccination and Non-Pharmaceutical Interventions on the Dynamics and Control of COVID-19: A Computational Epidemic Modeling Approach. *Mathematics* **2023**, *11*, 4253. [CrossRef]
52. Margenov, S.; Popivanov, N.; Ugrinova, I.; Hristov, T. Differential and Time-Discrete SEIRS Models with Vaccination: Local Stability, Validation and Sensitivity Analysis Using Bulgarian COVID-19 Data. *Mathematics* **2023**, *11*, 2238. [CrossRef]
53. Raizah, Z.; Zarin, R. Advancing COVID-19 Understanding: Simulating Omicron Variant Spread Using Fractional-Order Models and Haar Wavelet Collocation. *Mathematics* **2023**, *11*, 1925. [CrossRef]
54. Topîrceanu, A. On the Impact of Quarantine Policies and Recurrence Rate in Epidemic Spreading Using a Spatial Agent-Based Model. *Mathematics* **2023**, *11*, 1336. [CrossRef]
55. Al-Shbeil, I.; Djenina, N.; Jaradat, A.; Al-Husban, A.; Ouannas, A.; Grassi, G. A New COVID-19 Pandemic Model Including the Compartment of Vaccinated Individuals: Global Stability of the Disease-Free Fixed Point. *Mathematics* **2023**, *11*, 576. [CrossRef]
56. Zakharov, V.; Balykina, Y.; Ilin, I.; Tick, A. Forecasting a New Type of Virus Spread: A Case Study of COVID-19 with Stochastic Parameters. *Mathematics* **2022**, *10*, 3725. [CrossRef]
57. Akaev, A.; Zvyagintsev, A.I.; Sarygulov, A.; Devezas, T.; Tick, A.; Ichkitidze, Y. Growth Recovery and COVID-19 Pandemic Model: Comparative Analysis for Selected Emerging Economies. *Mathematics* **2022**, *10*, 3654. [CrossRef]
58. Ahmad, Z.; Almaspoor, Z.; Khan, F.; El-Morshedy, M. On predictive modeling using a new flexible Weibull distribution and machine learning approach: Analyzing the COVID-19 data. *Mathematics* **2022**, *10*, 1792. [CrossRef]
59. Verrelli, C.M.; Della Rossa, F. Two-age-structured COVID-19 epidemic model: Estimation of virulence parameters to interpret effects of national and regional feedback interventions and vaccination. *Mathematics* **2021**, *9*, 2414. [CrossRef]
60. Verrelli, C.M.; Della Rossa, F. Two-age-structured COVID-19 epidemic model: Estimation of virulence parameters through new data incorporation. *Mathematics* **2024**, *12*, 825. [CrossRef]

Disclaimer/Publisher's Note: The statements, opinions and data contained in all publications are solely those of the individual author(s) and contributor(s) and not of MDPI and/or the editor(s). MDPI and/or the editor(s) disclaim responsibility for any injury to people or property resulting from any ideas, methods, instructions or products referred to in the content.

Article

Two-Age-Structured COVID-19 Epidemic Model: Estimation of Virulence Parameters to Interpret Effects of National and Regional Feedback Interventions and Vaccination

Cristiano Maria Verrelli [1,*] and Fabio Della Rossa [2]

1 Electronic Engineering Department, University of Rome Tor Vergata, Via del Politecnico 1, 00133 Rome, Italy
2 Department of Electronic, Information and Biomedical Engineering, Politecnico di Milano, 20133 Milan, Italy; fabio.dellarossa@polimi.it
* Correspondence: verrelli@ing.uniroma2.it; Tel.: +39-(0)6-72597410

Abstract: The COVID-19 epidemic has recently led in Italy to the implementation of different external strategies in order to limit the spread of the disease in response to its transmission rate: strict national lockdown rules, followed first by a weakening of the social distancing and contact reduction feedback interventions and finally the implementation of coordinated intermittent regional actions, up to the application, in this last context, of an age-stratified vaccine prioritization strategy. This paper originally aims at identifying, starting from the available age-structured real data at the national level during the specific aforementioned scenarios, external-scenario-dependent sets of virulence parameters for a two-age-structured COVID-19 epidemic compartmental model, in order to provide an interpretation of how each external scenario modifies the age-dependent patterns of social contacts and the spread of COVID-19.

Keywords: COVID-19 epidemic; model identification; parameter estimation; compartmental model; national lockdown; regional action; vaccine prioritization strategy

1. Introduction

COVID-19 (SARS-CoV-2) is at the root of the recent economic and public health crisis worldwide. Since it was reported in December 2019 in China, the virus quickly took pandemic proportions throughout six continents and over 210 countries. Over 100 countries declared lockdowns and curfews, with an estimated global economic loss of one trillion US dollars in 2020 (see [1] and references therein). By October 2020, over 36 million people were definitely reported to be infected with COVID-19 and more than one million people had died from virus-related complications.

COVID-19 causes respiratory disease: common symptoms include fever, dry cough, fatigue, shortness of breath and loss of smell or taste, with possible complications including pneumonia and acute respiratory distress syndrome up to severe respiratory failures, septic shocks and death [2].

A huge amount of effort has been spent with the aim of finding novel methods for mathematical modeling and control of epidemics [3]. Mathematical models can in fact accurately portray the epidemic's dynamic spread [3–9]. Eight stages of infection, namely susceptible (S), infected (I), diagnosed (D), ailing (A), recognized (R), threatened (T), healed (H) and extinct (E), are presented in [8] to illustrate how restrictive social-distancing measures have to include a combination with widespread testing and contact tracing. A control-oriented SIR model that stresses the effects of delays and compares the outcomes of different containment policies is proposed in [5], whereas stochastic transmission models are considered in [3,9]. A method for the joint optimal lockdown and release design in a pandemic is proposed in [4] and then applied in a realistic simulation scenario based on the data of COVID-19's evolution in Italy. A dynamical model specifically designed for COVID-19 is used in [6] to describe the epidemic evolution in Italy, with different kinds of control actions (social, political, and medical) being explicitly modeled. A parameter-varying modification of the SIRD model is finally proposed in [7] for describing and predicting the behavior of the COVID-19 contagion in Italy through identification of model parameters, written as linear combinations of basis functions.

An important feature of COVID-19 is its highly non-uniform attack of different age strata of society [2]: the infection fatality ratio for individuals older than 80 is likely significantly higher than the infection fatality ratio for individuals younger than 50 years. In particular, [10], which fit an age-structured mathematical model to epidemic data from China, Italy, Japan, Singapore, Canada and South Korea, shows that susceptibility to infection in individuals under 20 years of age is approximately half that in adults aged over 20 years, and that clinical symptoms manifest in 21% of infections in 10- to 19-year-olds, rising to 69% of infections in people aged over 70 years. On the other hand, ref. [11] estimates an overall infection fatality rate of 1.29%, as well as large differences by age, with a low infection fatality rate of 0.05% for those under 60 years old and a substantially higher 4.25% for people above 60 years of age; even if only 10% of the population were infected, the infection fatality rate would not rise above 0.2% for people under 60.

Since social contacts are influenced by age structure of the population and the frequency of contacts across the population [12], the spread of the disease relies on contact patterns among different subjects in the infected population. Epidemic models that are stratified by age are therefore particularly relevant when the hospital load and fatalities related to COVID-19 are to be estimated, with age-dependent patterns of social contacts being incorporated. Such models, which take into account the mechanism of its transmission, including the (possibly heterogeneous) pattern of mixing among the population, the susceptibility within the population, the virulence of the infection, the probability of transmission per contact, and the changes in behavior in the affected population in response to an epidemic [13], constitute useful tools to understand and characterize the complex transmission dynamics acting among different groups of the population (see [13]), as well as to identify and predict the effects of different age-stratified intervention strategies in slowing the spread. They concurrently (i) provide valuable information for public-health policy makers, (ii) avoid health systems saturation, and (iii) mitigate the impact on costs.

In this respect, despite containing simplifying assumptions, common variants of SIR-type models, including age-dependent substructures, are of great help in characterizing epidemics (see [14] and references therein). The reader is referred to the very recent [1,12,14–18] for age-structured modeling of the COVID-19 epidemic and related age-dependent analyses. The reader is also referred to [19] for the latest study adopting a model with an age-dependent pre-pandemic contact matrix, reflecting the goal of a return to pre-pandemic routines once a vaccine is available, to compare five age-stratified vaccine prioritization strategies. A modified age-structured SIR model—based on known patterns of social contact and distancing measures within Washington, USA—is presented in [20]: population age-distribution has a significant effect on disease spread and mortality rate and contributes to the efficacy of age-specific contact and treatment measures. On the other hand, ref. [21] shows that if transmission rates' return to normal in the future and the epidemic ends only when population immunity is sufficient to survive reintroduction of infection, then age-targeted mitigations can still achieve a large mortality reduction.

This paper, unlike any other approach in the literature, aims to illustrate how six different diseases transmission scenarios and concurrently adopted social distancing and feedback interventions—including an age-stratified vaccine prioritization strategy—modify the age-dependent patterns of social contacts and the spread of COVID-19 disease. To this purpose, we use a (deterministic) two-age-structured COVID-19 epidemic compartmental model, in which two-age-classes (lower than 60 years old and not lower than 60 years old) are adopted, in order to comply with the adopted vaccination strategy while avoiding issues related to the lack of identifiability. In innovation is that identification of virulence parameters within the two groups is performed during the different phases. Real data, indeed, are taken from the Italian context, in which the implementation of the following subsequent-in-time different strategies has been carried out over time in response to different disease transmission scenarios:

(i) a strict national lockdown rule (scenario *a*), as necessary in the first place (in the presence of a relatively high estimate of the disease transmission rate) to remove social contacts in workplaces, schools, markets and other public areas;

(ii) a weakened feedback social distancing and contact reduction intervention (in the presence of a relatively low estimate of the disease transmission rate), which is composed of a weakened lockdown phase (scenario *b*), a low distancing phase (scenario *c*), a low distancing + workplace/school-contacts re-activation phase (scenario *d*), with a progressive release of the population back to their daily routine appearing;

(iii) a coordinated intermittent regional action (scenario *e*)—in the presence of a newly alarming increase in the estimated disease transmission rate—where social distancing measures are put in place or relaxed independently by each region according to the ratio between hospitalized individuals and the total capacity of the health system in that region; and

(iv) direct mRNA-vaccination of subjects—especially the elderly—(scenario *f*) at highest risk for severe outcomes, along with Vaxzevria-vaccination of young subjects belonging to crucial occupational categories, to indirectly protect subjects at highest risk for severe outcomes.

In particular, the recent [22] is at the foundation of the intermittent intervention, which is inspired by the regionalism as an integral part of the Italian constitution and involves regional feedback strategies, where each of the twenty regions strengthens or weakens local mitigating actions, namely social distancing, inflow–outflow control, as a function of the saturation of their hospital capacity. It is worth noticing that the regional action was actually intermittent in a strict sense, owing to the presence, over time, of Italian regions with actually weakened social distancing.

The resulting framework originally exploited in the paper thus enriches the one proposed in [19], while turning out to be useful to compare impacts of national or regional intervention contexts and age-stratified vaccine prioritization strategies on the virulence parameters (within the age-dependent groups) of the age-stratified model (though the effects of seasonality on COVID-19 remain unsettled, they might be represented by the estimation results.). The results of this paper thus move in the direction of understanding the impact of human contact networks and human behavior on the spread of infectious diseases (no prediction purpose is declared), while assessing the implications of this for the planning of public health policy. In fact, even though the simplest mathematical models assume that the population mixes homogeneously, such an assumption is often only sufficient to obtain general insights, with the pattern of contacts between different age groups playing an essential role in determining the spread of disease. Indeed, in a real epidemic, the behavioral changes will not only reduce the number of contacts and intensity but will even change the structure of the contact network.

2. The Model

The (deterministic) compartmental model used in this paper is reported in this section. It is a natural extension of the classical SIR model [23]. The compartments are subdivided into two different age groups: group of subjects with age lower than 60; group of subjects with age not lower than 60. There are many reasons for the choice to split the population into these two groups. The first one, which is more conceptual, is that this division more or less coincides with a division of active and retired populations, in which one may expect different patterns of social interactions, leading to different transmission dynamics. The second one is in accordance with the recent literature, where empirical estimates based on population-level data show a sharp difference in fatality rates between young and old people and firmly rule out overall fatality ratios below 0.5% in populations with more than 30% being over 60 years old [11]. The third reason relies on the fact that this choice is compatible with the vaccination strategy adopted in *scenario f*. Finally, a closer look at the Italian data reveals that this choice is the one that divides the number of COVID-19 cases in the most uniform way, allowing a better exploitation of the data information and reducing the bias due to non-uniform cardinality division between the two classes. The resulting model thus reads:

$$\begin{aligned}
S_y(t+1) &= S_y(t) - S_y(t)(v_{11}I_y(t) + v_{12}I_o(t))/N(t) \\
S_o(t+1) &= S_o(t) - S_o(t)(v_{21}I_y(t) + v_{22}I_o(t))/N(t) \\
I_y(t+1) &= (1 - \tau_1 - \gamma)I_y(t) + S_y(t)(v_{11}I_y(t) + v_{12}I_o(t))/N(t) \\
I_o(t+1) &= (1 - \tau_2 - \gamma)I_o(t) + S_o(t)(v_{21}I_y(t) + v_{22}I_o(t))/N(t) \\
C_y(t+1) &= C_y(t) + \tau_1 I_y(t) \\
C_o(t+1) &= C_o(t) + \tau_2 I_o(t)
\end{aligned} \quad (1)$$

where: t is the time, measured in days; S_i, I_i and C_i, $i = y, o$ are the numbers of susceptible, infected and reported cases for the two age classes, respectively; and $N(t)$ is the number of persons who are not quarantined, hospitalized or dead at time t. The parameters v_{ij}, $i, j = 1, 2$ represent the virulence of the virus among the different age classes, while $1/\tau_i$, $i = 1, 2$ is the average time for disease identification and $\gamma = 0.07$ is the rate of asymptomatic infected who recover without being reported [22,24]. Note that the model is not autonomous, since $N(t)$ cannot be reconstructed from the state variables. In other words, such a model does not count the number of quarantined, hospitalized, recovered or dead, and thus it needs this time-series in order to be simulated. Even though this model might be certainly extended so it can used for prediction, this would be out of the scope of this paper. As aforementioned, the goal here is to understand the demographical habits (how people interconnected) starting from the epidemiological data while estimating model parameters to overview people's reactions under the six different scenarios.

3. Estimation of Model Parameters

To fit the model, we first detected the six scenarios of Section 1 that characterized the COVID-19 pandemic in Italy:

a. From $t_0^a = 9$ March 2020 to $t_e^a = 28$ April 2020: strict national lockdown rule in which social contacts in workplaces, schools, markets and other public areas are removed;
b. From $t_0^b = 7$ May 2020 to $t_e^b = 3$ June 2020: weakened feedback social distancing and contact reduction intervention, with a slow release of the population back to their daily routine appearing (especially the elderly, as a psychological toll due to the suffered isolation);
c. From $t_0^c = 9$ June 2020 to $t_e^c = 8$ September 2020: low feedback social distancing and contact reduction intervention, due to a low ratio between hospitalized individuals and the total capacity of the national health system;
d. From $t_0^d = 15$ September 2020 to $t_e^d = 27$ October 2020: low feedback social distancing and contact reduction intervention, with social contacts in workplaces and schools being re-activated;
e. From $t_0^e = 7$ November 2020 to $t_e^e = 29$ December 2020: coordinated intermittent regional action, where social contacts in schools is decreased at national level and social distancing measures are put in place or relaxed independently by each region according to the ratio between hospitalized individuals and the total capacity of the health system in that region; and
f. from $t_0^f = 5$ January 2021 to $t_e^f = 15$ May 2021: direct mRNA-vaccination of subjects (the elderly) at highest risk for severe outcomes and indirect protection through Vaxzevria vaccination of young subjects belonging to crucial occupational categories.

Real data about the pandemic on each of these scenarios are taken from the official Ministerial website https://www.epicentro.iss.it/coronavirus/aggiornamenti (accessed on 1 June 2021). In particular, in the data centre, it is possible to find:

- The cumulative detected cases on a weekly scale $C(t)$ divided by age (so it is possible to compute $C_y(t)$ and $C_o(t)$);
- The number of recovered people (not divided by age) $R(t)$.

After noting that

$$N(t) = N(0) - (C_y(t) + C_o(t)) + R(t)$$

we fit the model to the real data by minimizing the squared relative error between the measured data of the detected cases and the ones predicted from the model. More precisely, starting from the time windows that identify *scenario a*, i.e., $t \in [t_0^i, t_e^i]$, $i = a$, we compute the trajectory of the model

$$\begin{aligned}
S_y^i(t+1) &= S_y^i(t) - S_y^i(t)(v_{11}^i I_y^i(t) + v_{12}^i I_o^i(t))/N(t) \\
S_o^i(t+1) &= S_o^i(t) - S_o^i(t)(v_{21}^i I_y^i(t) + v_{22}^i I_o^i(t))/N(t) \\
I_y^i(t+1) &= (1 - \tau_1^i - \gamma)I_y^i(t) + S_y^i(t)(v_{11}^i I_y^i(t) + v_{12}^i I_o^i(t))/N(t) \\
I_o^i(t+1) &= (1 - \tau_2^i - \gamma)I_o^i(t) + S_o^i(t)(v_{21}^i I_y^i(t) + v_{22}^i I_o^i(t))/N(t)
\end{aligned} \quad (2)$$

that starts from the initial condition

$$\begin{aligned}
I_y^i(t_0^i) &= I_{t_0y}^i, & S_y^i(t_0^i) &= N_y(0) - C_y(t_0^i) - I_{t_0y}^i, \\
I_o^i(t_0^i) &= I_{t_0o}^i, & S_o^i(t_0^i) &= N_o(0) - C_o(t_0^i) - I_{t_0o}^i
\end{aligned}$$

for a particular set of the parameters that we are identifying, which are

- v_{11}^i, characterizing the intra-juvenile virulence;
- v_{12}^i, characterizing the juvenile-elder virulence;
- v_{21}^i, characterizing the elder-juvenile virulence;
- v_{22}^i, characterizing the intra-elder virulence;
- $1/\tau_1^i$, denoting the average time for disease identification in young subjects;
- $1/\tau_2^i$, denoting the average time for disease identification in old subjects;
- $I_{t_0y}^i$, representing the young subjects infected at the beginning of the scenario time window;
- $I_{t_0o}^i$, representing the old subjects infected at the beginning of the scenario time window.

We then compute as cost the relative error between the predicted and the real new daily cases, i.e.,

$$J^i = \sum_{t=t_0^i}^{t_e^i} \left(\frac{C_y(t+1) - \left(C_y(t) + \tau_1^i I_y^i(t)\right)}{C_y(t)} \right)^2 + \left(\frac{C_o(t+1) - \left(C_o(t) + \tau_2^i I_o^i(t)\right)}{C_o(t)} \right)^2.$$

We repeat the same procedure for the following scenarios, i.e., $i = b, \ldots, f$, but, this time, to guarantee the continuity of the identified solution, we impose that the initial condition of the current scenario (parameters $I_{t_0y}^i$ and $I_{t_0o}^i$) is different from the final condition of the previous one of at most 10%. We then tune the 48 parameters (10 of them are constrained) in order to minimize the sum of the six cost functions:

$$J = \sum_{i \in \{a,\ldots,f\}} J^i$$

through the fmincon routine in Matlab©. The total number of data we use for our fitting procedure over the considered six scenarios is 128 [22, 10, 26, 14, 18, 38, respectively], with the problem of estimation needing at least 48 data points (one for each of the parameter we are estimating, eight per scenario). A practical identifiability analysis [25–27] of the parameters around the estimation point confirms that the values we obtained with this procedure can be locally determined from the data we used (the local minimum we have found has no directions on which the cost function does not significantly increase with respect to the parameter variations). This allows us to provide the picture of the age-dependent patterns of social contacts and the spread of COVID-19 disease in the Italian context, which is depicted in Figure 1.

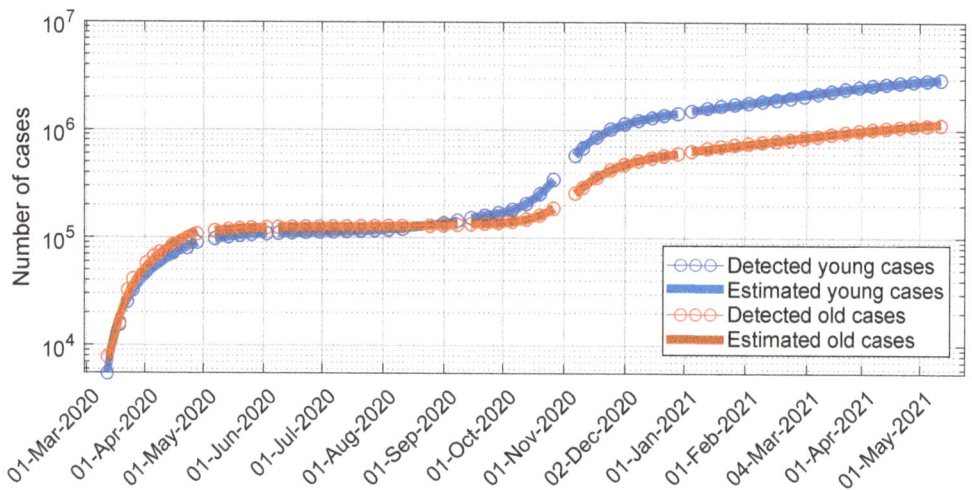

Figure 1. Data fitting for the compartmental model: actual and estimated cumulative profiles for young subjects infected and old subjects infected (logarithmic scale).

The estimated parameters (in the different scenarios $i \in \{a, \ldots, f\}$) $v_{kl}^i, k, l \in \{1, 2\}$ are reported in Table 1, while the estimated parameters $\tau_l^i, l \in \{1, 2\}$, and I_{t0y}^i, I_{t0o}^i appear in Table 2. The modulus and phase of the complex numbers

$$\begin{aligned} \lambda_1^i &= v_{11}^i + \mathbf{i} v_{12}^i \\ \lambda_2^i &= v_{22}^i + \mathbf{i} v_{21}^i \end{aligned} \tag{3}$$

are reported in Table 3 (in the different scenarios $a - f$), with $\mathbf{i}$ here representing the imaginary unit satisfying $\mathbf{i}^2 = -1$. Notice that a non-zero real number λ_1^i or λ_2^i thus represents the case in which only intra-group virulences appear. Conversely, a purely imaginary number λ_1^i or λ_2^i represents the case in which only cross-group virulences appear. On the other hand, the alternative option for coordinates in the complex plane constituted by the polar coordinate system—which uses the distance of a point from the origin and the angle subtended between the positive real axis and the line segment connecting the origin and the point in a counterclockwise way—allows us to directly visualize the complex number λ_1^i or λ_2^i through modulus and phase. Zero phase therefore means a real number, whereas $\pi/2$ phase means a purely imaginary number (according to the previously reported interpretation), with the modulus $|\lambda_1^i| = \sqrt{v_{11}^{i2} + v_{12}^{i2}}$ or $|\lambda_2^i| = \sqrt{v_{22}^{i2} + v_{21}^{i2}}$ constituting a measure of both the contributions coming from the involved virulence parameters.

Table 1. Estimated parameters v^i_{kl}, $k,l \in \{1,2\}$, in the different scenarios $i \in \{a,\ldots,f\}$ [v^i_{11} for intra-juvenile virulence; v^i_{12} for juvenile–elder virulence; v^i_{21} for elder–juvenile virulence; v^i_{22} for intra-elder virulence].

Scenario a			
v^a_{11}	v^a_{12}	v^a_{21}	v^a_{22}
0.7456	0.0000	1.3642	0.0001
Scenario b			
v^b_{11}	v^b_{12}	v^b_{21}	v^b_{22}
0.0266	0.5243	0.0356	0.3912
Scenario c			
v^c_{11}	v^c_{12}	v^c_{21}	v^c_{22}
0.4183	0.1728	0.0302	0.9567
Scenario d			
v^d_{11}	v^d_{12}	v^d_{21}	v^d_{22}
0.0582	4.1363	0.0001	1.2367
Scenario e			
v^e_{11}	v^e_{12}	v^e_{21}	v^e_{22}
0.3914	1.3690	0.1562	0.7877
Scenario f			
v^f_{11}	v^f_{12}	v^f_{21}	v^f_{22}
0.4583	2.1428	0.1856	0.7119

Table 2. Estimated parameters τ^i_l, $l \in \{1,2\}$, and I^i_{t0y}, I^i_{t0o} in the different scenarios $i \in \{a,\ldots,f\}$ [$1/\tau^i_1$ as average time for disease identification in young subjects; $1/\tau^i_2$ as average time for disease identification in old subjects; I^i_{t0y} for the initial young subjects infected; I^i_{t0o} for the initial old subjects infected].

Scenario a			
τ^a_1	τ^a_2	I^a_{t0y}	I^a_{t0o}
0.3448	0.2303	5.5966×10^3	0.7804×10^3
Scenario b			
τ^b_1	τ^b_2	I^b_{t0y}	I^b_{t0o}
0.1311	0.1944	4.9146×10^3	3.2516×10^3
Scenario c			
τ^c_1	τ^c_2	I^c_{t0y}	I^c_{t0o}
0.1319	0.7070	3.3592×10^2	6.5398×10^2
Scenario d			
τ^d_1	τ^d_2	I^d_{t0y}	I^d_{t0o}
0.1917	0.3867	8.7358×10^3	0.3079×10^3

Table 2. Cont.

Scenario e			
τ_1^e	τ_2^e	I_{t0y}^e	I_{t0o}^e
0.3360	0.6050	9.4860×10^4	1.3570×10^4
Scenario f			
τ_1^f	τ_2^f	I_{t0y}^f	I_{t0o}^f
0.3474	0.7364	2.2351×10^4	0.5169×10^4

4. Discussion

The following comments are in order. They provide a meaningful interpretation of the estimation results in accordance with Tables 1–3.

- All the estimates corresponding to the different scenarios, including the estimated I_{t0y}^i, I_{t0o}^i (initial young subjects infected; initial old subjects infected), allow the estimated profile to satisfactorily reproduce the actual one along the different scenarios, as shown by Figure 1.
- Comments for estimated $1/\tau_1^i$ (average time for disease identification in young subjects). This average time takes homogeneous values: it varies from 3 to 7 days weeks over all the scenarios, with about 7 days passing for scenarios b and c. Actually, after the lockdown period and the related concerns, young subjects paid much less attention to their symptoms (recall that scenarios b and c cover a period starting from 7 May 2020 up to 8 September 2020). In addition, recall that young subjects have a higher probability of being asymptomatic (or even weakly symptomatic), while old subjects have a lower probability of being asymptomatic. Asymptomatic subjects usually continue their social interactions, infecting many people before recognizing that they are sick, and are then isolated.
- Comments for estimated $1/\tau_2^i$ (average time for disease identification in old subjects). This average time varies from 1 to 5 days, with less than 3 days occurring in scenarios c–f in which the elderly paid a higher level of attention to symptoms, as a psychological toll due to the suffered isolation in scenarios a–b.
- Comments for estimated v_{kl}^i, $k, l \in \{1, 2\}$, (intra-juvenile virulence; juvenile–elder virulence; elder–juvenile virulence; intra-elder virulence) and related measures.
    - During scenario a, a very small intra-elder virulence appears due to the strict national lockdown rule, with an increase during scenario b, due to the weakened feedback social distancing and contact reduction intervention.
    - During scenarios c and d, a larger increase in the intra-elder virulence occurs, during summer holidays (as a consequence of the juvenile-elder virulence of scenario b) and owing to the re-activation of contacts in workplaces and schools. Recall that school closures during epidemics and pandemics aim to decrease transmission among children. They seemingly have whole-population effects, whenever children are major contributors to community transmission rates.
    - During scenarios e and f, a decrease in the intra-elder virulence is exhibited (when compared to scenarios c and d), as a consequence of an imposed decrease in social contacts in schools and in the direct mRNA vaccination of subjects (the elderly) at highest risk for severe outcomes, in spite of a re-activation of social contacts in schools and in Christmas-related activities. Notice that the intermittent intervention of scenarios e and f, in which each of the twenty regions strengthens or weakens local mitigating actions as a function of the saturation of their hospital capacity, has been largely lighter than the lockdown intervention of scenario a, leading to the possibility of reinvigorating economy and mitigating costs due to the epidemic's spread.

- Large intra-juvenile virulence (>0.39) is exhibited in scenarios *a, c, e* and *f*, i.e., during the strict lockdown (with the virus circulating within families), as well as during summer holidays and after the first days of November, whereas small values accordingly appear in scenarios *b* and *d*, in which more attention was paid by young subjects after the perceived social alarms coming after the end of the strict lockdown and the end of summer vacations.
- Large juvenile–elder virulence (>1.36) is exhibited in scenarios *d–f*, after 15 September 2020, owing to the (typically Italian) juvenile–elder contacts coming from school re-activation, with the smallest value actually occurring in scenario *e*, in which social contacts in schools are decreased at national level and social distancing measures are put in place or relaxed independently by each region according to the ratio between hospitalized individuals and the total capacity of the health system in that region. Nevertheless, a large phase of λ_1^b (with a rather small modulus of λ_1^b) is exhibited in scenario *b*, owing to a weakened feedback social distancing and contact reduction intervention after the strict lockdown.
- The elder–juvenile virulence appears to be relatively small (about zero) in all the scenarios, except for scenario *a*, in which the virus circulated within families (see also the phase of λ_2^a).
- The sum of the two λ_1 and λ_2 phases is small only in scenario *c*, i.e., during holiday vacations, in which a sort of decoupling between the two age classes appeared.

On the other hand, once the estimates of the model parameters have been obtained (Tables 1 and 2), the values of the *reproduction number* $R_t^i[m]$ associated with model (1) in each scenario *i* [*m* stands for model-based computation], as average number of new infections caused by an infected person, can be computed through the formula (adapted from [24])

$$R_t^i[m] = \frac{1}{N(t)} \sigma_1 \left(\begin{bmatrix} S_y^i(t) & 0 \\ 0 & S_o^i(t) \end{bmatrix} \left(\frac{1}{\tau_1^i + \gamma} \begin{bmatrix} v_{11}^i & v_{12}^i \\ v_{11}^i & v_{12}^i \end{bmatrix} + \frac{1}{\tau_2^i + \gamma} \begin{bmatrix} v_{21}^i & v_{22}^i \\ v_{21}^i & v_{22}^i \end{bmatrix} \right) \right)$$

where $\sigma_1(\cdot)$ denotes the biggest among the moduli of the eigenvalues of the matrix argument. The resulting mean reproduction numbers $R_t^i[m]$ over the scenarios $i \in \{a, \ldots, f\}$ read: 1.2, 0.7, 0.7, 2.8, 0.9, 1.1. They are compatible – excepting for a relatively slight overestimate of scenarios *f* – with the maximum likelihood values of the national reproduction number in Figure 2, computed from raw data through the EpiEstim toolbox [28] (a gamma distribution is used as prior distribution, with shape $\alpha = 1.87$, rate $\beta = 0.28$, whereas the R_t-maximum likelihood value is taken as $(\alpha - 1)/\beta$ regarding the analytically expressed posterior distributions), showing that model (1) – though relying on just two age-based subgroups and neglecting, for instance, gender-based different behaviours – is able to catch the main epidemic features along the considered scenarios.

Finally, estimating τ_1 and τ_2 in the six scenarios leads to the possibility of identifying the number of infected cases, in accordance to

$$\begin{aligned} \tau_1^{-1}(C_y(t+1) - C_y(t)) &= I_y(t) \\ \tau_2^{-1}(C_o(t+1) - C_o(t)) &= I_o(t) \end{aligned}$$

coming from the last two equations in (1). This is an advantageous feature of our approach. Looking at Table 2: τ_1^{-1} (young age class) is close to 3 in scenarios *a, e, f*, whereas it is larger than 5 in the least juvenile-action-restrictive scenarios *b-d*; τ_2^{-1} (old age class) is close to 1.5 in the most-decoupled or vaccine-characterized scenarios *c, e, f*, whereas it is close to 5 in the scenarios *a-b* in the middle of the pandemic wave.

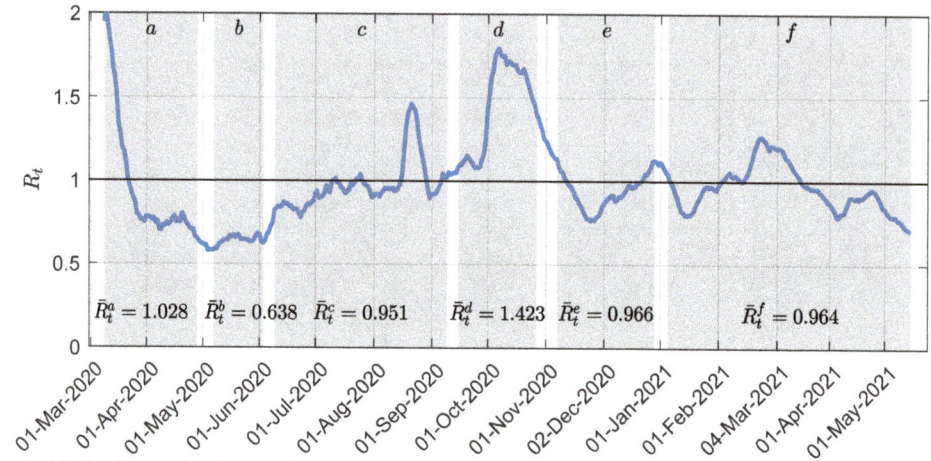

Figure 2. National reproduction number R_t within the considered time windows. Each shaded portion of the plane corresponds to a specific scenario. The mean value $\bar{R}_t^i$ (among the values corresponding to our sampling) within each time window $i \in \{a, \ldots, f\}$ is reported at the bottom of each shaded region.

Table 3. Modulus and phase of the complex numbers $\lambda_1^i = v_{11}^i + i v_{12}^i$ and $\lambda_2 = v_{22}^i + i v_{21}^i$ in the different scenarios $i \in \{a, \ldots, f\}$.

		Scenario a		
$\|\lambda_1^a\|$		$\angle(\lambda_1^a)$	$\|\lambda_2^a\|$	$\angle(\lambda_2^a)$
0.7456		0.0000	1.3642	1.5707
		Scenario b		
$\|\lambda_1^b\|$		$\angle(\lambda_1^b)$	$\|\lambda_2^b\|$	$\angle(\lambda_2^b)$
0.5250		1.5201	0.3928	0.0908
		Scenario c		
$\|\lambda_1^c\|$		$\angle(\lambda_1^c)$	$\|\lambda_2^c\|$	$\angle(\lambda_2^c)$
0.4256		0.3917	0.9572	0.0316
		Scenario d		
$\|\lambda_1^d\|$		$\angle(\lambda_1^d)$	$\|\lambda_2^d\|$	$\angle(\lambda_2^d)$
4.1367		1.5567	1.2367	0.0001
		Scenario e		
$\|\lambda_1^e\|$		$\angle(\lambda_1^e)$	$\|\lambda_2^e\|$	$\angle(\lambda_2^e)$
1.4239		1.2923	0.8030	0.1958
		Scenario f		
$\|\lambda_1^f\|$		$\angle(\lambda_1^f)$	$\|\lambda_2^f\|$	$\angle(\lambda_2^f)$
2.1913		1.3601	0.7357	0.2550

5. Conclusions

Starting from the available age-structured real data at national level (from 9 March 2020 up to 12 May 2021), the parameters of a two-age-structured COVID-19 epidemic compartmental model—with the same two-age-classes definition adopted in the implemented vaccination strategy and more or less coinciding with a division of active and retired populations, as well as with a division characterized by relatively large differences in fatality rates and by uniformity in subgroups cardinality—have been identified in the different scenarios reported in Section 3: the ways in which external scenarios have modified the age-dependent patterns of social contacts and the spread of COVID-19 disease has been assessed. In particular, an epidemiological model for Covid-19 has been developed, which considers the epidemic within the younger age group and older age group separately. Such a model provides insight, at national level, in the different evolution of the epidemic within these two interacting age groups, while simultaneously evaluating, through the estimation of model parameters along time, the impact of changes in social distancing measures and vaccination due to varying external strategies. The results of the present study exhibit some specific limits at their root, such as the inclusion of just two age-dependent subgroups or the absence of gender differences in the subgroups, which might be further investigated in future studies (see [29] for related results). In this respect, it is worth noticing that such a modeling choice is, however, motivated by the fact that adopting more complex SIR variants may fall into unidentifiability problems owing to insufficient data in the details of the many involved compartments, or because of their overly complex structure whose different features cannot be caught in the initial fast-increasing phases. Nevertheless, our study also possesses points of strength. It certainly gives a deep interpretative insight into the time-varying action of parameters within a well-defined COVID epidemics model structure, while providing useful information regarding not only the number of undetected infected cases but also the effects of strategic actions and behaviours. This is rather meaningful, especially on the eve of possible occurrences of variant-based epidemic waves in response to an exact duration of the immunity for the vaccines that is, at this moment, uncertain. The exact structure of the contact patterns in the general population is, in fact, still unknown to a large extent and merits specific research efforts, especially when model-based prediction has to be performed in the presence of political choices that change rules governing social distancing. Such political choices, when they vary along time and affect interactions between subgroups, influence variations of the internal parameters of deterministic models that should be suitably taken into consideration in the related context.

Author Contributions: Conceptualization: C.M.V., F.D.R.; Methodology: C.M.V., F.D.R.; Software and Resources: F.D.R.; Formal Analysis: C.M.V., F.D.R.; Validation, Investigation: C.M.V., F.D.R.; Writing—Original Draft: C.M.V.; Writing—Review and Editing: C.M.V., F.D.R. All authors have read and agreed to the published version of the manuscript.

Funding: This research received no external funding.

Institutional Review Board Statement: Not applicable.

Informed Consent Statement: Not applicable.

Data Availability Statement: The publicly archived datasets that were analyzed during the study are explicitly quoted in the paper.

Acknowledgments: The authors are grateful to E. Cottafava, in her quality of General Secretary of Fondazione GIMBE, Via Amendola, 2, 40121 Bologna, for her willingness to provide data concerning the profile of the Italian reproduction number along time. The authors are also indebted to M. di Bernardo for helpful preliminary discussions about the topics of this paper.

Conflicts of Interest: The authors declare no conflict of interest.

References

1. Mizrahi, L.; Shekhidem, H.A.; Stern, S. Age separation dramatically reduces COVID-19 mortality rate in a computational model of a large population. *Open Biol.* **2020**, *10*, 200213. [CrossRef]
2. Verity, R.; Okell, L.C.; Dorigatti, I.; Winskill, P.; Whittaker, C.; Imai, N.; Cuomo-Dannenburg, G.; Thompson, H.; Walker, P.G.T.; Fu, H.; et al. Ferguson. Estimates of the severity of coronavirus disease 2019: A model-based analysis. *Lancet Infect. Dis.* **2020**, *20*, 669–677. [CrossRef]
3. Borri, A.; Palumbo, P.; Papa, F. Spread/removal parameter identification in a SIR epidemic model. In Proceedings of the 2021 60th IEEE Conference on Decision and Control (CDC), Austin, TX, USA, 13–15 December 2021.

4. Borri, A.; Palumbo, P.; Papa, F.; Possieri, C. Optimal design of lock-down and reopening policies for early-stage epidemics through SIR-D models. *Annu. Rev. Control* **2021**, *51*, 511–524. [CrossRef]
5. Casella, F. Can the COVID-19 epidemic be controlled on the basis of daily test reports? *IEEE Control Syst. Lett.* **2021**, *5*, 1079–1084. [CrossRef]
6. Di Giamberardino, P.; Iacoviello, D.; Papa, F.; Sinisgalli, C. Dynamical evolution of COVID-19 in Italy with an evaluation of the size of the asymptomatic infective population. *IEEE J. Biomed. Health Inform.* **2021**, *25*, 1326–1332. [CrossRef]
7. Calafiore G.C.; Novara, C.; Possieri, C. A time-varying SIRD model for the COVID-19 contagion in Italy. *Annu. Rev. Control* **2020**, *50*, 361–372. [CrossRef]
8. Blanchini, G.G.F.; Bruno, R.; Colaneri, P.; di Filippo, A.; di Matteo, A.; Colaneri, M. Modelling the COVID-19 epidemic and implementation of population-wide interventions in Italy. *Nat. Med.* **2020**, *26*, 855–860.
9. Kucharski, A.J.; Russell, T.W.; Diamond, C.; Liu, Y.; Edmunds, J.; Funk, S.; Eggo, R.M. Early dynamics of transmission and control of COVID-19: A mathematical modelling study. *Lancet Infect. Dis.* **2020**, *20*, 553–558. [CrossRef]
10. Davies, N.G.; Klepac, P.; Liu, Y.; K; Prem; Jit, M.; Eggo, R.M. Age-dependent effects in the transmission and control of COVID-19 epidemics. *Nat. Med.* **2020**, *26*, 1205–1211. [CrossRef]
11. Paradisi, M.; Rinaldi, G. An empirical estimate of the infection fatality rate of COVID-19 from the first Italian outbreak. *The Lancet Infectious Diseases*. Available online: https://ssrn.com/abstract=3582811 (accessed on 15 September 2021).
12. Kimathi, M.; Mwalili, S.; Ojiambo, V.; Gathungu, D.K. Age-structured model for COVID-19: Effectiveness of social distancing and contact reduction in Kenya. *Infect. Dis. Model.* **2021**, *6*, 15–23. [CrossRef]
13. Del Valle, S.Y.; Hyman, J.M.; Chitnis, N. Mathematical models of contact patterns between age groups for predicting the spread of infectious diseases. *Math. Biosci. Eng.* **2013**, *10*, 2013.
14. Lyra, W., Jr.; Belkhiria, J.; de Almeida, L.; Chrispim, P.P.M.; de Andrade, I. COVID-19 pandemics modeling with modified determinist SEIR, social distancing, and age stratification. The effect of vertical confinement and release in Brazil. *PLoS ONE* **2020**, *15*, e0237627. [CrossRef]
15. Balabdaoui, F.; Mohr, D. Age-stratified discrete compartment model of the COVID-19 epidemic with application to Switzerland. *Sci. Rep.* **2020**, *10*, 21306. [CrossRef]
16. Bentout, S.; Tridane, A.; Djilali, S.; Touaoula, T.M. Age-structured modeling of COVID-19 epidemic in the USA, UAE and Algeria. *Alex. Eng. J.* **2021**, *60*, 401–411. [CrossRef]
17. Colombo, R.M.; Garavello, M.; Marcellini, F.; Rossi, E. An age and space structured SIR model describing the Covid-19 pandemic. *J. Math. Ind.* **2020**, *10*, 22. [CrossRef]
18. Dudel, C.; Riffe, T.; Acosta, E.; van Raalte, A.; Strozza, C.; Myrskylä, M. Monitoring trends and differences in COVID-19 case-fatality rates using decomposition methods: Contributions of age structure and age-specific fatality. *PLoS ONE* **2020**, *15*, e0238904. [CrossRef]
19. Bubar, K.M.; Reinholt, K.; Kissler, S.M.; Lipsitch, M.; Cobey, S.; Grad, Y.H.; Larremore, D.B. Model-informed COVID-19 vaccine prioritization strategies by age and serostatus. *Science* **2021**, *317*, 916–921. [CrossRef]
20. Ram, V.; Schaposnik, L.P. A modifed age-structured SIR model for COVID-19 type viruses. *Sci. Rep.* **2021**, *11*, 15194. [CrossRef]
21. Chikina, M.; Pegden, W. Modeling strict age-targeted mitigation strategies for COVID-19. *PLoS ONE* **2020**, *15*, e0236237. [CrossRef]
22. Della Rossa, F.; Salzano, D.; di Meglio, A.; de Lellis, F.; Coraggio, M.; Calabrese, C.; Guarino, A.; Cardona-Rivera, R.; de Lellis, P.; Liuzza, D.; et al. A network model of Italy shows that intermittent regional strategies can alleviate the COVID-19 epidemic. *Nat. Commun.* **2020**, *11*, 5106. [CrossRef]
23. Kermack, W.O.; McKendrick, A.G. A contribution to the mathematical theory of epidemics. *Proc. R. Soc. Lond. Ser. A* **1927**, *115*, 700–721.
24. Gatto, M.; Bertuzzo, E.; Mari, L.; Miccoli, S.; Carraro, L.; Casagrandi, R.; Rinaldo, A. Spread and dynamics of the COVID-19 epidemic in Italy: Effects of emergency containment measures. *Proc. Natl. Acad. Sci. USA* **2020**, *117*, 10484-10491. [CrossRef] [PubMed]
25. Gábor, A.; Villaverde, A.F.; Banga, J.R. Parameter identifiability analysis and visualization in large-scale kinetic models of biosystems. *BMC Syst. Biol.* **2017**, *11*, 1–16. [CrossRef] [PubMed]
26. Saltelli, A.; Tarantola, S.; Campolongo, F. Sensitivity analysis as an ingredient of modeling. *Stat. Sci.* **2000**, *15*, 377–395.
27. Turányi, T. Sensitivity analysis of complex kinetic systems. Tools and applications. *J. Math. Chem.* **1990**, *5*, 203–248. [CrossRef]
28. Cori, A.; Ferguson, N.M.; Fraser, C.; Cauchemez, S. A new framework and software to estimate time-varying reproduction numbers during epidemics. *Am. J. Epidemiol.* **2013**, *178*, 1505–1512. [CrossRef]
29. Caselli, G.; Egidi, V. Gender differences in COVID-19 cases and death rates in Italy. *Ital. J. Gend.-Specif. Med.* **2020**, *6*, 96–99.

Article

On Predictive Modeling Using a New Flexible Weibull Distribution and Machine Learning Approach: Analyzing the COVID-19 Data

Zubair Ahmad [1], Zahra Almaspoor [1,*], Faridoon Khan [2] and Mahmoud El-Morshedy [3,4]

1. Department of Statistics, Yazd University, Yazd P.O. Box 89175-741, Iran; zubair@stu.yazd.ac.ir
2. PIDE School of Economics, Islamabad 44000, Pakistan; faridoonkhan_18@pide.edu.pk
3. Department of Mathematics, College of Science and Humanities in Al-Kharj, Prince Sattam Bin Abdulaziz University, Al-Kharj 11942, Saudi Arabia; mah_elmorshedy@mans.edu.eg
4. Department of Mathematics, Faculty of Science, Mansoura University, Mansoura 35516, Egypt
* Correspondence: z.almaspoor@stu.yazd.ac.ir

Abstract: Predicting and modeling time-to-events data is a crucial and interesting research area. For modeling and predicting such types of data, numerous statistical models have been suggested and implemented. This study introduces a new statistical model, namely, a new modified flexible Weibull extension (NMFWE) distribution for modeling the mortality rate of COVID-19 patients. The introduced model is obtained by modifying the flexible Weibull extension model. The maximum likelihood estimators of the NMFWE model are obtained. The evaluation of the estimators of the NMFWE model is assessed in a simulation study. The flexibility and applicability of the NMFWE model are established by taking two datasets representing the mortality rates of COVID-19-infected persons in Mexico and Canada. For predictive modeling, we consider two pure statistical models and two machine learning (ML) algorithms. The pure statistical models include the autoregressive moving average (ARMA) and non-parametric autoregressive moving average (NP-ARMA), and the ML algorithms include neural network autoregression (NNAR) and support vector regression (SVR). To evaluate their forecasting performance, three standard measures of accuracy, namely, root mean square error (RMSE), mean absolute error (MAE), and mean absolute percentage error (MAPE) are calculated. The findings demonstrate that ML algorithms are very effective at predicting the mortality rate data.

Keywords: flexible Weibull extension; mortality rate; COVID-19 event; simulation; statistical modeling

MSC: 62N01; 62N02

1. Introduction

The coronavirus disease 2019 (COVID-19) pandemic has strongly affected the schedule of everyday life; particularly, it has created public health crises the likes of which we have never before faced. Biomedical researchers are constantly paying attention to estimating and predicting the average of new cases, the number or ratio of deaths, or the rate of recovery of the infected patients to make the appropriate arrangements (Hogan et al. [1]). In this regard, several studies on the COVID-19 pandemic have appeared. For example, Mizumoto et al. [2] estimated the asymptomatic proportion of COVID-19 cases in Japan. Ilyas et al. [3] studied the scenario of the COVID-19 pandemic in Pakistan. Rao et al. [4] investigated COVID-19 data using the Weibull distribution under indeterminacy. Up to 27 November 2021, 10:53 GMT, the total number of registered cases has reached 261 million, the total number of deaths around the globe has reached 5.2 million, and 235.86 million infected persons have recovered. Based on the latest updates about the COVID-19 pandemic, the United States of America is at the top of the list, having 49 million total cases and 799,138 deaths.

Several statistical models (SMs) have been implemented to describe, estimate, and predict the nature of the COVID-19 pandemic. For example, Singhal et al. [5] modeled and predicted the COVID-19 epidemic using the Gaussian model. Qin et al. [6] estimated the distribution of the incubation period of COVID-19 events. Almetwally et al. [7] implemented a new version of the inverted Topp–Leone (ITL) distribution to analyze the COVID-19 mortality rate. Almongy et al. [8] applied an extended version of the Rayleigh distribution to the COVID-19 data. Liu et al. [9] modeled the survival times of COVID-19-infected persons in China. El-Sagheer et al. [10] applied the mortality distribution to the COVID-19 data based on randomly censored observations.

In today's competitive era, the data generated from various fields are becoming increasingly more complex. As a result, in modeling such data, we need machine learning tools under probability distributions that are best suited for analytical studies of multidimensional and complex data. Machine learning algorithms are often expressed in terms of probability; most machine learning tools are based on inferential statistics, where the statistics are based on probability theory. In essence, probability theory mathematically expresses how likely something is, given our assumptions. There are now probabilistic interpretations of black-box algorithms such as deep learning. These interpretations help us understand how such algorithms work, and how to improve them. There are many researchers who base their entire models for computer learning on statistics. At a basic level, they think that the world, or at least their problem, is driven by or best represented by certain combinations of random variables, which are best expressed by statistics. These models are typically suited to very different types of problems than are multifactor models. Furthermore, forecasting stock prices is a good example, where statistical models can be used for machine learning. Thus, machine learning is more linked to statistics and probabilities. See, for example, Eliwa et al. [11], El-Morshedy et al. [12,13], Altun et al. [14,15], among others.

In the current scenario, the best description of the COVID-19 pandemic is a crucial research topic. Several SMs are available that can be used to describe the behavior of the COVID-19 pandemic adequately, in addition to machine learning tools. Among the available SMs, the two-parameter flexible Weibull extension (FWE) model holds a key place (see Bebbington et al. [16]). Different variants of the FWE model have been introduced and implemented for dealing with the data in numerous sectors; see El-Morshedy et al. [17], El-Morshedy et al. [18], and Abubakari et al. [19].

Let a random variable W have the FWE model with parameters $\sigma_1 > 0$ and $\sigma_2 > 0$; its cumulative distribution function (CDF) can be expressed as

$$K(w; \sigma_1, \sigma_2) = 1 - e^{-Y(w;\sigma_1,\sigma_2)}, \quad w \geq 0, \quad (1)$$

with the probability density function (PDF) given by

$$k_{FWE}(w; \sigma_1, \sigma_2) = \left(\sigma_1 + \frac{\sigma_2}{w^2}\right) Y(w; \sigma_1, \sigma_2) e^{-Y(w;\sigma_1,\sigma_2)}, \quad w > 0,$$

where $Y(w; \sigma_1, \sigma_2) = e^{\sigma_1 w - \frac{\sigma_2}{w}}$. To add further flexibility to the FWE model, El-Gohary et al. [20] proposed the exponentiated FWE (Exp-FWE) model with parameters $\sigma_1 > 0, \sigma_2 > 0$, and $\delta_1 > 0$. The CDF of the Exp-FWE model is given by

$$K(w; \sigma_1, \sigma_2, \delta_1) = \left(1 - e^{-Y(w;\sigma_1,\sigma_2)}\right)^{\delta_1}, \quad w \geq 0.$$

El-Damcese et al. [21] further modified the Exp-FWE model by introducing the Kumaramswamy FWE (Ku-FWE) model with parameters $\sigma_1 > 0, \sigma_2 > 0, \delta_1$, and $\delta_2 > 0$. The CDF $K(w; \sigma_1, \sigma_2, \delta_1, \delta_2)$ of the Ku-FWE model is given by

$$K(w; \sigma_1, \sigma_2, \delta_1, \delta_2) = 1 - \left(1 - \left[1 - e^{-Y(w;\sigma_1,\sigma_2)}\right]^{\delta_1}\right)^{\delta_2}, \quad w \geq 0.$$

Recently, Ahmad et al. [22] further contributed to this research area by proposing a new family of distributions with CDF, given by

$$M(w;\lambda,\boldsymbol{\vartheta}) = \frac{\lambda K(w;\boldsymbol{\vartheta})}{\lambda - 1 + K(w;\boldsymbol{\vartheta})}, \quad w \in \mathbb{R}, \lambda > 1, \quad (2)$$

where $K(w;\boldsymbol{\vartheta})$ is the CDF of the baseline model with parameter vector $\boldsymbol{\vartheta}$. The corresponding PDF, survival function (SF), and hazard function (HF) to Equation (2) are given by

$$m(w;\lambda,\boldsymbol{\vartheta}) = \frac{\lambda(\lambda-1)k(w;\boldsymbol{\vartheta})}{[\lambda - 1 + K(w;\boldsymbol{\vartheta})]^2}, \quad w \in \mathbb{R},$$

$$S(w;\lambda,\boldsymbol{\vartheta}) = 1 - \frac{\lambda K(w;\boldsymbol{\vartheta})}{\lambda - 1 + K(w;\boldsymbol{\vartheta})}, \quad w \in \mathbb{R},$$

and

$$h(w;\lambda,\boldsymbol{\vartheta}) = \frac{\lambda k(w;\boldsymbol{\vartheta})}{(1 - K(w;\boldsymbol{\vartheta}))[\lambda - 1 + K(w;\boldsymbol{\vartheta})]}, \quad w \in \mathbb{R},$$

respectively.

As we know, heavy-tailed (HT) distributions play a vital role in medical and other related sectors ((Gardiner et al. [23]), (Zhao et al. [24])). However, in the literature, there are only few distributions that possess the HT characteristics ((Bhati and Ravi, [25]), (Ahmad et al. [26]), (Ahmad et al. [27])). Keeping in view the importance of the HT distributions, we introduce a new HT distribution, namely, a new modified flexible Weibull extension (NMFWE) distribution. The HT characteristics of the NMFWE distributions are proved mathematically (see Section 3). The NMFWE distribution is introduced by incorporating $K(w;\sigma_1,\sigma_2) = 1 - e^{-Y(w;\sigma_1,\sigma_2)}$ in Equation (2).

2. A New Modified Flexible Weibull Extension

A random variable W has the NMFWE distribution with parameters $\lambda > 1, \sigma_1 > 0$, and $\sigma_2 > 0$, if its CDF can be formulated as

$$M(w;\lambda,\sigma_1,\sigma_2) = \frac{\lambda - \lambda e^{-Y(w;\sigma_1,\sigma_2)}}{\lambda - e^{-Y(w;\sigma_1,\sigma_2)}}, \quad w \geq 0. \quad (3)$$

In link to $M(w;\lambda,\sigma_1,\sigma_2)$, the PDF and HF can be expressed as

$$m(w;\lambda,\sigma_1,\sigma_2) = \frac{\lambda(\lambda-1)\left(\sigma_1 + \frac{\sigma_2}{w^2}\right)Y(w;\sigma_1,\sigma_2)e^{-Y(w;\sigma_1,\sigma_2)}}{\left[\lambda - e^{-Y(w;\sigma_1,\sigma_2)}\right]^2}, \quad w > 0, \quad (4)$$

and

$$h(w;\lambda,\sigma_1,\sigma_2) = \frac{\lambda\left(\sigma_1 + \frac{\sigma_2}{w^2}\right)Y(w;\sigma_1,\sigma_2)}{\lambda - e^{-Y(w;\sigma_1,\sigma_2)}}, \quad w > 0,$$

respectively.

For different values of λ, σ_1, and σ_2, visual illustrations of $m(w;\lambda,\sigma_1,\sigma_2)$, and $h(w;\lambda,\sigma_1,\sigma_2)$ are presented in Figure 1.

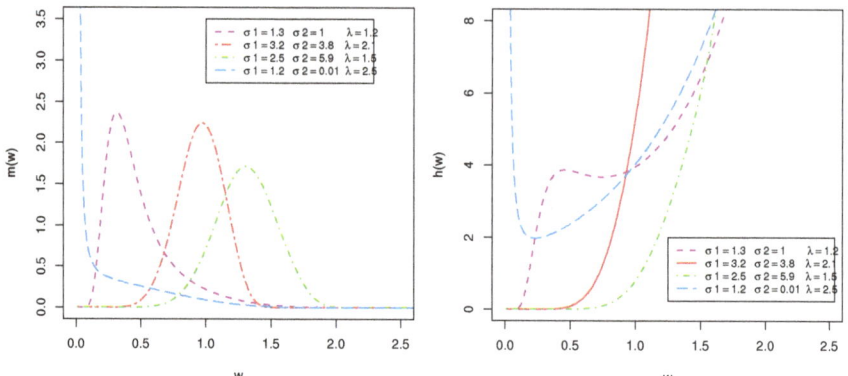

Figure 1. The PDF (**left panel**) and HRF (**right panel**) plots for the NMFWE distribution.

It is found that the proposed model can be used effectively in modeling symmetric and asymmetric data. Moreover, it can be utilized as a probability tool to discuss various kinds of failure rates.

3. The HT Characteristics

This section is devoted to proving the HT characteristics of the NMFWE distribution.

Regular Variational Property

Here, we prove the regular variational property of the NMFWE distribution. According to Seneta [28], in terms of SF $S(w; \boldsymbol{\vartheta}) = 1 - K(w; \boldsymbol{\vartheta})$, we have:

Theorem 1. *If $[1 - K(w; \boldsymbol{\vartheta})]$ is the SF of the regular varying distribution (RVD), then $[1 - M(w; \lambda, \sigma_1, \sigma_2)]$ is an RVD.*

Proof. Suppose $\lim_{w \to \infty} \frac{[1-K(aw;\boldsymbol{\vartheta})]}{[1-K(w;\boldsymbol{\vartheta})]} = f(w)$ is finite but nonzero for every $w > 0$. Incorporating Equation (5), we obtain

$$\lim_{w \to \infty} \frac{1 - M(aw; \lambda, \sigma_1, \sigma_2)}{1 - M(w; \lambda, \sigma_1, \sigma_2)} = \lim_{w \to \infty} \frac{(\lambda - 1)[1 - K(aw; \boldsymbol{\vartheta})]}{\lambda - 1 + K(aw; \boldsymbol{\vartheta})} \times \frac{\lambda - 1 + K(w; \boldsymbol{\vartheta})}{(\lambda - 1)[1 - K(w; \boldsymbol{\vartheta})]},$$

$$\lim_{w \to \infty} \frac{1 - M(aw; \lambda, \sigma_1, \sigma_2)}{1 - M(w; \lambda, \sigma_1, \sigma_2)} = \lim_{w \to \infty} \frac{[1 - K(aw; \boldsymbol{\vartheta})]}{[1 - K(w; \boldsymbol{\vartheta})]} \times \frac{\lambda - 1 + K(w; \boldsymbol{\vartheta})}{\lambda - 1 + K(aw; \boldsymbol{\vartheta})},$$

$$\lim_{w \to \infty} \frac{1 - M(aw; \lambda, \sigma_1, \sigma_2)}{1 - M(w; \lambda, \sigma_1, \sigma_2)} = \lim_{w \to \infty} f(w) \times \frac{\lambda - 1 + K(w; \boldsymbol{\vartheta})}{\lambda - 1 + K(aw; \boldsymbol{\vartheta})}. \qquad (5)$$

Using Equation (3) in Equation (5), we obtain

$$\lim_{w \to \infty} \frac{1 - M(aw; \lambda, \sigma_1, \sigma_2)}{1 - M(w; \lambda, \sigma_1, \sigma_2)} = \lim_{w \to \infty} f(w) \times \frac{\lambda - 1 + \left(1 - e^{-e^{\sigma_1 w} - \frac{\sigma_2}{w}}\right)}{\lambda - 1 + \left(1 - e^{-e^{\sigma_1 (aw)} - \frac{\sigma_2}{(aw)}}\right)},$$

$$\lim_{w \to \infty} \frac{1 - M(aw; \lambda, \sigma_1, \sigma_2)}{1 - M(w; \lambda, \sigma_1, \sigma_2)} = \lim_{w \to \infty} f(w) \times \frac{\lambda - 1 + \left(1 - e^{-e^{\sigma_1 \times \infty} - \frac{\sigma_2}{\infty}}\right)}{\lambda - 1 + \left(1 - e^{-e^{\sigma_1 (a \times \infty)} - \frac{\sigma_2}{(a \times \infty)}}\right)},$$

$$\lim_{w\to\infty}\frac{1-M(aw;\lambda,\sigma_1,\sigma_2)}{1-M(w;\lambda,\sigma_1,\sigma_2)}=\lim_{w\to\infty}f(w)\times\frac{\lambda-1+\left(1-e^{-e^\infty}\right)}{\lambda-1+(1-e^{-e^\infty})},$$

$$\lim_{w\to\infty}\frac{1-M(aw;\lambda,\sigma_1,\sigma_2)}{1-M(w;\lambda,\sigma_1,\sigma_2)}=\lim_{w\to\infty}f(w)\times\frac{\lambda-1+(1-e^{-\infty})}{\lambda-1+(1-e^{-\infty})},$$

$$\lim_{w\to\infty}\frac{1-M(aw;\lambda,\sigma_1,\sigma_2)}{1-M(w;\lambda,\sigma_1,\sigma_2)}=\lim_{w\to\infty}f(w)\times\frac{\lambda-1+\left(1-\frac{1}{e^\infty}\right)}{\lambda-1+\left(1-\frac{1}{e^\infty}\right)},$$

$$\lim_{w\to\infty}\frac{1-M(aw;\lambda,\sigma_1,\sigma_2)}{1-M(w;\lambda,\sigma_1,\sigma_2)}=\lim_{w\to\infty}f(w)\times\frac{\lambda-1+1}{\lambda-1+1},$$

$$\lim_{w\to\infty}\frac{1-M(aw;\lambda,\sigma_1,\sigma_2)}{1-M(w;\lambda,\sigma_1,\sigma_2)}=\lim_{w\to\infty}f(w). \tag{6}$$

Since Equation (6) is nonzero for every $w>0$, $[1-M(w;\lambda,\sigma_1,\sigma_2)]$ is the SF of the RVD. □

A Supportive Example of RVP

Suppose W follows a power-law behavior; then, as per the definition of the HT property, we have

$$1-K(w;\boldsymbol{\vartheta})=\mathbb{P}(W>w)\sim w^{-\beta}.$$

By implementing Karamata's characterization theorem (Seneta, [28]), we can write the expression $[1-M(aw;\lambda,\sigma_1,\sigma_2)]$ as

$$1-M(w;\lambda,\sigma_1,\sigma_2)=w^{-\beta}L(w),$$

where the quantity $L(w)$ represents the slowly varying function (SVF). From Equation (5), we have

$$1-M(w;\lambda,\sigma_1,\sigma_2)=\frac{[1-K(w;\boldsymbol{\vartheta})](\lambda-1)}{\lambda-1+K(w;\boldsymbol{\vartheta})},$$

$$1-M(w;\lambda,\sigma_1,\sigma_2)=\frac{w^{-\beta}(\lambda-1)}{\lambda-1+K(w;\boldsymbol{\vartheta})},$$

$$1-M(w;\lambda,\sigma_1,\sigma_2)=w^{-\beta}L(w), \tag{7}$$

where $L(w)=\frac{(\lambda-1)}{\lambda-1+K(w;\boldsymbol{\vartheta})}$. If we can show that $L(w)$ is an SVF, then the result obtained in Equation (7) is true. To show $L(w)$ is an SVF, we must satisfy

$$\lim_{z\to\infty}\frac{L(aw)}{L(w)}=1.$$

So,

$$\frac{L(aw)}{L(w)}=\frac{\frac{(\lambda-1)}{\lambda-1+K(aw;\boldsymbol{\vartheta})}}{\frac{(\lambda-1)}{\lambda-1+K(w;\boldsymbol{\vartheta})}},$$

$$\frac{L(aw)}{L(w)}=\frac{\lambda-1+K(w;\boldsymbol{\vartheta})}{\lambda-1+K(aw;\boldsymbol{\vartheta})},$$

$$\frac{L(aw)}{L(w)}=\frac{\lambda-1+\left(1-e^{-e^{\sigma_1\times w-\frac{\sigma_2}{w}}}\right)}{\lambda-1+\left(1-e^{-e^{\sigma_1(a\times w)-\frac{\sigma_2}{a\times w}}}\right)}.$$

Applying the limit, we obtain

$$\lim_{w\to\infty}\frac{L(aw)}{L(w)} = \frac{\lambda - 1 + \left(1 - e^{-e^{\sigma_1 \times \infty} - \frac{\sigma_2}{\infty}}\right)}{\lambda - 1 + \left(1 - e^{-e^{\sigma_1(a\times\infty)} - \frac{\sigma_2}{a\times\infty}}\right)},$$

$$\lim_{w\to\infty}\frac{L(aw)}{L(w)} = \frac{\lambda - 1 + 1}{\lambda - 1 + 1},$$

$$\lim_{w\to\infty}\frac{L(aw)}{L(w)} = 1.$$

4. Estimation and Simulation

In this section, we adopt a known estimation procedure to obtain the maximum likelihood estimators (MLEs) $(\hat{\sigma}_1, \hat{\sigma}_2, \hat{\lambda})$ of the parameters $(\sigma_1, \sigma_2, \lambda)$. After obtaining the MLEs of the parameters, we conduct a simulation study (SimS) to assess the performances of the estimators.

Let $W_1, W_2, \cdots, W_n$ be an observed random sample (RS) of size n, taken from $m(w; \lambda, \sigma_1, \sigma_2)$. In link to $m(w; \lambda, \sigma_1, \sigma_2)$, the likelihood function (LiF), say $\Delta(\lambda, \sigma_1, \sigma_2 | w_1, w_2, \cdots, w_n)$, is given by

$$\Delta(\lambda, \sigma_1, \sigma_2 | w_1, w_2, \cdots, w_n) = \prod_{a=1}^{n} \frac{\lambda(\lambda - 1)\left(\sigma_1 + \frac{\sigma_2}{w_a^2}\right) Y_a(w_a; \sigma_1, \sigma_2) e^{-Y_a(w_a; \sigma_1, \sigma_2)}}{\left[\lambda - e^{-Y_a(w_a; \sigma_1, \sigma_2)}\right]^2}, \quad (8)$$

where $Y_a(w_a; \sigma_1, \sigma_2) = e^{\sigma_1 w_a - \frac{\sigma_2}{w_a}}$. The corresponding log LiF to $\Delta(\lambda, \sigma_1, \sigma_2 | w_1, w_2, \cdots, w_n)$ can be formulated as

$$\delta(\lambda, \sigma_1, \sigma_2 | w_1, w_2, \cdots, w_n) = n \log \lambda + n \log(1 - \lambda) + \sum_{a=1}^{n} \log\left(\sigma_1 + \frac{\sigma_2}{w_a^2}\right) + \sum_{a=1}^{n} \sigma_1 w_a$$

$$- \sum_{a=1}^{n} \frac{\sigma_2}{w_a} - \sum_{a=1}^{n} Y_a(w_a; \sigma_1, \sigma_2) - 2 \sum_{a=1}^{n} \log\left[\lambda - e^{-Y_a(w_a; \sigma_1, \sigma_2)}\right].$$

Based on $\delta(\lambda, \sigma_1, \sigma_2 | w_1, w_2, \cdots, w_n)$, the partial derivatives are given by

$$\frac{\partial}{\partial \sigma_1}\delta(\lambda, \sigma_1, \sigma_2 | w_1, w_2, \cdots, w_n) = \sum_{a=1}^{n} \frac{1}{\left(\sigma_1 + \frac{\sigma_2}{w_a^2}\right)} + \sum_{a=1}^{n} w_a - \sum_{a=1}^{n} w_a Y_a(w_a; \sigma_1, \sigma_2)$$

$$- 2 \sum_{a=1}^{n} \frac{w_a Y_a(w_a; \sigma_1, \sigma_2) e^{-Y_a(w_a; \sigma_1, \sigma_2)}}{\left[\lambda - e^{-Y_a(w_a; \sigma_1, \sigma_2)}\right]},$$

$$\frac{\partial}{\partial \sigma_2}\delta(\lambda, \sigma_1, \sigma_2 | w_1, w_2, \cdots, w_n) = \sum_{a=1}^{n} \frac{\frac{1}{w_a}}{\left(\sigma_1 + \frac{\sigma_2}{w_a^2}\right)} - \sum_{a=1}^{n} \frac{1}{w_a} + \sum_{a=1}^{n} \frac{1}{w_a} Y_a(w_a; \sigma_1, \sigma_2)$$

$$+ 2 \sum_{a=1}^{n} \frac{\frac{1}{w_a} Y_a(w_a; \sigma_1, \sigma_2) e^{-Y_a(w_a; \sigma_1, \sigma_2)}}{\left[\lambda - e^{-Y_a(w_a; \sigma_1, \sigma_2)}\right]},$$

and

$$\frac{\partial}{\partial \lambda}\delta(\lambda, \sigma_1, \sigma_2 | w_1, w_2, \cdots, w_n) = \frac{n}{\lambda} + \frac{n}{(1 - \lambda)} - 2 \sum_{a=1}^{n} \frac{1}{\left[\lambda - e^{-Y_a(w_a; \sigma_1, \sigma_2)}\right]},$$

respectively.

Solving $\frac{\partial}{\partial \sigma_1}\delta(\lambda,\sigma_1,\sigma_2|w_1,w_2,\cdots,w_n) = 0$, $\frac{\partial}{\partial \sigma_2}\delta(\lambda,\sigma_1,\sigma_2|w_1,w_2,\cdots,w_n) = 0$, and $\frac{\partial}{\partial \lambda}\delta(\lambda,\sigma_1,\sigma_2|w_1,w_2,\cdots,w_n) = 0$ yields $\hat{\sigma}_1, \hat{\sigma}_2$, and $\hat{\lambda}$, respectively.

Next, we assess the performances of $\hat{\sigma}_1, \hat{\sigma}_2$, and $\hat{\lambda}$ via an SimS. For carrying out the SimS, an RS, say $n = 25, 50, \cdots, 500$, was obtained from the NMFWE model. The SimS was performed for two schemes as follows: scheme I: $\sigma_1 = 0.6, \sigma_2 = 1.4, \lambda = 1.1$); scheme II: $\sigma_1 = 1.1, \sigma_2 = 1.6, \lambda = 1.4$. Furthermore, two evaluation criteria, bias and mean square error (MSE), were considered for assessing $\hat{\sigma}_1, \hat{\sigma}_2$, and $\hat{\lambda}$. These criteria were, respectively, computed using the below expressions:

$$Bias(\hat{\Theta}) = \frac{1}{n}\sum_{a=1}^{n}\left(\hat{\Theta} - \Theta\right),$$

and

$$MSE(\hat{\Theta}) = \frac{1}{n}\sum_{a=1}^{n}\left(\hat{\Theta} - \Theta\right)^2,$$

where $\Theta = (\sigma_1, \sigma_2, \lambda)$.

Corresponding to scheme I, the results of the SimS are provided in Table 1 and presented visually in Figure 2, whereas Table 2 (numerical illustration) and Figure 3 (visual illustration) offer the results of the SimS for schema II. The SimS was performed with the objective that (i) as the value of n increases, the values of $\hat{\sigma}_1, \hat{\sigma}_2$, and $\hat{\lambda}$ tend to stability, and (ii), the biases and mean square errors tend to zero as the sample size grows; this proves the consistency property for the estimators. Thus, we can conclude that the maximum likelihood approach works quite well in estimating the model parameters under various sample sizes.

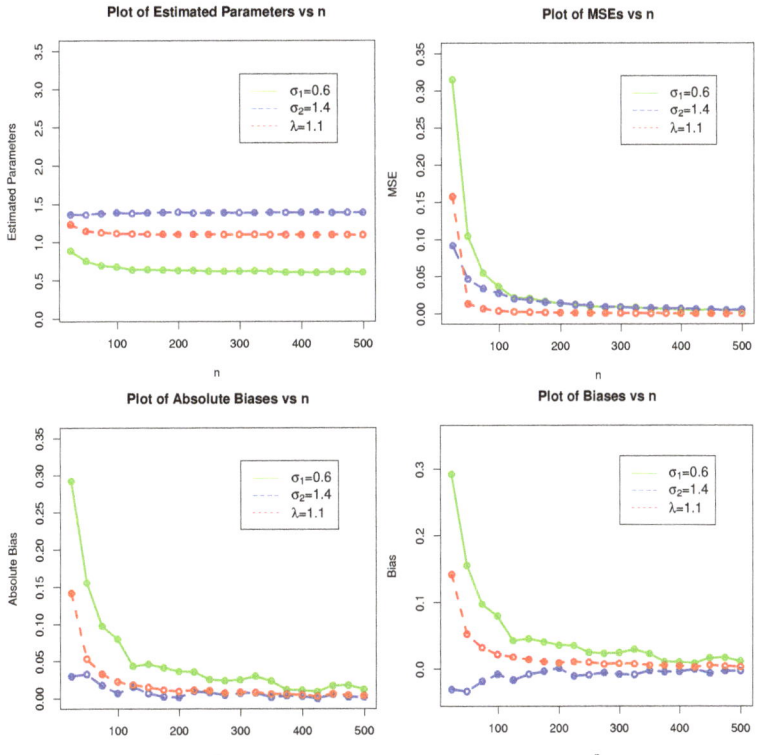

Figure 2. A visual display of the results of the SimS of the NMFWE model for $\sigma_1 = 0.6, \sigma_2 = 1.4$, and $\lambda = 1.1$.

Table 1. The results of the SimS of the NMFWE model for $\sigma_1 = 0.6, \sigma_2 = 1.4$, and $\lambda = 1.1$.

n	Parameters	MLEs	MSEs	Biases
25	σ_1	0.89302560	0.315274822	0.293025560
	σ_2	1.36985000	0.092232346	−0.03015010
	λ	1.24237000	0.158247051	0.142369775
50	σ_1	0.75604460	0.105204950	0.156044587
	σ_2	1.36711800	0.047163494	−0.03288202
	λ	1.15347600	0.013885900	0.053475874
75	σ_1	0.69801090	0.055053130	0.098010942
	σ_2	1.38220000	0.034336224	−0.01779995
	λ	1.13308800	0.007274670	0.033088078
100	σ_1	0.68048360	0.037073941	0.080483598
	σ_2	1.39280200	0.028143866	−0.00719760
	λ	1.12277300	0.004239282	0.022773367
150	σ_1	0.64644740	0.021180684	0.046447381
	σ_2	1.39273800	0.018909237	−0.00726232
	λ	1.11521500	0.002497769	0.015214547
200	σ_1	0.63668930	0.014568689	0.036689263
	σ_2	1.40180900	0.014734874	0.001809424
	λ	1.10985700	0.001831123	0.009857116
250	σ_1	0.62596330	0.010368965	0.025963329
	σ_2	1.39177300	0.011995360	−0.00822708
	λ	1.11065800	0.001393079	0.010658350
300	σ_1	0.62550910	0.009587663	0.025509126
	σ_2	1.39270200	0.009345810	−0.00729828
	λ	1.10861100	0.001066879	0.008610622
350	σ_1	0.62391800	0.006948299	0.023917996
	σ_2	1.39811500	0.008009759	−0.00188529
	λ	1.10609300	0.000808322	0.006093125
400	σ_1	0.61132940	0.005626040	0.011329370
	σ_2	1.39673000	0.007298053	−0.00327033
	λ	1.10493100	0.000693269	0.004930592
450	σ_1	0.61765410	0.005751539	0.017654106
	σ_2	1.39418200	0.006140556	−0.00581764
	λ	1.10637100	0.000606152	0.006371330
500	σ_1	0.61250040	0.004853576	0.012500388
	σ_2	1.39779400	0.006180866	−0.00220605
	λ	1.10388200	0.000549189	0.003882268

Table 2. The results of the SimS of the NMFWE model for $\sigma_1 = 1.1, \sigma_2 = 1.6$, and $\lambda = 1.4$.

n	Parameters	MLEs	MSEs	Biases
25	σ_1	1.21214000	0.109696162	0.112139576
	σ_2	1.65247100	0.211066950	0.052471477
	λ	2.22540700	2.677630360	0.825406630
50	σ_1	1.16302800	0.046138277	0.063028001
	σ_2	1.60940600	0.127701240	0.009405599
	λ	1.96382400	1.630116700	0.563823750
75	σ_1	1.14084100	0.029142804	0.040841203
	σ_2	1.61535700	0.080794640	0.015357220
	λ	1.70831400	0.785862990	0.308313730

Table 2. Cont.

n	Parameters	MLEs	MSEs	Biases
100	σ_1	1.12400600	0.018225475	0.024006192
	σ_2	1.60450400	0.065363500	0.004504030
	λ	1.63171300	0.494613440	0.231713380
150	σ_1	1.11539700	0.012912002	0.015397237
	σ_2	1.59474700	0.048261320	−0.005252810
	λ	1.55643900	0.268352060	0.156439100
200	σ_1	1.12505500	0.009975488	0.025054780
	σ_2	1.59002600	0.031762970	−0.00997431
	λ	1.53205300	0.172517940	0.132052860
250	σ_1	1.10665200	0.007469320	0.006651779
	σ_2	1.61439400	0.028265080	0.014393549
	λ	1.45660500	0.079107080	0.056605210
300	σ_1	1.10702000	0.005739174	0.007019600
	σ_2	1.60981400	0.022193800	0.009813938
	λ	1.43945700	0.037317540	0.039456630
350	σ_1	1.10944500	0.005131958	0.009445283
	σ_2	1.60127800	0.019648730	0.001277727
	λ	1.45528500	0.056621610	0.055285400
400	σ_1	1.10729900	0.004561712	0.007299329
	σ_2	1.60803700	0.017440070	0.008037465
	λ	1.43753900	0.036744330	0.037538550
450	σ_1	1.10532200	0.003908403	0.005322438
	σ_2	1.60144000	0.016748580	0.001439921
	λ	1.44244400	0.033483900	0.042444350
500	σ_1	1.10438500	0.003518714	0.004384936
	σ_2	1.60492800	0.013190150	0.004928443
	λ	1.43178200	0.024440440	0.031781690

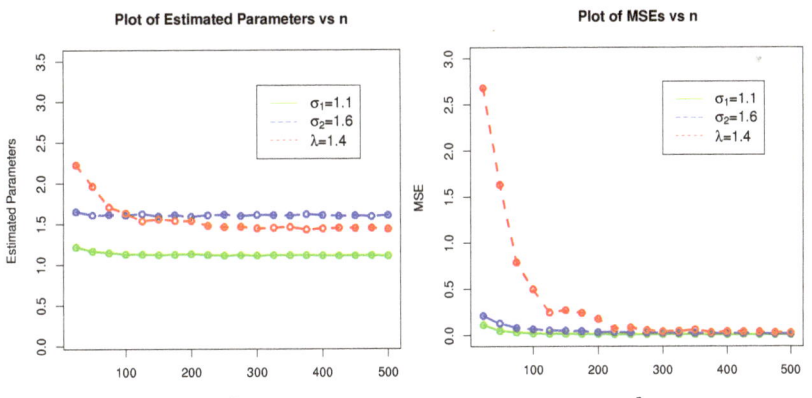

Figure 3. Cont.

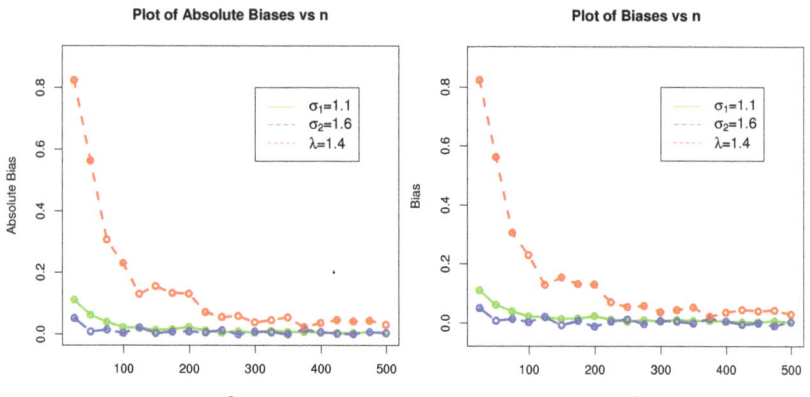

Figure 3. A visual display of the results of the SimS of the NMFWE model for $\sigma_1 = 1.1, \sigma_2 = 1.6$, and $\lambda = 1.4$.

5. Data Analysis

This section deals with data analysis to illustrate the crucial and important role of the NMFWE model in real life data modeling. To show the applicability of the NMFWE model and to carry out its illustration, two datasets from the health sector are considered. The first dataset (Data 1) consists of 106 observations and represents the mortality rate of patients during the COVID-19 pandemic in Mexico. The second dataset (Data 2) consists of 224 observations and represents the mortality rate of patients during the COVID-19 pandemic in Canada. Both datasets are provided in Table 3.

Table 3. The COVID-19 datasets.

Data 1	1.7652, 1.2210, 1.8782, 2.9924, 2.0766, 1.4534, 2.6440, 3.2996, 2.3330, 1.2030, 2.1710, 1.2244, 1.3312, 0.6880, 1.1708, 2.1370, 2.0070, 1.0484, 0.8688, 1.0286, 1.5260, 2.9208, 1.5806, 1.2740, 0.7074, 1.2654, 0.9460, 0.6430, 1.8568, 2.5756, 1.7626, 2.0086, 1.4520, 1.1970, 1.2824, 0.6790, 0.8848, 1.9870, 1.5680, 1.9100, 0.6998, 0.7502, 1.3936, 0.6572, 2.0316, 1.6216, 1.3394, 1.4302, 1.3120, 0.4154, 0.7556, 0.5976, 0.6672, 1.3628, 1.6650, 1.5708, 1.7102, 0.6456, 1.4972, 1.3250, 1.2280, 0.9818, 0.9322, 1.0784, 2.4084, 1.7392, 0.3630, 0.6654, 1.0812, 1.2364, 0.2082, 0.3600, 0.9898, 0.8178, 0.6718, 0.4140, 0.6596, 1.0634, 1.0884, 0.9114, 0.8584, 0.5000, 1.3070, 0.9296, 0.9394, 1.0918, 0.8240, 0.7844, 0.6438, 0.2804, 0.4876, 0.6514, 0.7264, 0.6466, 0.6054, 0.4704, 0.2410, 0.6436, 0.5852, 0.5202, 0.4130, 0.6058, 0.4116, 0.4652, 0.5012, 0.3846
Data 2	0.9636, 2.7852, 3.8628, 2.6436, 3.0120, 2.1780, 1.7952, 1.9236, 1.0176, 1.3272, 2.9796, 2.3520, 2.8644, 1.0488, 1.1244, 2.0904, 0.9852, 3.0468, 2.4324, 2.0088, 2.1444, 1.9680, 0.6228, 1.1328, 0.8964, 1.0008, 2.0436, 2.4972, 2.3556, 2.5644, 0.9684, 2.2452, 1.9872, 1.8420, 1.4724, 1.3980, 1.6176, 3.6120, 2.6088, 0.5436, 0.9972, 1.6212, 1.8540, 0.3120, 0.5400, 1.4844, 1.2264, 1.0068, 0.6204, 0.9888, 1.5948, 1.6320, 1.3668, 1.2876, 0.7500, 1.9596, 1.3944, 1.4088, 1.6368, 1.2360, 1.1760, 0.9648, 0.4200, 0.7308, 0.9768, 1.0896, 0.9696, 0.9072, 0.7056, 0.3612, 0.9648, 0.8772, 0.7800, 0.6192, 0.9084, 0.6168, 0.6972, 0.7512, 0.5760, 5.2956, 3.6624, 5.6340, 8.9772, 6.2292, 4.3596, 7.9320, 9.8988, 6.9984, 3.6084, 6.5124, 3.6732, 3.9936, 2.0640, 3.5124, 6.4104, 6.0204, 3.1452, 2.6064, 3.0852, 4.5780, 8.7624, 4.7412, 3.8220, 2.1216, 3.7956, 2.8380, 1.9284, 5.5704, 7.7268, 5.2872, 6.0252, 4.3560, 3.5904, 3.8472, 2.0364, 2.6544, 5.9604, 4.7040, 5.7300, 2.0988, 2.2500, 4.1808, 1.9716, 6.0948, 4.8648, 4.0176, 5.1300, 1.9368, 4.4916, 3.9744, 3.6840, 2.9448, 2.7960, 3.2352, 7.2252, 5.2176, 1.0884, 1.9956, 3.2436, 3.7092, 0.6240, 1.0800, 2.9688, 2.4528, 2.0148, 1.2420, 1.9788, 3.1896, 3.2652, 2.7336, 2.5752, 1.5000, 3.9204, 2.7888, 2.8176, 3.2748, 2.4720, 2.3532, 1.9308, 0.8412, 1.4628, 1.9536, 2.1792, 1.9392, 1.8156, 1.4112, 0.7224, 1.9308, 1.7556, 1.5600, 1.2384, 1.8168, 1.2348, 1.3956, 1.5036, 1.1532, 4.2360, 2.9304, 4.5072, 7.1808, 4.9836, 3.4872, 6.3456, 7.9188, 5.5992, 2.8872, 5.2104, 2.9376, 3.1944, 1.6512, 2.8092, 5.1288, 4.8168, 2.5152, 2.0844, 2.4684, 3.6624, 7.0092, 3.7932, 3.0576, 1.6968, 3.0360, 2.2704, 1.5432, 4.4556, 6.1812, 4.6764, 1.3188, 3.7068, 6.6516, 3.8244, 3.1848, 3.7476, 4.5180, 5.4912, 7.3872, 3.4908, 3.0804, 3.3684, 4.1184, 3.0912, 1.3176, 3.4884, 4.9176

Corresponding to Data 1, the initial density shape is reported using the non-parametric kernel density estimation (KDE) approach in Figure 4, and it is noted that the density is asymmetric and unimodal. The normality condition is checked via the quantile–quantile (Q–Q) plot in Figure 4. The extremes are spotted using the box plot in Figure 4, and it is showed that some extreme observations were listed. Moreover, Figure 4 indicates that Data 1 has an increasing failure shape, based on the total time test (TTT) plot. For Data 2, the initial density shape, KDE, Q–Q plot, box plot, and TTT plot are presented in Figure 5. From the plots in Figure 5, we can see that the second dataset is unimodal, skewed to the right, and has an increasing failure shape.

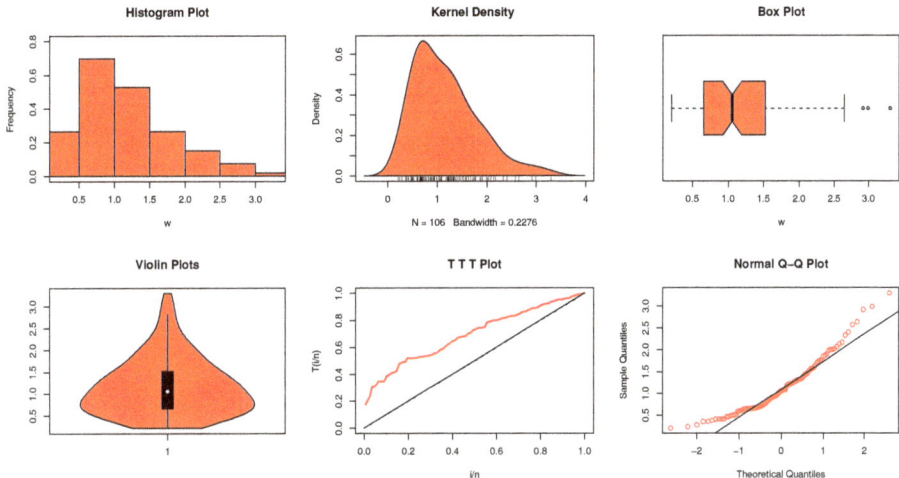

Figure 4. Nonparametric plots for Data 1.

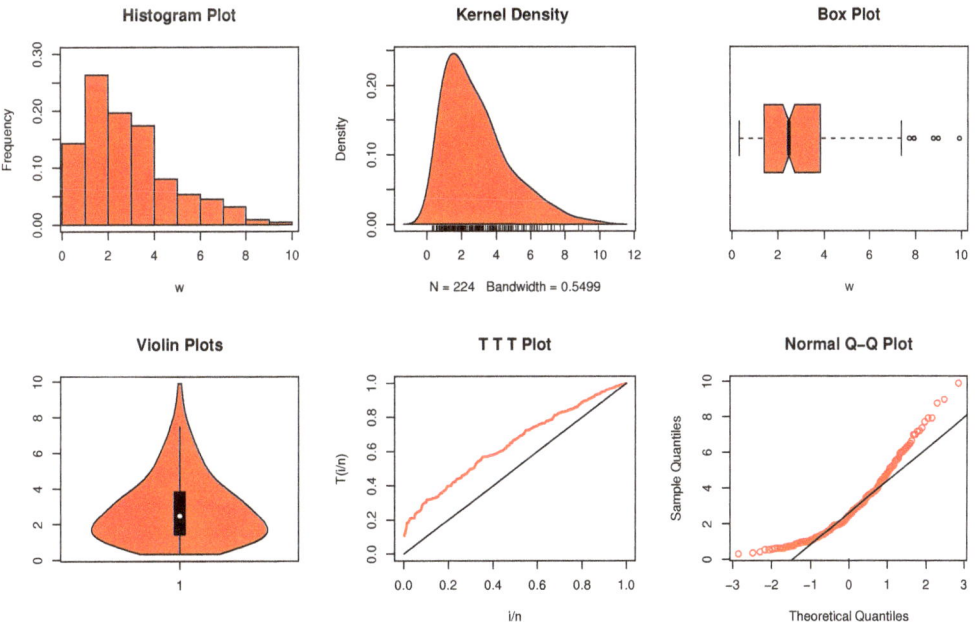

Figure 5. Nonparametric plots for Data 2.

Using the above mortality rate datasets, we show the applicability and best fitting capability of the NMFWE distribution. For this purpose, the comparison of the performance of the NMFWE distribution is made with the baseline FWE model, an exponentiated version of the FWE distribution, namely, an exponentiated FWE (E-FWE), the Weibull model, a generalization of the Weibull model, namely, the exponentiated Weibull (E-Weibull), and another famous extension of the Weibull model called the Kumaraswamy Weibull (K-Weibull) distribution. The SFs of the selected models are

- FWE:
$$M(w; \sigma_1, \sigma_2) = e^{-e^{Y(w;\sigma_1,\sigma_2)}}, \quad w \geq 0,$$
where $\sigma_1 > 0$ and $\sigma_2 > 0$;

- E-FWE:
$$M(w; \beta_1, \sigma_1, \sigma_2) = \left(1 - e^{-Y(w;\sigma_1,\sigma_2)}\right)^{\beta_1}, \quad w \geq 0,$$
where $\sigma_1 > 0$, $\sigma_2 > 0$, and $\beta_1 > 0$;

- Weibull:
$$M(w; \sigma_1, \sigma_2) = 1 - e^{-\sigma_2 w^{\sigma_1}}, \quad w \geq 0,$$
where $\sigma_1 > 0$ and $\sigma_2 > 0$;

- E-Weibull:
$$M(w; \beta_1, \sigma_1, \sigma_2) = \left(1 - e^{-\sigma_2 w^{\sigma_1}}\right)^{\beta_1}, \quad w \geq 0,$$
where $\sigma_1 > 0$, $\sigma_2 > 0$, and $\beta_1 > 0$;

- K-Weibull:
$$M(w; \beta_1, \beta_2, \sigma_1, \sigma_2) = 1 - \left[1 - \left(1 - e^{-\sigma_2 y^{\sigma_1}}\right)^{\beta_1}\right]^{\beta_2}, \quad w \geq 0,$$
where $\sigma_1 > 0$, $\sigma_2 > 0$, β_1, and $\beta_2, > 0$.

After choosing the competing models for comparative purposes, the very next step is to select the statistical tools to judge the performances of the fitted models. For the illustration and evaluation of these distributions, certain statistical tools and tests were selected and computed. These tools are given by

- AIC (Akaike information criterion), obtained as
$$2k - 2\delta(\lambda, \sigma_1, \sigma_2 | w_1, w_2, \cdots, w_n);$$

- CAIC (corrected Akaike information criterion), calculated by
$$\frac{2nk}{n-k-1} - 2\delta(\lambda, \sigma_1, \sigma_2 | w_1, w_2, \cdots, w_n);$$

- BIC (Bayesian information criterion), computed as
$$k \log(n) - 2\delta(\lambda, \sigma_1, \sigma_2 | w_1, w_2, \cdots, w_n);$$

- HQIC (Hannan–Quinn information criterion), obtained using the formula
$$2k \log(\log(n)) - 2\delta(\lambda, \sigma_1, \sigma_2 | w_1, w_2, \cdots, w_n);$$

- AD (Anderson–Darling) test, having a mathematical expression given by
$$-n - \frac{1}{n} \sum_{a=1}^{n} (2a-1)[\log M(w_a) + \log\{1 - M(w_{n-a+1})\}];$$

- CM (Cramér-von Mises) test, obtained using the formula

$$\frac{1}{12n} + \sum_{a=1}^{n}\left[\frac{2a-1}{2n} - M(w_a)\right]^2;$$

- KS (Kolmogorov–Smirnov) test, whose value is computed using the expression

$$sup_w[M_n(w) - M(w)].$$

From the expressions of the MLEs obtained in the previous section, we can observe that these expressions

$$\frac{\partial}{\partial \sigma_1}\delta(\lambda, \sigma_1, \sigma_2|w_1, w_2, \cdots, w_n),$$

$$\frac{\partial}{\partial \sigma_2}\delta(\lambda, \sigma_1, \sigma_2|w_1, w_2, \cdots, w_n),$$

and

$$\frac{\partial}{\partial \lambda}\delta(\lambda, \sigma_1, \sigma_2|w_1, w_2, \cdots, w_n),$$

are not in simple forms. Therefore, we have to adopt an optimization procedure to obtain the numerical values of $\hat{\sigma}_1, \hat{\sigma}_2$, and $\hat{\lambda}$.

For Data 1, the numerical values of $\hat{\sigma}_1, \hat{\sigma}_2, \hat{\lambda}, \hat{\beta}_1$, and $\hat{\beta}_2$ are presented in Table 4. The values of the comparative tools are provided in Tables 5 and 6. From the numerical illustrations of the fitted models in Tables 5 and 6, we observe that the NMFWE model is the best one for modeling the mortality rate data. For the NMFWE distribution, the numerical values of the selected statistical measures are AIC = 186.12600, CAIC = 186.36130, BIC = 194.11630, HQIC = 189.36450, CM = 0.03276, AD = 0.20485, and KS = 0.05085, with p-value = 0.94680. Based on the KS criterion with the p-value, the FWE is the second-best model, with the respective values given by 0.05313 and 0.92580, whereas, by considering the AD and CM tools, the E-FWE is the best model. For the E-FWE model, these values are given by AD = 0.03866 and CM = 0.25671. From Tables 5 and 6, it is now obvious that the NMFWE model is the best choice to apply for modeling the mortality rate data.

Furthermore, a visual illustration to support the numerical results is provided in Figure 6. For a visual illustration of the NMFWE distribution, the plots of the fitted PDF, PP, CDF, HF, CHF, and SF functions were obtained. These plots visually confirm the best fitting of the NMFWE distribution.

Table 4. The numerical values of $\hat{\sigma}_1, \hat{\sigma}_2, \hat{\lambda}, \hat{\beta}_1$, and $\hat{\beta}_2$, using the first COVID-19 dataset.

Model	$\hat{\sigma}_1$	$\hat{\sigma}_2$	$\hat{\lambda}$	$\hat{\beta}_1$	$\hat{\beta}_2$
NMFWE	0.61568 (0.06012)	1.33469 (0.20807)	2.86140 (1.76820)	-	-
FWE	0.64201 (0.05039)	1.11759 (0.10961)	-	-	-
E-FWE	0.65089 (0.11588)	1.35112 (2.14650)	-	0.82677 (1.39368)	-
Weibull	1.92159 (0.14090)	0.58694 (0.07121)	-	-	-
E-Weibull	1.00398 (0.32020)	1.78865 (0.75560)	-	4.02508 (3.10070)	-
K-Weibull	1.44294 (0.14370)	3.76192 (NaN)	-	3.13605 (1.63131)	0.24665 (NaN)

Table 5. The values of AIC, CAIC, BIC, and HQIC of the fitted models, using the first COVID-19 dataset.

Model	AIC	CAIC	BIC	HQIC
NMFWE	186.12600	186.36130	194.11630	189.36450
FWE	189.04580	189.16230	196.37270	191.20480
E-FWE	187.01970	187.25500	195.01000	190.25820
Weibull	191.38590	191.50240	196.71280	193.54490
E-Weibull	188.2469 0	188.48220	196.23720	191.48540
K-Weibull	189.18680	189.58290	199.84060	193.50490

Table 6. The values of CM, AD and KS of the fitted models, using the first COVID-19 dataset.

Model	CM	AD	KS	p-Value
NMFWE	0.03276	0.20485	0.05085	0.94680
FWE	0.03963	0.26343	0.05313	0.92580
E-FWE	0.03866	0.25671	0.05589	0.89500
Weibull	0.10233	0.65790	0.06967	0.68220
E-Weibull	0.05380	0.29853	0.06758	0.71820
K-Weibull	0.04335	0.24179	0.06477	0.76540

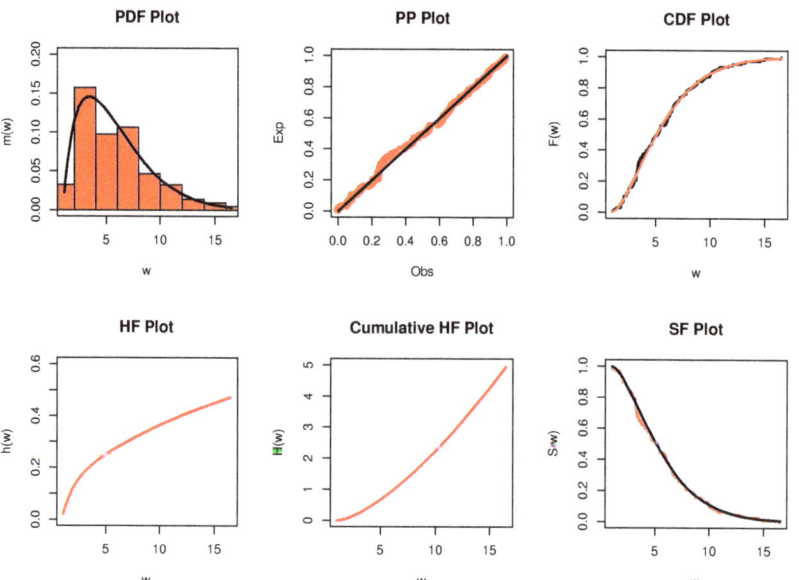

Figure 6. A visual illustration of the NMFWE model using Data 1.

For Data 2, the numerical values of $\hat{\sigma}_1, \hat{\sigma}_2, \hat{\lambda}, \hat{\beta}_1$, and $\hat{\beta}_2$ are provided in Table 7. For this data, the values of the analytical tools are presented in Tables 8 and 9. From the numerical comparison of the competing distributions in Tables 8 and 9, we observe that the proposed NMFWE model is the best choice to implement for dealing with the mortality rate data. For the NMFWE distribution, the values of the analytical measures are AIC = 848.33910, CAIC = 848.44800, BIC = 858.57400, HQIC = 852.47040, CM = 0.03762, AD = 0.21668, and KS = 0.04217, with p-value = 0.82040. The second-best model, based on the KS test with p-value, is the K-Weibull distribution, with the respective values given by 0.04345 and 0.79130. By considering the other analytical tools, we observe that the E-FWE model is the second-best model.

To support the best fitting power of the NMFWE model, a visual illustration is provided in Figure 7. From the visual illustration in Figure 7, we can see that the NMFWE distribution follows the fitted PDF, CDF, and SF very closely.

Table 7. The numerical values of $\hat{\sigma}_1, \hat{\sigma}_2, \hat{\lambda}, \hat{\beta}_1$, and $\hat{\beta}_2$, using the second COVID-19 dataset.

Model	$\hat{\sigma}_1$	$\hat{\sigma}_2$	$\hat{\lambda}$	$\hat{\beta}_1$	$\hat{\beta}_2$
NMFWE	0.21080 (0.02141)	2.14767 (0.01293)	10.42059 (2.72864)	-	-
FWE	0.64201 (0.05039)	1.11759 (0.10961)	-	-	-
E-FWE	0.21612 (0.01809)	3.86071 (1.84962)	-	0.54707 (0.27101)	-
Weibull	1.61908 (0.08247)	0.14782 (0.02037)	-	-	-
E-Weibull	0.89210 (0.69863)	0.81533 (0.69098)	-	3.72610 (2.65132)	-
K-Weibull	1.20004 (NaN)	2.22247 (NaN)	-	4.63532 (0.13187)	0.14624 (0.01021)

Table 8. The values of AIC, CAIC, BIC, and HQIC of the fitted models, using the second COVID-19 dataset.

Model	AIC	CAIC	BIC	HQIC
NMFWE	848.33910	848.44800	858.57400	852.47040
FWE	851.87659	851.90876	863.87654	856.75648
E-FWE	850.06480	850.17658	861.29978	854.19629
Weibull	859.51210	859.56640	866.33540	862.26630
E-Weibull	855.16730	855.27630	865.40220	859.29860
K-Weibull	852.71950	852.90220	866.36610	858.22800

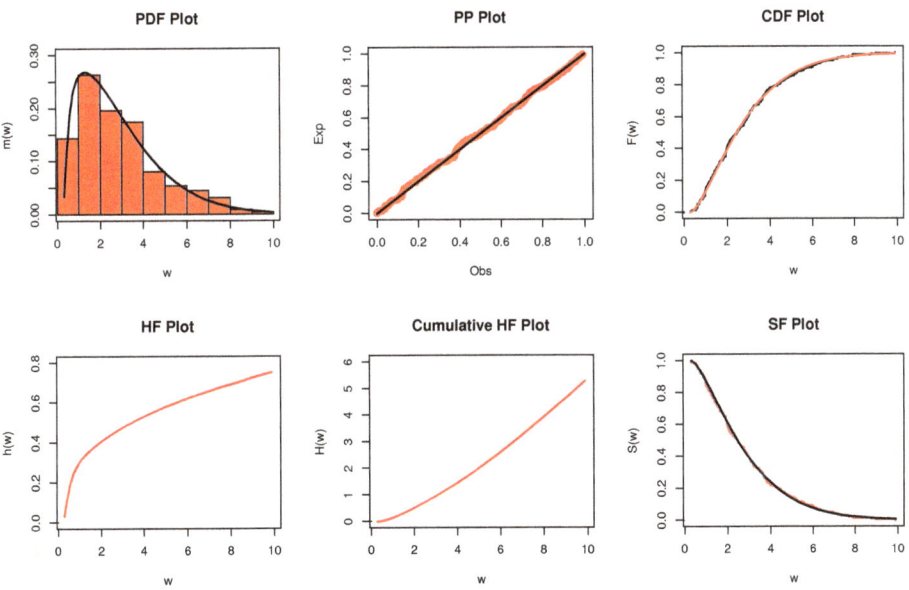

Figure 7. A visual illustration of the NMFWE model using Data 2.

Table 9. The values of CM, AD, and KS of the fitted models, using the second COVID-19 dataset.

Model	CM	AD	KS	p-Value
NMFWE	0.03762	0.21668	0.04217	0.82040
FWE	0.04076	0.25807	0.04746	0.71927
E-FWE	0.03975	0.24540	0.04819	0.67560
Weibull	0.13201	0.92260	0.05400	0.53080
E-Weibull	0.05951	0.40453	0.04895	0.65640
K-Weibull	0.04285	0.26641	0.04345	0.79130

Based on the obtained results in Tables 4–9, we observe that the NMFWE model works quite well for analyzing the COVID-19 datasets. Therefore, it can be considered the best model among all competitive distributions, and we can utilize it as an alternative probability tool in prediction, rather than of recording data for a long period of time.

6. An Econometric Approach

In the previous section, the modified distribution is compared with numerous existing distributions under simulation and real data related to the mortality rate caused by the COVID-19 epidemic in Mexico and Canada. In this section, some pure statistical models are compared with machine learning algorithms via forecasting on the same datasets. The parametric autoregressive moving average (ARMA) and non-parametric autoregressive integrated moving average (NP-ARMA) are pure time series models, while neural network autoregression (NNAR) and support vector regression (SVR) are machine learning algorithms. Data splitting is needed to segment the data into two parts, in the form of training data and testing data, in order to obtain forecast errors. Therefore, 80 percent of the data is provided for model fitting, and 20 percent is preserved for the models' comparison, following (Qi and Zhang, [29]). Details regarding each technique used for the modeling are given below.

6.1. The ARMA Model

In the time series forecasting literature, the ARMA is a powerful tool for univariate modeling. In the last few decades, ARMA has found successful applications in different areas such as economics, finance, engineering, and so forth (Khashei and Bijari, [30]). Generally, ARMA is a combination of autoregressive (AR) and moving average (MA) models. Mathematically, the ARMA can be written as

$$\pi_t = \mu + \sum_{a=1}^{m} \delta_a \pi_{t-a} + \sum_{b=0}^{n} \zeta_b \aleph_{t-b}, \qquad (9)$$

where μ indicates an intercept term, $\delta_a (a = 1, 2, \cdots, m)$ and $\zeta_b (b = 1, 2, \cdots, m)$ represent the coefficients of AR and MA, respectively, and $\aleph_{t-b}$ represents the white noise term with zero mean and variance $\sigma_\aleph^2$. The order of m and n is often determined by an autocorrelation function (ACF) and by partial autocorrelation (PACF); see Bibi et al. [31]. In our case, we fit an ARMA (2, 1) model to the underlying time series π_t.

6.2. The NP-ARMA Method

The additive non-parametric counterpart of the ARMA process leads to an additive model (NP-ARMA), where the association between π_t and its lagged variables do not have any specific known functional form. Probably, for any sort of non-linear form which is stated as

$$\pi_t = g_1(\pi_{t-1}) + g_2(\pi_{t-2}) + ,...,+ g_k(\pi_{t-m}) + \aleph_t, \qquad (10)$$

where g_i $(i = 1, 2, \cdots, k)$ show the smoothing functions which describe the association between π_t and its own lagged variables, the functions g_i represent the cubic regression splines (Shah et al., [32]). In the recent case, we incorporate four lags while estimating the model.

6.3. The NNAR Method

Customarily, a network or circuit of neurons leads to a neural network (NN). If the neurons or nodes are artificial, it leads to an artificial neural network. Neural network models have the potential to capture the complex non-linear nexus between an outcome variable and its covariates. A feedback NN is built with lagged time series variables as a covariate and hidden layer(s) with dimension nodes. NNAR consists of at least three layers of nodes: an output layer, a hidden layer, and an input layer. The outputs of a single layer are utilized as inputs to the succeeding one. A nonlinear NNAR model can be fitted/trained to predict a series by using its lagged variables as inputs $\pi_t, \pi_{t-1}, \cdots, \pi_{t-m}$; this process entails "so-called" feedback delays, where t represents the time delay parameter. The expression NNAR (h, w) shows that there are h delay inputs and w nodes in the hidden layers. The NNAR is the same as ARMA $(h, 0)$, conditionally, if there are zero nodes, i.e., NNAR $(h, 0)$. However, here, the parameter which ensures stationarity is not incorporated (Bibi et al., [31]). The nonlinear NNAR equation can be expressed as

$$\pi_t = \Omega_0 + \sum_{c=1}^{h} \Omega_c \zeta \left(\phi_c + \sum_{w=1}^{z} \phi_{cw} \pi_{t-w} \right) + \aleph_t, \quad (11)$$

where $\phi_c (c = 1, 2, \cdots, h, \ w = 1, 2, \cdots, z)$ and $\Omega_c (c = 1, 2, \cdots, h)$ indicate the weights of interconnection, h shows the length of the hidden layers with activation function ζ, and z shows the length of input layers. In our study, NNAR (6, 2) is utilized, which reveals six lagged variables, which are used as inputs, and two hidden layers. The input and hidden layers are selected for the model estimation through a trial-and-error approach, following (Khashei and Hajirahimi, [33]).

6.4. The SVR Method

Support vector regression is an alternative tool for solving regression issues such as nonlinearity and complexity in the data by introducing an alternative loss function (Vapnik et al. [34]; Vapnik [35]). SVR is based on the same principles as support vector machines (SVMs). It is an effective tool and has shown remarkable forecasting performance in many practical applications. The SVR utilizes different kernel functions to compute the resemblance between two data points to overcome the non-linearity. The core benefit of SVR lies in its capability to capture the covariate nonlinearity and then utilize it to boost the forecasting situations. It helps researchers discover a model's acceptable margin of error (Bibi et al. [31]; Ribeiro et al. [36]). The mathematical form of SVR with kernel function can be described as

$$\pi_t = \sum_{c=1}^{h} (\gamma_c - \gamma_c^*) M(u_c, u) + \varphi, \quad (12)$$

where the kernel function $M(u_c, u)$ refers to the inner product, φ is adjusted within the kernel function, and $\sum_{c=1}^{h} (\gamma_c - \gamma_c^*)$ is a constraint. Among numerous kernel functions, radial basis function (RBF) is commonly used, which can be described as

$$M(u_c, u) = exp\left(-\frac{||u_c - u_m||^2}{2\sigma^2} \right),$$

where $||u_c - u_m||^2$ represents the Euclidean distance amid the two covariate vectors squared and σ_2 shows the width of RBF (Lu et al., [37]). Our study proceeds with the RBF kernel function.

The predictive potential of all econometric models is evaluated by utilizing standard accuracy measures computed from a testing dataset. Statistically, the forecast errors are a more suitable criterion for assessing forecasting capability and for choosing the best tool.

The widely used principles are mean absolute error (MAE), root mean square error (RMSE), and mean absolute percentage error (MAPE). Hence, our study adopts these three criteria to judge the models' prediction performance. Their mathematical forms can be written as

$$MAE = mean(|\pi_t - \hat{\pi}_t|),$$

$$RMSE = \sqrt{mean(\pi_t - \hat{\pi}_t)^2},$$

and

$$MAPE = mean\left(\left|\frac{\pi_t - \hat{\pi}_t}{\pi_t}\right|\right) \times 100,$$

respectively.

6.5. Empirical Results

This section presents the findings of the forecasting experiments and some graphical representations. In this paper, we use the mortality rate of COVID-19 patients in Mexico and Canada, respectively, in order to quantify the predictability of the pure statistical and ML models. We split the data into two parts, intending to facilitate the out-of-sample prediction accuracy. For estimation, we use 80 percent of the data, and the remaining 20 percent of the data is used for checking the models' multistep-ahead out-of-sample forecasting accuracy.

6.5.1. Analyzing the COVID-19 Data Taken from Mexico

In Figure 8, the mortality rates of the COVID-19 patient data are divided by a vertical blue dotted line, where the training part is used for model estimation and the second part (testing data) is used for out-of-sample prediction.

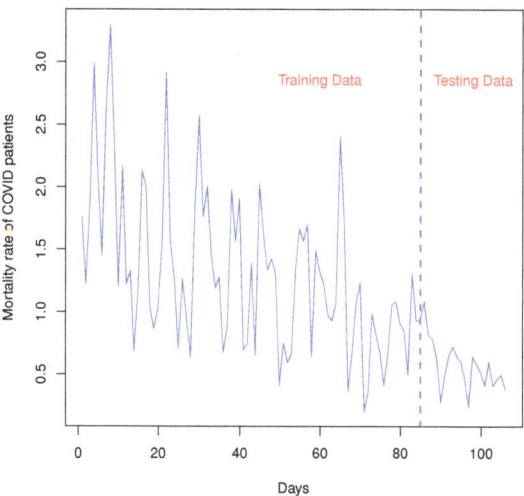

Figure 8. Divison of mortality rate of the COVID-19 patients data taken from Mexico.

The data pattern in Figure 9 shows non-constancy in the mean, variance, and covariance over time, which provides a piece of evidence about the unit root problem. Similarly, ACF and PACF also illustrate that the original data of the mortality rates of COVID-19 patients is non-stationary; see Figure 10. In general, time series models such as ARMA require stationary series for modeling; thus, to achieve stationarity, we adopted a differencing approach. Post-differencing, ACF and PACF confirmed that the transformed series is stationary.

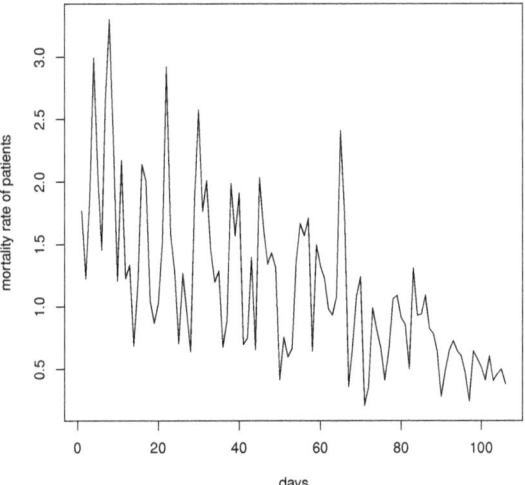

Figure 9. Trend of mortality rate data taken from Mexico.

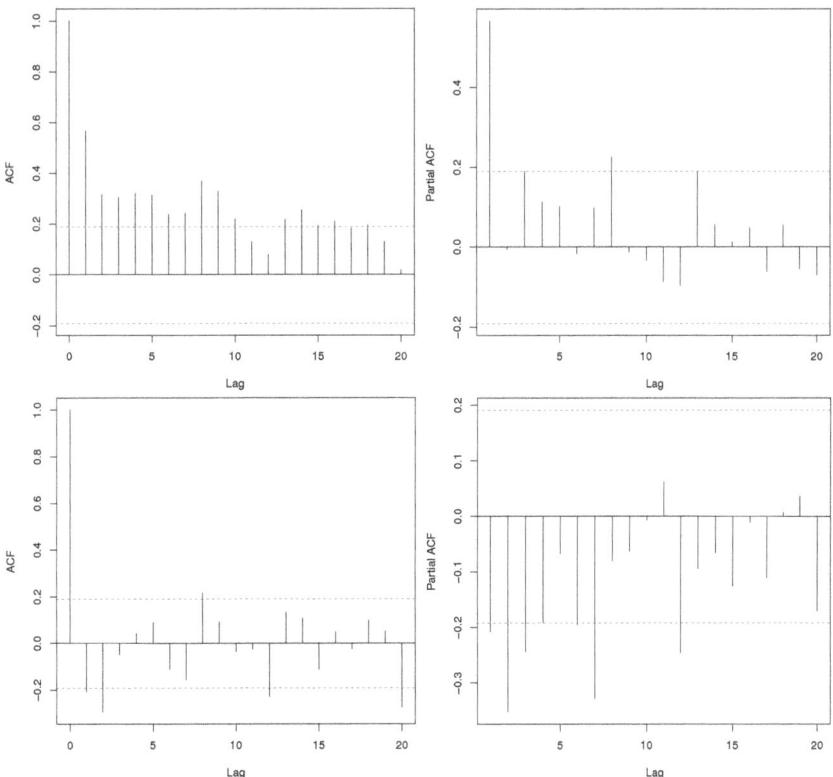

Figure 10. ACF and PACF for level (**first row**) and differenced data (**second row**).

Table 10 presents the results for the Jarque–Bera and Box–Ljung tests. The corresponding p-values exceed a five percent significance level; hence, we cannot reject the null hypothesis of random and normally distributed residuals of an estimated model. To be specific, it is declared that the residuals of the fitted model are uncorrelated and normally distributed. Thus, the ARMA model can be used for prediction.

Table 10. The results of the Box–Ljung and Q-statistics tests.

Test	χ^2	p-Value
Box–Ljung test	23.697	0.10
JB test	1.932	0.38

Alternatively, to identify the normality and randomness of the fitted models' residuals is to consider the graphs of the ACF, the Box–Ljung test, and the Q–Q plot of the residuals; see Figure 11. The plots in Figure 11 demonstrate that the residuals of the estimated model are random and normally distributed.

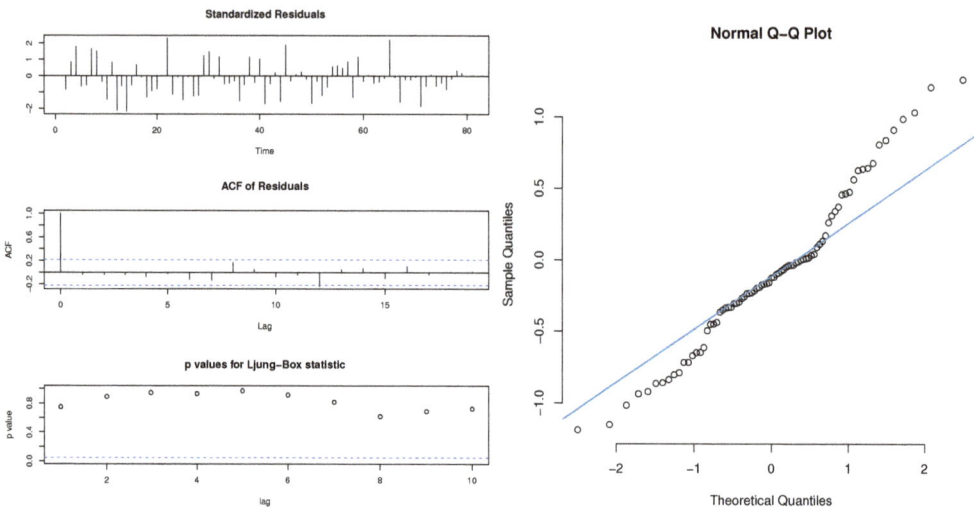

Figure 11. Diagnostic check.

The three standard measures of accuracy under the COVID-19 dataset are reported in Table 11. We can notice that RMSE, MAE, and MAPE computed for machine learning (ML) tools such as NNAR and SVR are substantially smaller than their pure statistical counterparts. Therefore, it can be concluded that predictions via ML tools tend to perform better than the rival statistical counterparts in terms of forecasting.

Table 11. The error metrics using the COVID-19 dataset taken from Mexico.

Criteria	ARIMA	NP-ARMA	NNAR	SVR
RMSE	0.359	0.576	0.230	0.073
MAE	0.320	0.525	0.169	0.043
MAPE	0.696	1.127	0.431	0.104

Furthermore, amid ML tools, the SVR outperforms the NNAR. A flowchart of forecast comparison is also presented in Figure 12. The plots in Figure 12 illustrate that ML tools, particularly SVR, remain effective tools for predicting the COVID-19 patient mortality rate trend. Moreover, Figure 13 also shows the performance of all models, and supports the output of Figure 12.

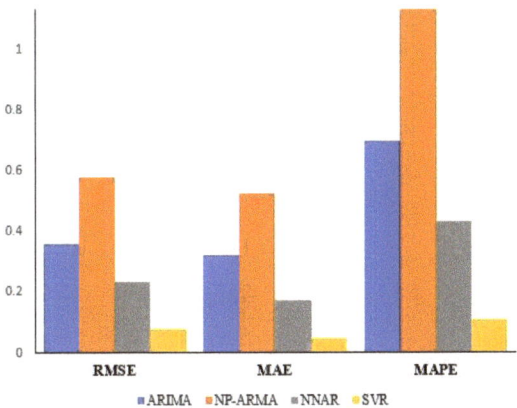

Figure 12. Forecasts comparsion for the COVID-19 dataset taken from Mexico.

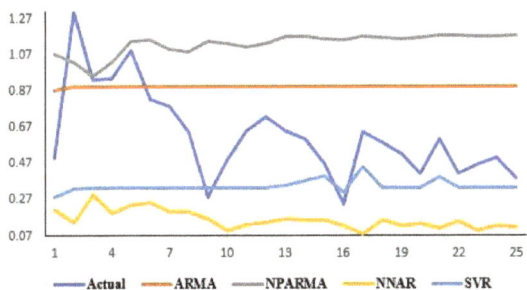

Figure 13. Forecasting performance of the models.

6.5.2. Analyzing the COVID-19 Data Taken from Canada

For estimating the models, we utilize 80 percent of the data, and the remaining 20 percent of the data is used for assessing the models' multistep-ahead post-sample predictive power. In Figure 14, the mortality rate of the COVID-19 patients' data is halved by a vertical blue dotted line, where the training part is utilized for model estimation and the second part (testing data) is utilized for post-sample forecasting.

The data pattern in Figure 14 shows non-constancy in the mean, variance, and covariance over time, which reflects the problem of a unit root. Likewise, the steady decline in the ACF plot reveals that the original data on the mortality rate of COVID-19 patients follow a random walk; see Figure 15. It is a fact that ARMA modeling requires stationary series; therefore, we take the first difference to make the underlying series stationary. Post-differencing, the ACF confirmed that the transformed series is stationary.

The numerical resutls of the Jarque–Bera and Box–Ljung tests are presented in Table 12. We can observe that the corresponding p-values exceed the five percent significance level; hence, we cannot reject the null hypothesis of random and normally distributed residuals. To be more specific, it is declared that the residuals of the fitted model are independent and follow a normal distribution. Therefore, the ARMA model can be used for prediction.

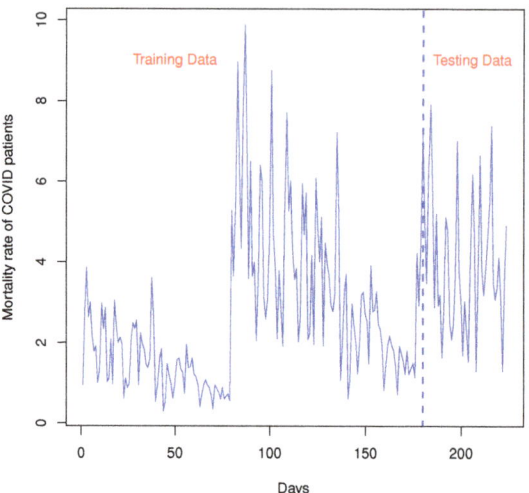

Figure 14. The mortality rate of the COVID-19 patients in Canada.

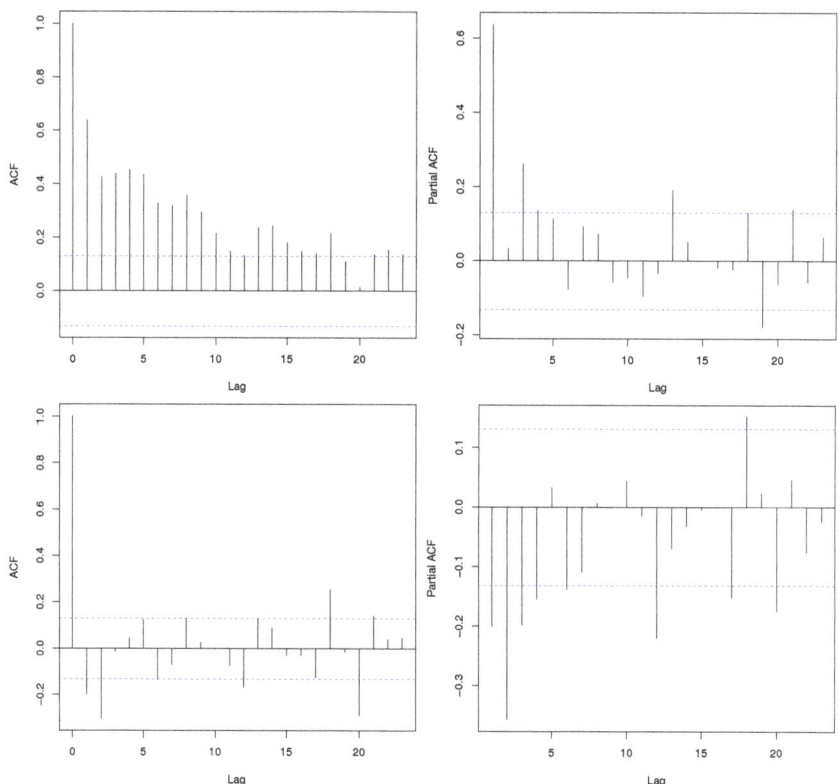

Figure 15. ACF and PACF for level (**first row**) and differenced data (**second row**).

Table 12. The resutls of the Box–Ljung and Q-statistics tests.

Test	χ^2	p-Value
Box–Ljung test	17.95	0.59
JB test	0.172	0.91

Alternatively, to identify the normality and randomness of the fitted models, we consider the graphs of the ACF, the Box–Ljung test, and the Q–Q plot of the residuals; see Figure 16. The plots in Figure 16 reveal that the residuals are random and normally distributed.

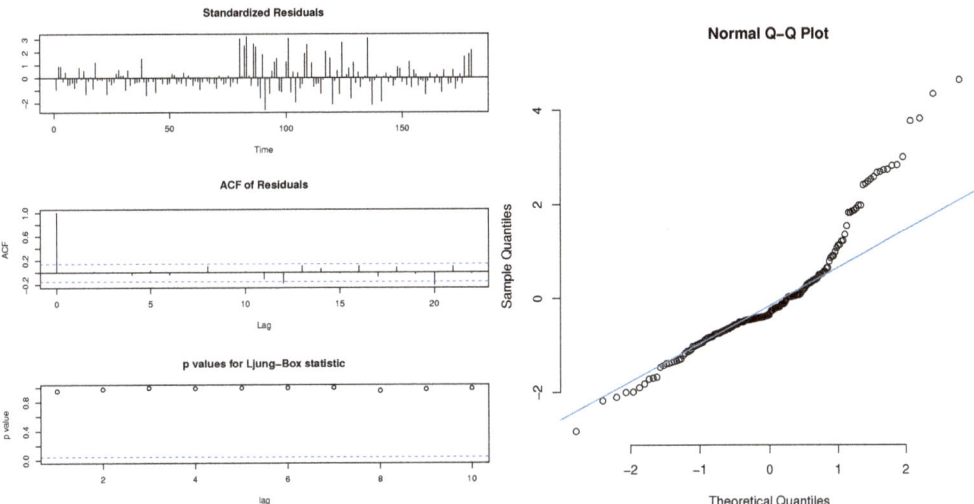

Figure 16. Diagnostic check.

For the COVID-19 dataset of Canada, the same three standard measures of accuracy are reported in Table 13. From the numerical results in Table 13, it is clear that the RMSE, MAE, and MAPE computed for the ML tools are substantially smaller. Hence, it can be inferred that predictions via the ML tools tend to perform better than their rival statistical counterparts in terms of forecasting. A flowchart of forecast comparison is also depicted in Figure 17.

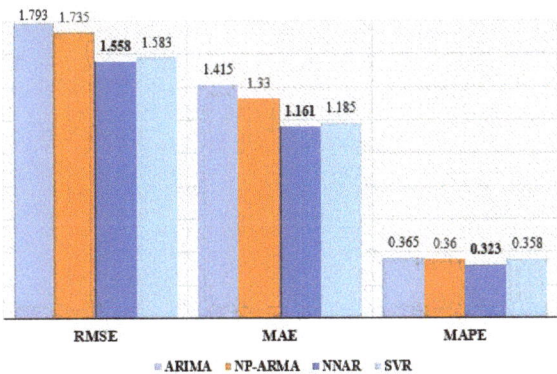

Figure 17. Flowchart of forecast errors.

Table 13. The error metrics.

Criteria	ARIMA	NP-ARMA	NNAR	SVR
RMSE	1.793	1.735	1.558	1.583
MAE	1.415	1.330	1.161	1.185
MAPE	0.365	0.360	0.323	0.358

The plots in Figure 18 reveal that the ML algorithms, particularly NNAR, remain a more effective tool in capturing the pattern of the mortality rate of COVID-19 patients in Canada. In addition, Figure 18 also portrays the performance of all models and supports the output of Figure 17.

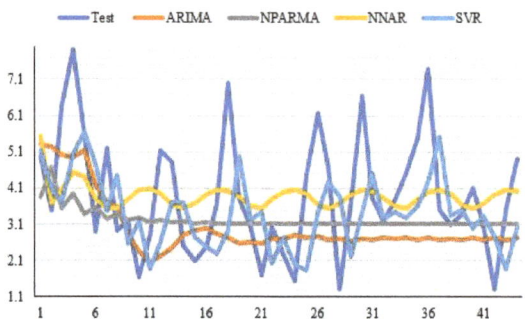

Figure 18. Forecasting performance of models.

7. Final Remarks

The COVID-19 epidemic has highly affected the business, trade, education, economy, and health sectors, etc. Among the affected areas, the health sector is one of the most-affected sectors. To have the best description and knowledge of the COVID-19 epidemic, many statistical studies have been carried out. This paper has added some further contributions towards the literature on COVID-19 data modeling. This paper suggested a new statistical model for analyzing the mortality rate of the COVID-19 pandemic in Mexico and Canada. The new model was named the NMFWE distribution and was applied to COVID-19 data in comparison with other statistical models. Based on seven statistical quantities, it is observed that the NMFWE model was the best competitor for dealing with mortality rate data. In addition, the COVID-19 datasets were also modeled through pure statistical models including ARMA, NP-ARMA, and two ML algorithms, including NNAR and SVR. The RMSE, MAE, and MAPE are utilized to evaluate the effectiveness of the underlying models. The findings illustrate that ML algorithms are successful at predicting the mortality rate of COVID-19 patients. The results also suggested that SVR provides a better forecast than NNAR in the case of Mexico. On the other hand, in the case of Canada, the NNAR outperforms the SVR, showing clearly that increasing the number of observations improves the NNAR forecasting performance, as compared to SVR. In other words, it can be inferred that NNAR requires more data for accurate predictions, in contrast to SVR.

In the future, we are committed to employing the proposed model in the machine learning field. We are also motivated to introduce the bivariate extension of the proposed model for analyzing bivariate data in the health sector.

Author Contributions: Formal analysis, Z.A. (Zubair Ahmad), Z.A. (Zahra Almaspoor), F.K. and M.E.-M.; Funding acquisition, Z.A. (Zubair Ahmad); Methodology, Z.A. (Zubair Ahmad) and Z.A. (Zahra Almaspoor); Software, Z.A. (Zahra Almaspoor), F.K. and M.E.-M.; Supervision, Z.A. (Zubair Ahmad); Validation, M.E.-M.; Writing—original draft, Z.A. (Zubair Ahmad), F.K. and M.E.-M. All authors have read and agreed to the published version of the manuscript.

Funding: This research received no external funding.

Institutional Review Board Statement: Not applicable.

Informed Consent Statement: Not applicable.

Data Availability Statement: The data sets are provided within the main body of the paper.

Acknowledgments: The authors are so grateful to the two anonymous reviewers for their constructive comments, which greatly improved the quality of the paper.

Conflicts of Interest: The authors declare no conflict of interest.

References

1. Hogan, C.A.; Sahoo, M.K.; Pinsky, B.A. Sample pooling as a strategy to detect community transmission of SARS-CoV-2. *JAMA* **2020**, *323*, 1967–1969. [CrossRef] [PubMed]
2. Mizumoto, K.; Kagaya, K.; Zarebski, A.; Chowell, G. Estimating the asymptomatic proportion of coronavirus disease 2019 (COVID-19) cases on board the Diamond Princess cruise ship, Yokohama, Japan, 2020. *Eurosurveillance* **2020**, *10*, 2000180. [CrossRef] [PubMed]
3. Ilyas, N.; Azuine, R.E.; Tamiz, A. COVID-19 pandemic in Pakistan. *Int. J. Transl. Med. Res. Public Health* **2020**, *4*, 37–49. [CrossRef]
4. Rao, G.S.; Aslam, M. Inspection plan for COVID-19 patients for Weibull distribution using repetitive sampling under indeterminacy. *BMC Med. Res. Methodol.* **2020**, *21*, 229. [CrossRef] [PubMed]
5. Singhal, A.; Singh, P.; Lall, B.; Joshi, S.D. Modeling and prediction of COVID-19 pandemic using Gaussian mixture model. *Chaos Solitons Fractals* **2020**, *138*, 110023. [CrossRef]
6. Qin, J.; You, C.; Lin, Q.; Hu, T.; Yu, S.; Zhou, X.H. Estimation of incubation period distribution of COVID-19 using disease onset forward time: A novel cross-sectional and forward follow-up study. *Sci. Adv.* **2020**, *6*, eabc1202. [CrossRef]
7. Almetwally, E.M.; Alharbi, R.; Alnagar, D.; Hafez, E.H. A new inverted topp-leone distribution: applications to the COVID-19 mortality rate in two different countries. *Axioms* **2021**, *10*, 25. [CrossRef]
8. Almongy, H.M.; Almetwally, E.M.; Aljohani, H.M.; Alghamdi, A.S.; Hafez, E.H. A new extended Rayleigh distribution with applications of COVID-19 data. *Results Phys.* **2021**, *23*, 104012. [CrossRef]
9. Liu, X.; Ahmad, Z.; KKhosa, S.; Yusuf, M.; Alamri, O.A.; Emam, W. A New Flexible Statistical Model: Simulating and Modeling the Survival Times of COVID-19 Patients in China. *Complexity* **2021**, *2021*, 6915742. [CrossRef]
10. EL-Sagheer, R.M.; Eliwa, M.S.; Alqahtani, K.M.; EL-Morshedy, M. Asymmetric randomly censored mortality distribution: Bayesian framework and parametric bootstrap with application to COVID-19 data. *J. Math.* **2022**, *2022*, 8300753. [CrossRef]
11. Eliwa, M.S.; Altun, E.; El-Dawoody, M.; El-Morshedy, M. A new three-parameter discrete distribution with associated INAR (1) process and applications. *IEEE Access* **2020**, *8*, 91150–91162. [CrossRef]
12. El-Morshedy, M.; Eliwa, M.S.; Altun, E. Discrete Burr-Hatke distribution with properties, estimation methods and regression model. *IEEE Access* **2020**, *8*, 74359–74370. [CrossRef]
13. El-Morshedy, M.; Altun, E.; Eliwa, M.S. A new statistical approach to model the counts of novel coronavirus cases. *Math. Sci.* **2020**, *16*, 37–50. [CrossRef]
14. Altun, H.K.; Ermumcu, M.S.K.; Kurklu, N.S. Evaluation of dietary supplement, functional food and herbal medicine use by dietitians during the COVID-19 pandemic. *Public Health Nutr.* **2021**, *24*, 861–869. [CrossRef]
15. Altun, E.; El-Morshedy, M.; Eliwa, M.S. A new regression model for bounded response variable: An alternative to the beta and unit-Lindley regression models. *PLoS ONE* **2021**, *16*, e0245627.
16. Bebbington, M.; Lai, C.D.; Zitikis, R. A flexible Weibull extension. *Reliab. Eng. Syst. Saf.* **2007**, *92*, 719–726. [CrossRef]
17. El-Morshedy, M.; El-Bassiouny, A.H.; El-Gohary, A. Exponentiated inverse flexible Weibull extension distribution. *J. Stat. Appl. Probab.* **2017**, *6*, 169–183. [CrossRef]
18. El-Morshedy, M.; Eliwa, M.S.; El-Gohary, A.; Almetwally, E.M.; EL-Desokey, R. Exponentiated Generalized Inverse Flexible Weibull Distribution: Bayesian and Non-Bayesian Estimation Under Complete and Type II Censored Samples with Applications. *Commun. Math. Stat.* **2021**, 1–22. [CrossRef]
19. Abubakari, A.G.; Kandza-Tadi, C.C.; Moyo, E. Modified Beta Inverse Flexible Weibull Extension Distribution. *Ann. Data Sci.* **2021**, *7*, 1–29. [CrossRef]
20. El-Gohary, A.; El-Bassiouny, A.H.; El-Morshedy, M. Exponentiated flexible Weibull extension distribution. *Int. J. Math. Its Appl.* **2015**, *3*, 1–12.
21. El-Damcese, M.A.; Mustafa, A.; El-Desouky, B.S.; Mustafa, M.E. The Kumaraswamy flexible Weibull extension. *Int. J. Math. Its Appl.* **2016**, *4*, 1–14.
22. Ahmad, Z.; Mahmoudi, E.; Dey, S. A new family of heavy tailed distributions with an application to the heavy tailed insurance loss data. *Commun. Stat.-Simul. Comput.* **2020**, *49*, 1–24. [CrossRef]
23. Gardiner, J.C.; Luo, Z.; Tang, X.; Ramamoorthi, R.V. Fitting heavy-tailed distributions to health care data by parametric and Bayesian methods. *J. Stat. Theory Pract.* **2014**, *8*, 619–652. [CrossRef]

24. Zhao, W.; Khosa, S.K.; Ahmad, Z.; Aslam, M.; Afify, A.Z. Type-I heavy tailed family with applications in medicine, engineering and insurance. *PLoS ONE* **2020**, *15*, e0237462. [CrossRef]
25. Bhati, D.; Ravi, S. On generalized log-Moyal distribution: A new heavy tailed size distribution. *Insur. Math. Econ.* **2018**, *79*, 247–259. [CrossRef]
26. Ahmad, Z.; Mahmoudi, E.; Hamedani, G.G.; Kharazmi, O. New methods to define heavy-tailed distributions with applications to insurance data. *J. Taibah Univ. Sci.* **2020**, *14*, 359–382. [CrossRef]
27. Ahmad, Z.; Mahmoudi, E.; Alizadeh, M.; Roozegar, R.; Afify, A.Z. The exponential TX family of distributions: Properties and an application to insurance data. *J. Math.* **2021**, *2021*, 3058170. [CrossRef]
28. Seneta, E. Karamata's characterization theorem, feller and regular variation in probability theory. *Publications de l'Institut Mathématique* **2002**, *71*, 79–89. [CrossRef]
29. Qi, M.; Zhang, G.P. An investigation of model selection criteria for neural network time series forecasting. *Eur. J. Oper. Res.* **2001**, *132*, 666–680. [CrossRef]
30. Khashei, M.; Bijari, M. An artificial neural network (p, d, q) model for timeseries forecasting. *Expert Syst. Appl.* **2010**, *37*, 479-489. [CrossRef]
31. Bibi, N.; Shah, I.; Alsubie, A.; Ali, S.; Lone, S.A. Electricity Spot Prices Forecasting Based on Ensemble Learning. *IEEE Access* **2021**, *9*, 150984–150992. [CrossRef]
32. Shah, I.; Iftikhar, H.; Ali, S. Modeling and forecasting medium-term electricity consumption using component estimation technique. *Forecasting* **2020**, *2*, 163–179. [CrossRef]
33. Khashei, M.; Hajirahimi, Z. A comparative study of series arima/mlp hybrid models for stock price forecasting. *Commun. Stat.-Simul. Comput.* **2019**, *48*, 2625–2640. [CrossRef]
34. Vapnik, V.; Golowich, S.; Smola, A. Support vector method for function approximation, regression estimation and signal processing. In *Advance in Neural Information Processing System*; Mozer, M., Jordan, M., Petsche, T., Eds.; MIT Press: Cambridge, MA, USA, 1997; Volume 9, pp. 281–287.
35. Vapnik, V. *Statistical Learning Theory*; Wiley: New York, NY, USA, 1998
36. Ribeiro MH, D.M.; da Silva, R.G.; Mariani, V.C.; dos Santos Coelho, L. Short-term forecasting COVID-19 cumulative confirmed cases: Perspectives for Brazil. *Chaos Solitons Fractals* **2020**, *135*, 109853. [CrossRef]
37. Lu, C.J.; Lee, T.S.; Chiu, C.C. Financial time series forecasting using independent component analysis and support vector regression. *Decis. Support Syst.* **2009**, *47*, 115–125. [CrossRef]

Article

Growth Recovery and COVID-19 Pandemic Model: Comparative Analysis for Selected Emerging Economies

Askar Akaev [1], Alexander I. Zvyagintsev [2], Askar Sarygulov [3], Tessaleno Devezas [4], Andrea Tick [5,*] and Yuri Ichkitidze [6]

1. Institute of Complex Systems Mathematical Research, Moscow State University, 119991 Moscow, Russia
2. Mikhailovskaya Military Artillery Academy, 195009 St. Petersburg, Russia
3. Center for Fundamental Research, St. Petersburg State University of Economics, 191023 St. Petersburg, Russia
4. Engineering Faculty, Atlântica Instituto Universitário, 2730-036 Barcarena, Portugal
5. Keleti Károly Faculty of Business and Management, Óbuda University, 1084 Budapest, Hungary
6. Department of Finance, HSE University, 101000 St. Petersburg, Russia
* Correspondence: tick.andrea@kgk.uni-obuda.hu

Citation: Akaev, A.; Zvyagintsev, A.I.; Sarygulov, A.; Devezas, T.; Tick, A.; Ichkitidze, Y. Growth Recovery and COVID-19 Pandemic Model: Comparative Analysis for Selected Emerging Economies. *Mathematics* **2022**, *10*, 3654. https://doi.org/10.3390/math10193654

Academic Editors: Cristiano Maria Verrelli and Fabio Della Rossa

Received: 30 August 2022
Accepted: 30 September 2022
Published: 5 October 2022

Copyright: © 2022 by the authors. Licensee MDPI, Basel, Switzerland. This article is an open access article distributed under the terms and conditions of the Creative Commons Attribution (CC BY) license (https://creativecommons.org/licenses/by/4.0/).

Abstract: The outburst of the COVID-19 pandemic and its rapid spread throughout the world in 2020 shed a new light on mathematic models describing the nature of epidemics. However, as the pandemic shocked economies to a much greater extent than earlier epidemics, the recovery potential of economies was emphasized and its inclusion in epidemic models is becoming more important. The present paper deals with the issues of modeling the recovery of economic systems that have undergone severe medical shocks, such as COVID-19. The proposed mathematical model considers the close relationship between the dynamics of pandemics and economic development. This distinguishes it from purely "medical" models, which are used exclusively to study the dynamics of the spread of the COVID-19 pandemic. Unlike standard SIR models, the present approach involves the introduction of the "vaccine" equation to the SIR model and introduces correction components that include the possibility of re-infection and other nuances such as the number of people at risk of infection (not sick with COVID but not vaccinated); sick with COVID; recovered; fully vaccinated (two doses) citizens; the rate of COVID infection; the rate of recovery of infected individuals; the vaccination coefficients, respectively, for those who have not been ill and recovered from COVID; the coefficient of revaccination; the COVID re-infection rate; and the population fluctuation coefficient, which takes into account the effect of population change as a result of births and deaths and due to the departure and return of citizens. The present model contains governance so that it not only generates scenario projections but also models specific governance measures as well to include the pandemic and restore economic growth. The model also adds management issues, so that it not only generates scenario forecasts but simultaneously models specific management measures as well, aiming to suppress the pandemic and restoring economic growth. The model was implemented on specific data on the dynamics of the spread of the COVID-19 pandemic in selected developing economies.

Keywords: COVID-19; economic systems; governance models; economic recovery; SIR model; Sanderson model

MSC: 00A72; 03C98

1. Introduction

Mathematical models describing the spread of the COVID-19 pandemic began to be developed almost simultaneously with the first outbreak in China in January 2020. These models are based on various approaches and aim to describe the continuous spread of COVID-19 adequately and accurately. Wang et al. [1] used a combination of a logistic model describing the rapid growth of the infected population and a model based on machine learning (FbProphet) to construct the epidemiological curve. Some other authors have proposed solutions based on the classical SIR-scheme models [2–4]. Finally, Contreras et al. [5] proposed a multi-group SEIRA model through populations with heterogeneous characteristics, taking into account the geographical features of the territory, behavioral features, and differences between social classes in a city, country, or region. However, an important feature of the COVID-19 pandemic is its systemic impact on the economy: temporary unemployment, the threat of bankruptcies, shocks in the anticipation of a decrease in production and demand, disruption of supply chains, and a sharp increase in sick pay [6–8]. All this forces change in economic policies and the development of mechanisms for an unprecedented amount of budget financing, including support for self-quarantine, as a means of flattening the epidemiological curve [9,10]. Another aspect of the impact of COVID-19 on the economy is the increasing income inequality through accumulating losses, mainly for workers with less education but not for those with an advanced degree [11]. In this regard, measures related to government support for specific segments of the economy, primarily employment systems, have been approved by the scientific community. For example, the EU's decision to adopt "temporary support to reduce the risks of emergency unemployment" (SURE), under which up to 100 billion euros in loans to maintain employment, either in the form of short-term work (STW) or similar schemes, has been allocated to EU member states [12], or addressing specific economic challenges such as supply chain resilience (SCR) in manufacturing and service operations in sectors such as the automotive industry and air transportation. The recommendations in such studies boil down to a series of management decisions that include steps such as developing localized sources of supply and using advanced Industry 4.0 technologies for the automotive industry, the need to ensure business continuity based on the correct definition of operations both at airports and on the flights of the airlines themselves, or the widespread use of big data analytics (BDA) [13]. The tourism sector has become a separate object of research in terms of developing the necessary management decisions. Škare, Soriano, and Porada-Rochon [14] showed that the recovery of the global tourism industry will take longer than the average expected recovery period of 10 months, requiring the development of new risk management practices, including rethinking the impact of epidemics on the industry and coordination measures of private and public policy for the survival of the industry infrastructure during such crises.

A similar nature of the impact of a medical disease on the economy that consists of stimulating productive managerial decisions by the epidemic has not yet been seen in modern history. Devezas [15,16] characterizes COVID-19 as a "provocative" innovation that has the potential to trigger a profound global transformation, causing a sharp surge in digitalization, accelerated development of artificial intelligence, the dominance of remote forms of work, and an e-commerce boom. Furthermore, Stiglitz [17] draws conclusions about the inefficiency of traditional market mechanisms in the context of the global COVID-19 pandemic and points to the dependence of further economic growth on the active participation of governments in overcoming the pandemic and its consequences. These two features of the COVID-19 pandemic, which characterize its impact on economic development, are still insufficiently reflected in the existing mathematical models describing the spread of the pandemic.

In this paper, such an approach is proposed that allows the combination of purely medical aspects of COVID-19 such as vaccination for those who have not been sick or recovered from COVID, revaccination, re-infections with COVID, population fluctuations, etc., with its economic aspects, thus building cyclical forecasts and possible scenarios of pandemics.

Earlier research proved that the COVID-19 pandemic has influenced various aspects of business and daily life, including healthcare, economic, and social aspects of life [18]. Specifically, the economic consequences include the disruption of supply chains, slowing of manufacturing, and significant slowing down in revenue growth, which results in a decreasing GDP.

Various mathematical models have been developed to model the dynamics of the pandemic such as an updated SEIR model [19] that uses the ABC-fractional operator [20]. This model proposes a time-fractional model, and next to the four components of the SEIR model—the susceptible, exposed, infected, and recovered population—it splits the infected population into asymptomatic and symptomatic and considers the hospitalized population as well. The model uses the fixed point theory, claiming the existence and uniqueness of the solutions via this specific theory. The fractional-order model developed in [20] helps to analyze the dynamics of even a new virus, and the model results are claimed to be critical in understanding the dynamics of an epidemic.

The present model, which uses various theories, such as the abovementioned fixed point theory or chaos control, has novelty in its nature of linking the features and dynamics of pandemics, the effects of vaccination, and its potential effect on restoring economies. The novel modeling links the first, most widely used classic SIR Kermack–McKendrick model [21] and the Sanderson model [22,23] for the first time to describe the interaction between the pandemic, the economy, and the economic recovery.

The presented model contributes to the literature and the research field by combining an epidemiological and an economic model to trace the trajectories for economic recovery in which the level of people protected against COVID-19 reaches saturation.

Five emerging economies were selected to verify the presented model: India, Brazil, Indonesia, South Africa, and Kazakhstan. To a large extent, the choice was dictated by the desire to cover the geography as widely as possible, since COVID-19 spread very quickly across all continents. Secondly, developing economies could not allocate funds to fight the pandemic in such a volume as the developed economies of the world; it was the vaccination of the entire population and economic measures to support it in the face of severe restrictions on economic activity that made it possible to stabilize the epidemiological curve in different countries [24–27]. Thirdly, economies of different sizes and populations were chosen, allowing for different economic opportunities to deal with the pandemic. Finally, these countries have different health care systems and, accordingly, different potential opportunities to fight the pandemic and restore their economies.

This paper is structured as follows: after the theoretical framework, the model building, and definition, the related literature is discussed, and the data and model estimates are presented. This paper finishes with the conclusion section, including the discussion of the implications of the developed model.

2. Theoretical Framework and Mathematical Modeling

This work is devoted to the mathematical modeling of the processes of overcoming the pandemic and the restoration of economic growth, and modeling of the required volume of anti-crisis measures and the effective timing of their implementation. The section includes the related literature to support the theoretical framework and the mathematical modeling. As a basis for our modeling, the classic SIR Kermack–McKendrick model [21] and the Sanderson model [22,23] are proposed, which is used for the first time to link the interaction between the pandemic and the economy.

2.1. The Kermack–McKendrick Model

The Kermack-McKendrick model considers three groups of individuals: susceptible to the disease (Susceptible), infected (Infected), and recovered (Recovered). Transmission of infection occurs from infected individuals to susceptible individuals. The SIR model is described by a system of three differential equations:

$$\begin{cases} \frac{dS}{dt} = -rSI \\ \frac{dI}{dt} = rSI - vI \\ \frac{dR}{dt} = vI \end{cases} \quad (1)$$

where $S(t)$, $I(t)$, and $R(t)$ are the numbers of susceptible, infected, and recovering individuals at time t, respectively; r is the rate of infection transmission; and v is the rate of recovery of infected individuals.

Some experts consider it wrong to use SIR models to describe the coronavirus pandemic [28], but if appropriate changes are made, then such models can be used to describe the coronavirus pandemic. The main drawback of the SIR model is that it is not capable of generating scenarios for the periodic undulating spread of COVID-19. This disadvantage is eliminated by the modification carried out in this paper. The possibility of such a modification was proved in [29]. Thus, our work, based on statistical data on the coronavirus pandemic in Russia [30], contains monthly indicators of the incidence and recovery of Russian citizens during the coronavirus pandemic and shows that the dynamics of incidence and recovery has waves with peaks repeated with a frequency of 7 months. The proposed model adds a "vaccine" equation and transforms the differential equations to difference equations, thus creating an extended model. Zvyagintsev [29] has shown that such a model allows the generation of cyclic trajectories with the required periodicity, which is a period of 7 months in the present case. Verification of the periodic orbit with specific initial conditions based on the statistical data on COVID-19 for Russia was conducted in Sadovnichiy et al. [30]. The proposed model is then combined with the "Wonderland model" [22,23] and then, to overcome the chaotic dynamics and the random characteristics of morbidity and recovery, the modern theory of chaos control is used [31,32] and a controlling function is determined. Finally, the constructed system makes the modeling of the stabilization process possible both in order to overcome the pandemic and achieve economic recovery and support governmental and management decision-making processes.

2.2. The Wonderland Model

The discrete model proposed by Sanderson and Lutz [22,23] is called the "Wonderland model". It models interrelated economic, demographic, and ecological processes and has the following form:

$$\begin{cases} x_{j+1} = x_j\left(1 + \beta_0\left(\beta_1 - \frac{e^{\beta y_j}}{1+e^{\beta y_j}}\right) - \alpha_0\left(\alpha_1 - \frac{e^{\alpha y_j}}{1+e^{\alpha y_j}}\right)\left(\alpha_2(1-z_j)^\theta + 1\right)\right) \\ y_{j+1} = y_j\left(1 + \gamma - (\gamma + \eta)(1-z_j)^\lambda\right) \\ z_{j+1} = \frac{z_j e^{\delta z_j^\rho - \omega x_j y_j p_j}}{1 - z_j + z_j e^{\delta z_j^\rho - \omega x_j y_j p_j}} \\ p_{j+1} = (1-\chi)p_j \end{cases} \quad (2)$$

where $\alpha_0, \alpha_1, \alpha_2\alpha, \beta_0, \beta_1, \beta, \theta, \gamma, \eta, \lambda, \delta, \rho, \omega, \chi$, are positive constants; $j = 0,1,2,\ldots$; x_j is the population of a selected country $x_j \geq 0$; y_j is the volume of GDP per capita; $y_j \geq 0$; z_j is a numerical characteristic of the ecological level (quality of the environment)—natural capital $0 \leq z_j \leq 1$; and p_j is a numerical characteristic of the degree of environmental pollution in the course of production, which is pollution per unit of production $0 \leq p_j \leq 1$.

In the "Wonderland model", the first equation defines the algorithm for the rate of population change in terms of the difference between birth and death rates. The second equation reflects the dynamics of the level of output per capita, taking into account the stock of natural capital. The third equation characterizes the quality of the environment and sets the growth of natural capital according to the logistic law. The fourth equation indicates the technological level of production and performs the function of controlling the intensity of environmental pollution.

In case of $z_j = 1$, it is believed that the ecology is in perfect condition and there is no environmental pollution at all. The case of $z_j = 0$ expresses the opposite limiting case when environmental pollution is so great that there is a maximum threat to human health and the economy. For variable p_j, the situation is the opposite. The value of $p_j = 1$ corresponds to the highest degree of pollution per unit of production. If $p_j = 0$, then there is no pollution per unit of production, which indicates the highest technological level of production.

2.3. The Proposed Model
2.3.1. Extension of the SIR Model

In the SIR model (1), it is assumed that the recovered individuals acquire immunity and cannot be re-infected. For the COVID-19 pandemic, this assumption does not hold. In addition, model (1) does not take into account the process of vaccination against a viral infection. Let us add a "vaccine" equation to the SIR model and introduce correction components that consider the possibility of re-infection and other nuances. Equation (3) presents the extended model:

$$\begin{cases} \frac{dS}{dt} = -rSI - qS + aS \\ \frac{dI}{dt} = rSI - vI + cR \\ \frac{dR}{dt} = vI - cR + dV \\ \frac{dV}{dt} = qS - dV + bR \end{cases} \quad (3)$$

where $S(t)$ is the number of people at risk of infection (not sick with COVID but not vaccinated), $I(t)$ is the number of people who are sick with COVID, $R(t)$ is the number of recovered, $V(t)$ is the number of fully vaccinated (two doses) citizens; r is the rate of COVID infection; v is the rate of recovery of infected individuals; q and b are the vaccination coefficients for those who have not been ill and recovered from COVID, respectively; d is the coefficient of revaccination; c is the COVID re-infection rate; and a is the population fluctuation coefficient, which takes into account the effect of population change as a result of births and deaths and due to the departure and return of citizens.

According to Comunian et al. [28], SIR models do not fit to describe the coronavirus pandemic; however, with appropriate changes such as in Equation (3), modified models can be used to describe the coronavirus pandemic.

Since we operate with discrete values of coronavirus statistics, it is expedient for modeling to switch from differential equations to difference equations. Then, the epidemiological-mathematical model (3) is transformed into a discrete system of equations:

$$\begin{cases} S_{j+1} = S_j(1 - rI_j - q + a) \\ I_{j+1} = I_j(1 + rS_j - v) + cR_j \\ R_{j+1} = R_j(1 - c) + vI_j + dV_j \\ V_{j+1} = V_j(1 - d) + qS_j + bR_j \end{cases} \quad (4)$$

where $j = 0,1,2, \ldots$, r is the rate of COVID infection; v is the rate of recovery of infected individuals; q and b are the vaccination coefficients for those who have not been ill and recovered from COVID, respectively; d is the coefficient of revaccination; c is the COVID re-infection rate; and a is the population fluctuation coefficient as a result of births and deaths and due to the migration of citizens. Actual coronavirus statistics are provided by official sources as specific discrete points in time. In this regard, in order to adapt the differential model (3) to actual realities, it is advisable to make the transition to a discrete model (4). Earlier, Zvyagintsev [26] showed that model (4) allows the generation of cyclic trajectories with the required periodicity. From the above, it follows that cycles with a period of 7 months are of interest. As a result of the computer experiment, the following numerical values of the system parameters (4) were selected:

$r = 2.14970486321458$; $v = 1.80089137686731$; $q = 1.25507485843425$;
$\alpha = 1.94597445654447$; $b = 0.0247039444045711$;
$c = 0.0619984179153856$; $d = 0.00807089279296773$.

With the help of approximation methods, the following initial conditions were found under which the nonlinear system (4) models a periodic orbit with a period of 7 months:

$$\tilde{S}_0 = 0.539753228690255; \tilde{I}_0 = 0.946101872256248; \tilde{R}_0 = 2.23329875094284; \tilde{V}_0 = 3.89439991609052. \quad (5)$$

These are the initial conditions, for which the solution of system (4) allows modeling of cyclical forecasts with a period of 7 in pandemic scenarios. These initial conditions (5) are universal for modeling 7-period cycles while the units of measurements are conditional individuals.

It should be noted that to find these 7-period cycles, it was necessary to calculate all the initial coordinates (5) up to the 14th decimal place, since the solutions of system (4) are very sensitive to changes in both the coefficients and the initial data. Furthermore, it should be noted that since real data are not always available for future forecasted time periods, the error cannot be established.

The periodic orbit modeled by system (4) with the initial conditions given in (5) based on the statistical data on COVID-19 for Russia was verified in Sadovnichiy, Akaev, Zvyagintsev, and Sarygulov [30]. In this case, the normalization coefficients B_1, C_1, B_2, C_2 were chosen in such a way that the trajectories $B_1 \widetilde{I}_j + C_1$ and $B_2 \widetilde{R}_j + C_2$ fluctuated, respectively, from $min I_j$ to $max I_j$ and from $min R_j$ to $max R_j$. As a result of the calculations, the following was obtained:

$$B_1 = \frac{max I_j - min I_j}{max \widetilde{I}_j - min \widetilde{I}_j} = 0.0024; \; C_1 = max I_j - B_1 \, max \widetilde{I}_j = 0.0036$$

$$B_2 = \frac{max R_j - min R_j}{max \widetilde{R}_j - min \widetilde{R}_j} = 0.0021; \; C_2 = max R_j - B_2 \, max \widetilde{R}_j = -0.0046$$

Therefore, the formulae in (6):

$$I_{prog} = 0.0024 \widetilde{I}_j + 0.0036 \text{ and } R_{prog} = 0.0021 \widetilde{R}_j - 0.0046 \tag{6}$$

help to model COVID-19 predictive cycles that are adapted to epidemiological realities.

Since the solutions of system (4) are hypersensitive to changes in the initial data, then, due to the variation of the initial conditions, model (4) is able to simulate various scenarios that may well correspond to real epidemiological processes. As an example, let us take such initial conditions that are very close to the values in (5):

$$S_0 = 0.54; \; I_0 = 0.95; \; R_0 = 2.23; \; V_0 = 3.89 \tag{7}$$

By solving system (4) with the initial conditions in (7) and applying the normalization coefficients from (6), forecasts with a decreasing amplitude of fluctuations in the coronavirus epidemic can be made.

For the mathematical interpretation of the mutual influence of the pandemic and the economy, the equations from system (2) are used. Since every month the sum of susceptible (S), infected (I), recovered (R), and vaccinated (V) individuals is the total population of a country, then:

$$x_j = S_j + I_j + R_j + V_j$$

holds.

2.3.2. Applying the "Wonderland Model"

Through z_j, we denote the epidemiological level of infection with coronavirus, and p_j is considered as an indicator of the effectiveness of medicine and healthcare in the fight against COVID. In the case of $z_j = 1$, it is considered that the epidemiological situation is in an ideal state and there is no infection with the coronavirus. The value $z_j = 0$ expresses the opposite extreme case, where the size of the pandemic is so large that there is a maximum threat to human health and the economy. For p_j, the situation is reversed. The value $p_j = 1$ corresponds to a catastrophic case of the highest degree of coronavirus infection. If $p_j = 0$, then there is no COVID infection, which indicates a very high level of efficiency in medicine and healthcare.

The value of p_j is calculated by the formula:

$$p_j = \frac{I_j}{x_j}$$

In the case of $p_j = 1$, the entire population is affected by the coronavirus $I_j = x_j$. If $p_j = 0$, there are no COVID patients $I_j = 0$. Then, the epidemiological situation is characterized by the equation (see the third equation in the "Wonderland model" (2)):

$$z_{j+1} = \frac{z_j e^{\delta z_j^\rho - \omega y_j l_j}}{1 - z_j + z_j e^{\delta z_j^\rho - \omega y_j l_j}}$$

This equation defines a logistic law such that $0 \leq z_{j+1} \leq 1$. If, as a result of modeling, $z_{j+1} \to 1$, the epidemiological situation improves, and the pandemic is overcome. If, as a result of modeling, $z_{j+1} \to 0$, the epidemiological situation worsens, and the pandemic becomes catastrophic. The dynamics of an economy, taking into account the epidemiological situation, is reflected by the equation (see the second equation in model (2)):

$$y_{j+1} = y_j\left(1 + \gamma - (\gamma + \eta)(1 - z_j)^\lambda\right)$$

where coefficients $\gamma, \eta, \lambda, \delta, \rho, \omega$ are constants and y_j is the volume of GDP per capita.

These equations demonstrate the dependence of an economy on the epidemiological situation. If, as a result of modeling, $z_j \to 1$, GDP is growing while, on the other hand, if, as a result of modeling, $z_j \to 0$, there is a decrease in GDP.

As a result, based on the aforementioned, the following mathematical model of the mutual influence of the pandemic and the economy is developed:

$$\begin{cases} S_{j+1} = S_j(1 - rI_j - q + a) \\ I_{j+1} = I_j(1 + rS_j - v) + cR_j \\ R_{j+1} = R_j(1 - c) + vI_j + dV_j \\ V_{j+1} = V_j(1 - d) + qS_j + bR_j \\ z_{j+1} = \frac{z_j e^{\delta z_j^\rho - \omega y_j l_j}}{1 - z_j + z_j e^{\delta z_j^\rho - \omega y_j l_j}} \\ y_{j+1} = y_j\left(1 + \gamma - (\gamma + \eta)(1 - z_j)^\lambda\right) \end{cases} \quad (8)$$

Governments are interested in overcoming the pandemic, restoring economic growth, and returning the socio-economic situation to a sustainable regime of a country. From a mathematical point of view, a stationary (fixed) point of system (8) corresponds to a stable regime. To find a fixed point, it is necessary to calculate such constants—$S_\#, I_\#, R_\#, V_\#, z_\#, y_\#$— so that for all values of the indices $j = 0,1,2,\ldots$, the identities $S_j = S_\#, I_j = I_\#, R_j = R_\#, V_j = V_\#, z_j = z_\#, y_j = y_\#$ are performed. Therefore, it is necessary to find a solution to the algebraic system of equations:

$$\begin{cases} S_\#(1 - rI_\# - q + a) = S_\# \\ I_\#(1 + rS_\# - v) + cR_\# = I_\# \\ R_\#(1 - c) + vI_\# + dV_\# = R_\# \\ V_\#(1 - d) + qS_\# + bR_\# = V_\# \\ \frac{z_\# e^{\delta z_\#^\rho - \omega y_\# l_\#}}{1 - z_\# + z_\# e^{\delta z_\#^\rho - \omega y_\# l_\#}} = z_\# \\ y_\#\left(1 + \gamma - (\gamma + \eta)(1 - z_\#)^\lambda\right) = y_\# \end{cases} \quad (9)$$

To find the coordinates of the fixed point of system (8), it is required to solve system (9). The solution of the algebraic system (9) has the form (10) and can be easily calculated by the standard method. The values from (10) are the coordinates of the fixed point of system (8). As a result of solving system (9), the fixed point coordinates for system (8) are the following:

$$S_\# = \frac{bv(q-a)}{r(ac - b(a-q))}; \quad I_\# = \frac{a-q}{r}; \quad R_\# = \frac{av(a-q)}{r(ac - b(a-q))}; \quad V_\# = \frac{bv(a-q)^2}{dr(ac - b(a-q))}; \quad z_\# = 1 - \left(\frac{\gamma}{\gamma + \eta}\right)^{\frac{1}{\lambda}}; \quad y_\# = \frac{\delta r}{\omega(a-q)}\left[1 - \left(\frac{\gamma}{\gamma + \eta}\right)^{\frac{1}{\lambda}}\right]^\rho. \quad (10)$$

Since the indicators of morbidity and recovery are characterized by randomness, in order to overcome the chaotic dynamics, model (8) is modified using the results from the modern theory of chaos control [28,29]. Let us introduce the notation:

$$w_1(j) = S_j; w_2(j) = I_j; w_3(j) = R_j; w_4(j) = V_j; w_5(j) = z_j; w_6(j) = y_j$$

and rewrite system (8) in the vector form:

$$w(j+1) = F(w(j)) ; j \in \{0, 1, 2, \ldots\} \quad (11)$$

where:

$$(j) = \begin{pmatrix} w_1(j) \\ w_2(j) \\ w_3(j) \\ w_4(j) \\ w_5(j) \\ w_6(j) \end{pmatrix} ; F(w(j)) = \begin{pmatrix} w_1(j)(1 - rw_2(j) - q + a) \\ w_2(j)(1 + rw_1(j) - v) + cw_3(j) \\ w_3(j)(1 - c) + vw_2(j) + dw_4(j) \\ w_4(j)(1 - d) + qw_1(j) + bw_3(j) \\ \frac{w_5(j)e^{\delta w_5^2(j) - \omega \cdot w_6(j)w_2(j)}}{1 - w_5(j) + w_5(j)e^{\delta w_5^2(j) - \omega \cdot w_6(j)w_2(j)}} \\ w_6(j)\left(1 + \gamma - (\gamma + \eta)(1 - w_5(j))^\lambda\right) \end{pmatrix}$$

For the vector function F, the Jacobi matrix has the following form:

$$A(w) = \begin{pmatrix} a_{11} & -rw_1 & 0 & 0 & 0 & 0 \\ rw_2 & a_{22} & c & 0 & 0 & 0 \\ 0 & v & 1-c & d & 0 & 0 \\ q & 0 & b & 1-d & 0 & 0 \\ 0 & a_{52} & 0 & 0 & a_{55} & a_{56} \\ 0 & 0 & 0 & 0 & a_{65} & a_{66} \end{pmatrix}$$

where:

$$a_{11} = 1 - rw_2 - q + a, \quad a_{22} = 1 + rw_1 - v, \quad a_{66} = 1 + \gamma - (\gamma + \eta)(1 - w_5)^\lambda,$$

$$a_{65} = \lambda(\gamma + \eta)w_6(1 - w_5)^{\lambda - 1}, \quad a_{52} = \frac{\omega w_6 w_5 (w_5 - 1)e^{\delta w_5^2 - \omega \cdot w_6 w_2}}{\left(1 - w_5 + w_5 e^{\delta w_5^2 - \omega \cdot w_6 w_2}\right)^2},$$

$$a_{55} = \frac{(1 + \delta \rho (1 - w_5)w_5^\rho)e^{\delta w_5^2 - \omega \cdot w_6 w_2}}{\left(1 - w_5 + w_5 e^{\delta w_5^2 - \omega \cdot w_6 w_2}\right)^2}, \text{ and } a_{56} = \frac{\omega w_2 w_5 (w_5 - 1)e^{\delta w_5^2 - \omega \cdot w_6 w_2}}{\left(1 - w_5 + w_5 e^{\delta w_5^2 - \omega \cdot w_6 w_2}\right)^2}.$$

After linearizing system (11) in the vicinity of the found fixed point $w_\#$ with the coordinates $w_{1\#} = S_\#, w_{2\#} = I_\#, w_{3\#} = R_\#, w_{4\#} = V_\#, w_{5\#} = z_\#, w_{6\#} = y_\#$, we have:

$$w(j+1) = F(w_\#) - F(w(j)) + A(w_\#)[w(j) - w_\#],$$

Then, applying the Pyragas method [33], we obtain:

$$w(j+1) = F(w_\#) - F(w(j)) + A(w_\#)[w(j) - w_\#] + P(j)[w(j) - w(j-1)],$$

As a result, a modified system is achieved:

$$w(j+1) = F(w(j)) + U(j) ; j \in \{0, 1, 2, \ldots\}, \quad (12)$$

where $U(j)$ is a control function designed to stabilize the behavior of the system's decisions. Based on the results of [28,29] on the stabilization of discrete systems, the control function of the following form is obtained:

$$U(j) = F(w_\#) - F(w(j)) + A(w_\#)[w(j) - w_\#] + P(j)[w(j) - w(j-1)]. \quad (13)$$

where *P(j)* is a periodic matrix of the following form:

$$P(j) = \begin{cases} (kE - A^2(w_\#))(A(w_\#) - E)^{-1}, & j = 2n \\ O, & j \neq 2n, \ n \in \{1, 2, \ldots\} \end{cases},$$

where $-1 < k < 1$, E is the identity matrix, O is the zero matrix, $A(w_\#)$ is the Jacobian matrix, and $w_\#$ is the fixed point.

The discrete system (12) is characterized by ultra-high sensitivity to parameter changes. Even minor changes in the coefficients lead to significant changes in the behavior of the solutions. Thus, due to the variation of the coefficients, model (12) is able to simulate various scenarios that are quite consistent with real epidemiological and economic processes.

Governments are interested in overcoming the pandemic and restoring economic growth. The ideal scenario is considered to be the case when, thanks to the development of antiviral medicine and the implementation of mass vaccination, the incidence of coronavirus, and hence recovery, drops to zero. According to most medical experts, in order to develop herd immunity against coronavirus, it is necessary that the vaccinated proportion of the population is 70–80 percent. Israel's positive experience shows that 100% vaccination is required to deal with the pandemic most effectively.

The derived Equations (12) and (13) are not only tools for efficient and adequate forecasting; the high efficiency is achieved as a result of finetuning of the model due to careful selection of the model parameters and their high-precision variation. A further advantage of this model is its ability to generate mathematically based management measures to stabilize socio-economic development in countries, which makes it possible to practically manage the dynamics of a pandemic and GDP. The constructed system (12) allows the modeling of the stabilization process due to the control function given in (13). The analytical Equation (13) obtained explicitly makes it possible to determine the size and time of proactive adjustments.

3. Data

Data from five emerging economies—Brazil, South Africa, Kazakhstan, India, and Indonesia—were collected and analyzed. The model considers each country independently. Data on incidence (Cases), the number of vaccinated (Vaccinations), mortality (Deaths), and recovered were taken from official websites [34,35] (https://ourworldindata.org/coronavirus (accessed on 27 August 2022)), https://www.worldometers.info/coronavirus/ (accessed on 5 May 2022)). The GDP forecast for 2022 was taken from the IMF review [36]. The number of employees and the dynamics of GDP in constant prices in 2015 were taken from the World Bank data (https://data.worldbank.org/indicator (accessed on 1 May 2022)) [37,38]. Data on investment (financial) multiples were used from the following sources [39–41].

4. Model-Based Estimates

Upon analyzing the epidemic waves and the virus variants, the highest peak of incidence fell on the Omicron strain, and the incidence dynamics had a "sawtooth" configuration. Figure 1 presents the daily rates of new coronavirus infections in the population of the selected countries during this period. The incidence dynamics has a clearly traceable cyclicity with a period of 7 days. Model (4) makes it possible to generate cyclic trajectories with the required frequency and configuration. As Figure 1 shows, the short-term cycles observed during the period of the Omicron strain are identical in these countries, and the frequency of these short-term cycles is 7 days for all the countries considered. Therefore, system (4) with the initial conditions given in (5) could be used again to model short-term cyclical forecasts during the period of the maximum COVID outbreak.

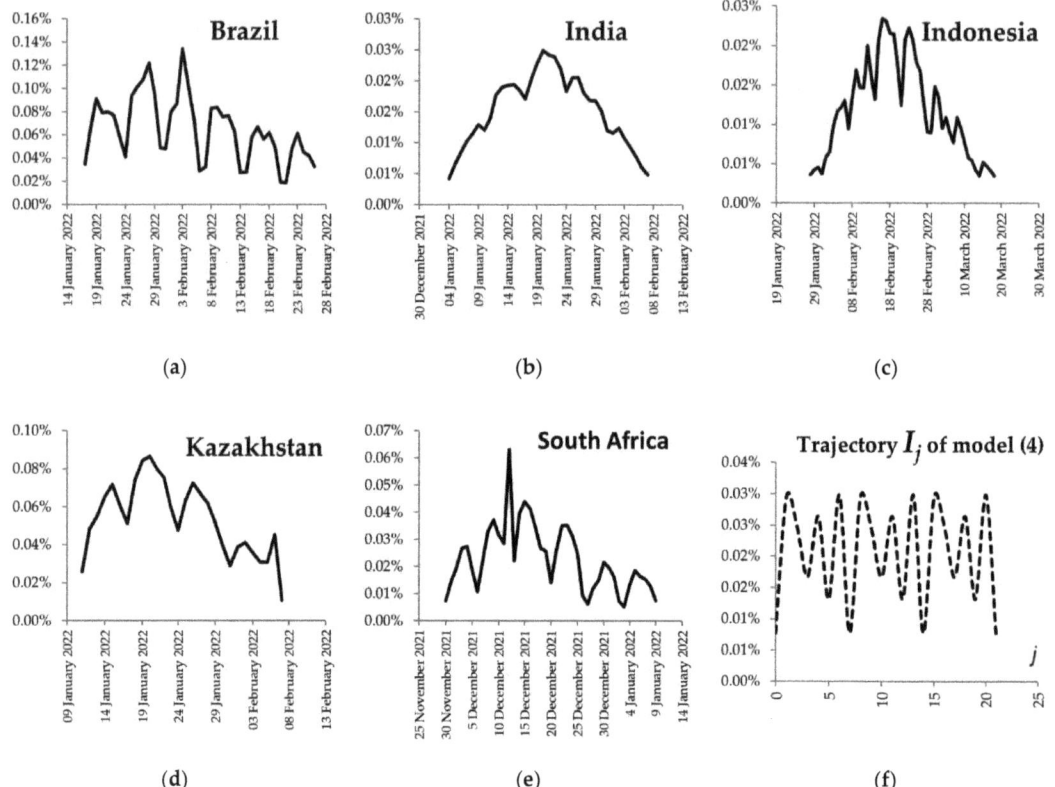

Figure 1. Infected proportion of the population and modeled 7-period morbidity trajectory of (**a**) Brazil, (**b**) India, (**c**) Indonesia, (**d**) Kazakhstan, (**e**) and South Africa, (**f**) trajectory I_j of model (4) (developed by authors).

In order to model the forecast for overcoming the pandemic and developing an economy, the system defined in (12) with initial conditions that correspond to actual realities is solved. Table 1 shows the initial indicators computed based on official statistical data as of 30 June 2022, collected from the sources given in Section 3 [34–41]. Data in Table 1 are computed based on statistics from the official sources indicated in Section 3. Units of measurement are shares of the population of countries, rate between 0 and 1, and USD for GDP per capita.

Table 1. Initial data for system (12).

	Brazil	India	Indonesia	Kazakhstan	South Africa
$w_1(0)$ (Susceptible, %)	0.198118	0.343269	0.390949	0.509977	0.68141
$w_2(0)$ (Infected, %)	0.006257	0.000221	0.000121	0.000052	0.00059
$w_3(0)$ (Recovered, %)	0.006235	0.00022	0.00012	0.000051	0.00058
$w_4(0)$ (Vaccinated, %)	0.78939	0.65629	0.60881	0.48992	0.31742
$w_5(0)$ (Epidemiological level of infection)	0.63818	0.6251	0.58678	0.41644	0.25091
$w_6(0)$ (GDP per capita, USD)	8549.62	1953.94	3855.9	11269.4	5864.0

Note that it is logical to assess the current level of overcoming the pandemic through vaccination and incidence rates. Therefore, it is assumed that:

$z_0 =$ [percentage of fully vaccinated population] $-$ [the share of the recovered population].

The coefficients of the system (12) are tools for setting up the developed model (Table 2). Table 2 displays the numerical values of the parameters that were carefully selected for the model (12) and (13) as a result of a computer experiment. Since the numerical values of the parameters for the model were selected solely on the basis of statistical data from official sources, the experiment is fully consistent with the real data. The main criterion for compliance with real data was the full compliance with the actual indicators of the selected countries in terms of incidence, vaccination, and GDP per capita.

Table 2. Coefficients of the model (12) and (13).

	Brazil	India	Indonesia	Kazakhstan	South Africa
a	0.001	0.001	0.001	0.001	0.001
b	0.5	5	10	50	5
c	0.3	0.3	0.3	0.3	0.3
d	0.000005263157895	0.000005263157895	0.000005263157895	0.000005263157895	0.000005263157895
r	1	10	20	100	10
v	0.35	0.8	1.3	5.3	0.8
q	0.0011	0.0011	0.0011	0.0011	0.0011
γ	0.00004	0.00004	0.00004	0.00004	0.00004
η	0.4	0.4	0.4	0.4	0.4
λ	2	2	2	2	2
δ	0.190487385822133	0.00458002457714418	0.00447593310948182	0.00247737693036435	0.013011433457796
ω	−0.2	−0.2	−0.2	−0.2	−0.2
ρ	3	3	3	3	3
k	0.8	0.8	0.8	0.8	0.8

Having solved system (12) with the initial conditions from Table 1, the predictive trajectories of the pandemic in the selected countries, presented in Figures 2–4, were obtained. Figure 2 clearly demonstrates that the countries under consideration have different medium-term cycles and different periodicities, it displays the COVID infected ratio, the recovery ratio, and the trajectories for the modeled morbidity and recovery ratios in Brazil, India, Indonesia, Kazakhstan, and South Africa.

Figure 3 includes vaccination in the model and presents the vaccination and COVID susceptible ratios in the selected countries and the forecasts for vaccination, susceptibility, and, for each observed country, the modeled level to overcome the pandemic.

Figures 2 and 3 clearly demonstrate that the vaccination of the population is the main tool in the fight against COVID-19, since the faster the population was vaccinated (slope of the curve of vaccinated ratio), the steeper the drop in the COVID susceptible ratio in each selected country (Figure 3). As the case of South Africa and Kazakhstan shown in Figure 3d,e, if vaccination stops, the rate of COVID susceptible stops decreasing and starts stagnating. Figure 3 shows that a higher vaccination rate results in a lower susceptible ratio and the trend of overcoming the pandemic shows a similar flattening, a slowing down tendency and converging to saturation level 1.

Figure 4 displays the modeled GDP per capita growth in the selected countries using systems (11) and (12), showing a definite recovery of economic growth even in countries experiencing a decreasing economic development trend. Figures 3 and 4 assume a significant relationship between overcoming the pandemic and the recovery of economic growth; therefore, the correlation was checked between the vaccination rate and economic recovery. A strong positive correlation was found between the obtained values of $w_5(j)$ and $w_6(j)$, equaling 0.9, which proves this assumption. As a result of modeling, the values of these indicators show to stationary trajectories (Figures 3 and 4). Since the indicators

$w_5(j) = z_j$ and $w_6(j) = y_j$ were defined before Equation (11), the correlation coefficient is quite informative.

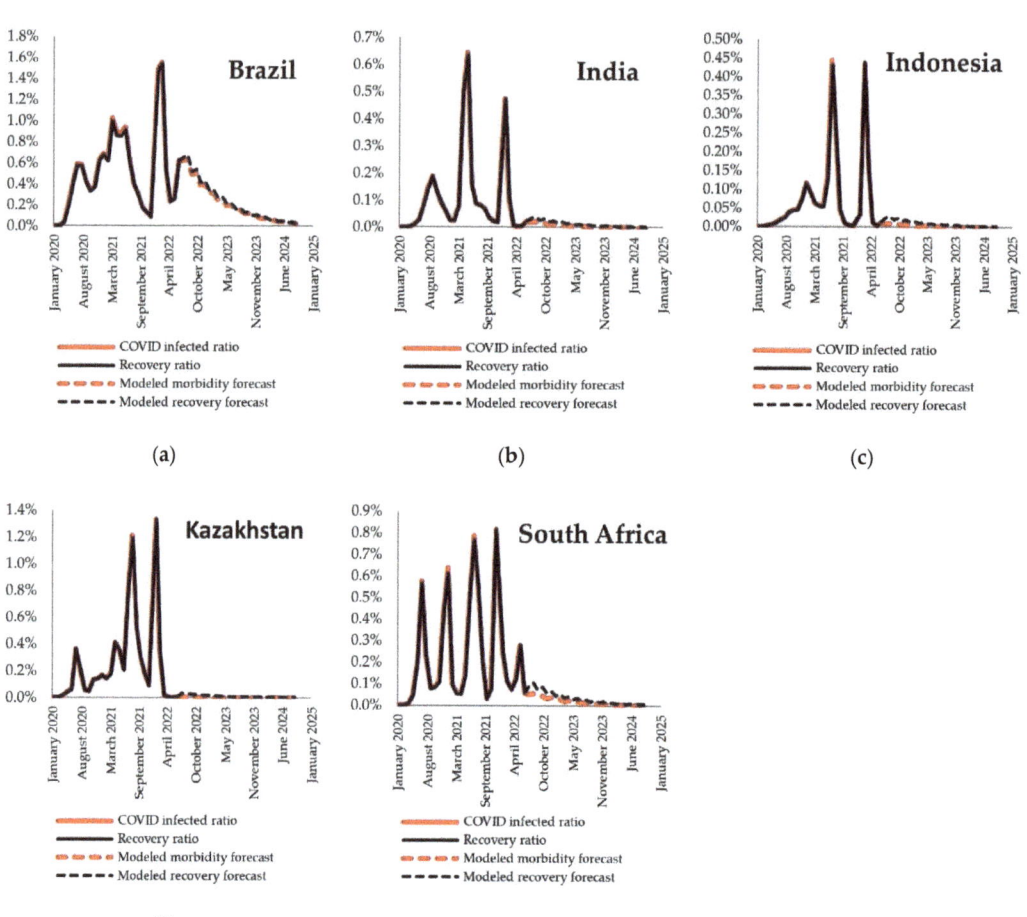

Figure 2. Modeled forecasts of the dynamics of morbidity and recovery in (**a**) Brazil, (**b**) India, (**c**) Indonesia, (**d**) Kazakhstan, and (**e**) South Africa (developed by authors).

As a result of the calculations and the verification carried out on the basis of the results of the simulation, and taking into account investment multipliers, the monthly values for the scale of vaccination and the volume of additional investments necessary to overcome the pandemic and ensure economic growth—displayed in Table 3—were identified. The values are summarized in Table 3, presenting the results of calculations by Equations (12) and (13), which, in fact, is a schedule for the implementation of preventive anti-crisis measures.

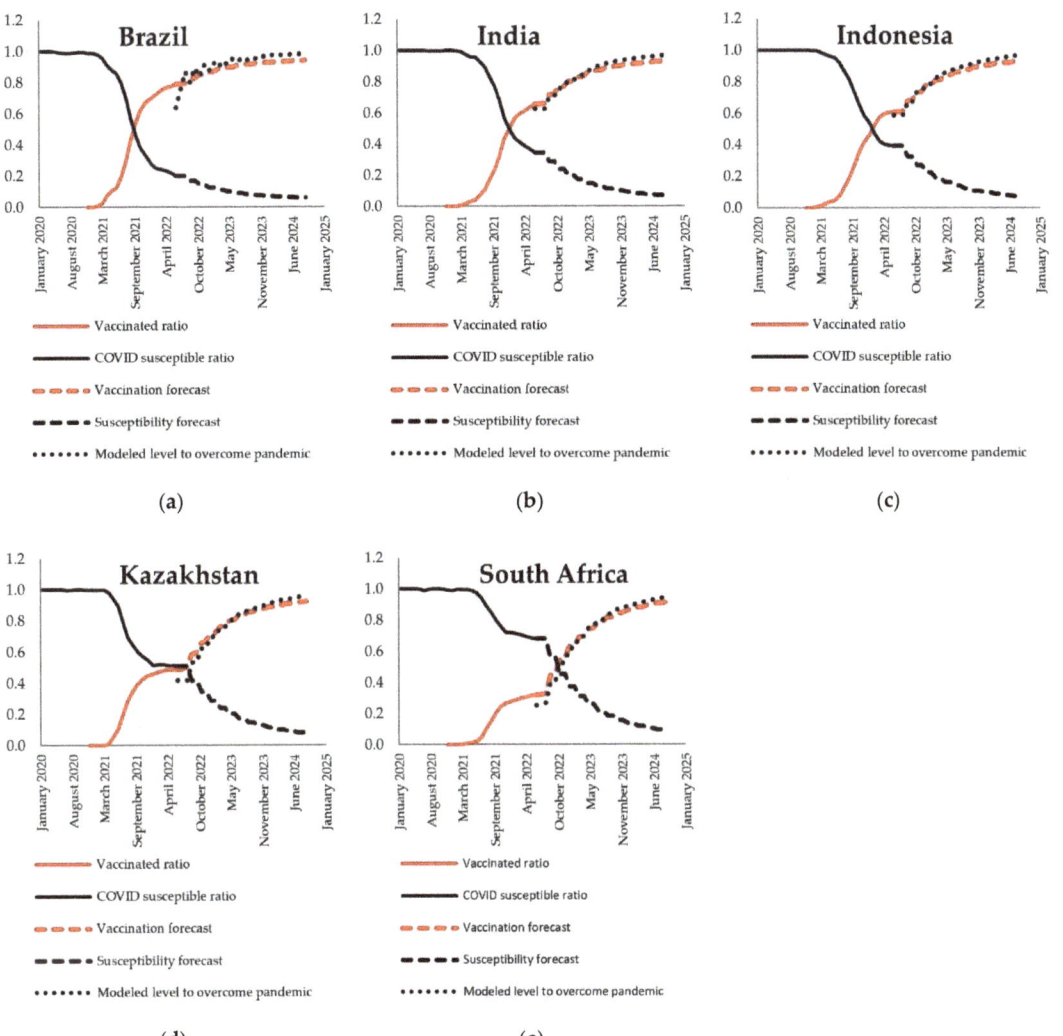

Figure 3. Modeled forecast of vaccination dynamics in (**a**) Brazil, (**b**) India, (**c**) Indonesia, (**d**) Kazakhstan, and (**e**) South Africa (developed by authors).

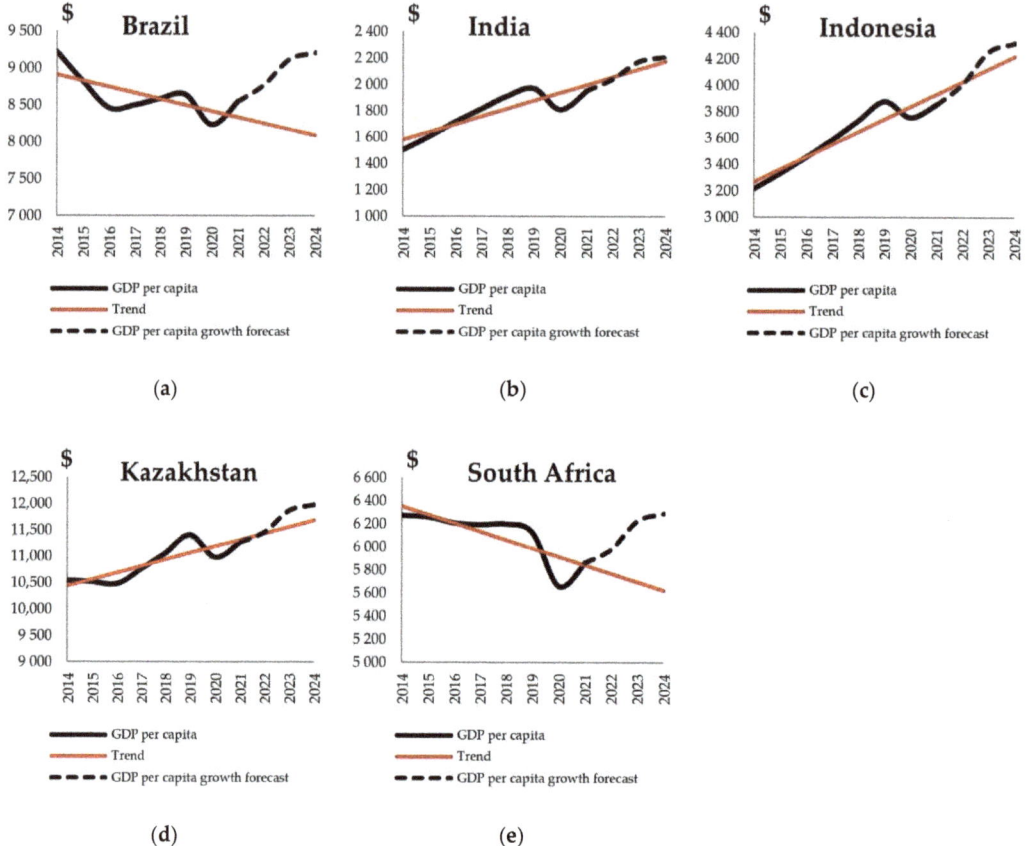

Figure 4. Modeled GDP per capita growth forecast in (**a**) Brazil, (**b**) India, (**c**) Indonesia, (**d**) Kazakhstan, and (**e**) South Africa (developed by authors).

Table 3. Schedule of preventive anti-crisis measures.

Date	Scale of Vaccination (% of Country's Population)					Additional Investments in the Economy (bln USD)				
	BRA	IND	IDN	KAZ	ZAF	BRA	IND	IDN	KAZ	ZAF
07.22	0.0	0.0	0.0	0.0	0.0	32.70	21.71	10.15	1.14	3.34
08.22	0.0	0.0	0.0	0.0	0.0	14.80	21.52	10.05	1.13	3.25
09.22	3.4	6.2	7.2	9.7	13.4	17.12	35.60	15.20	1.33	3.62
10.22	0.0	0.0	0.0	0.0	0.0	9.63	14.16	6.60	0.73	2.11
11.22	2.7	5.0	5.8	7.8	10.7	13.11	25.46	10.73	0.90	2.42
12.22	0.0	0.0	0.0	0.0	0.0	6.25	9.26	4.31	0.47	1.37
01.23	2.2	4.0	4.6	6.2	8.6	10.10	18.33	7.63	0.61	1.63
02.23	0.0	0.0	0.0	0.0	0.0	4.04	6.03	2.80	0.31	0.89
03.23	1.7	3.2	3.7	5.0	6.9	7.83	13.29	5.46	0.41	1.10
04.23	0.0	0.0	0.0	0.0	0.0	2.61	3.91	1.82	0.20	0.57

Table 3. Cont.

Date	Scale of Vaccination (% of Country's Population)					Additional Investments in the Economy (bln USD)				
	BRA	IND	IDN	KAZ	ZAF	BRA	IND	IDN	KAZ	ZAF
05.23	1.4	2.5	3.0	4.0	5.5	6.10	9.73	3.94	0.28	0.74
06.23	0.0	0.0	0.0	0.0	0.0	1.68	2.53	1.17	0.13	0.37
07.23	1.1	2.0	2.4	3.2	4.4	4.77	7.19	2.88	0.20	0.51
08.23	0.0	0.0	0.0	0.0	0.0	1.08	1.63	0.76	0.08	0.24
09.23	0.9	1.6	1.9	2.6	3.5	3.75	5.36	2.12	0.14	0.35
10.23	0.0	0.0	0.0	0.0	0.0	0.69	1.05	0.49	0.05	0.15
11.23	0.7	1.3	1.5	2.0	2.8	2.95	4.04	1.58	0.10	0.24
12.23	0.0	0.0	0.0	0.0	0.0	0.45	0.68	0.31	0.03	0.10
01.24	0.6	1.0	1.2	1.6	2.2	2.33	3.07	1.19	0.07	0.17
02.24	0.0	0.0	0.0	0.0	0.0	0.29	0.44	0.20	0.02	0.06
03.24	0.5	0.8	1.0	1.3	1.8	1.85	2.35	0.90	0.05	0.12
04.24	0.0	0.0	0.0	0.0	0.0	0.18	0.28	0.13	0.01	0.04
05.24	0.4	0.7	0.8	1.0	1.4	1.47	1.81	0.69	0.04	0.09
06.24	0.0	0.0	0.0	0.0	0.0	0.12	0.18	0.08	0.01	0.03
07.24	0.3	0.5	0.6	0.8	1.1	1.17	1.40	0.53	0.03	0.06
08.24	0.0	0.0	0.0	0.0	0.0	0.08	0.12	0.05	0.01	0.02
09.24	0.2	0.4	0.5	0.7	0.9	0.93	1.09	0.41	0.02	0.05

5. Conclusions

Based on the results obtained, it can be concluded that in order to overcome the pandemic and restore economic growth, it is necessary to implement proactive corrective measures in a timely manner. The proposed model (12) and (13) can be used as an auxiliary tool for stabilizing socio-economic development. Equation (13) allows accurate calculation of the size of and time required for the necessary proactive adjustments. Thus, a toolkit was developed for modeling management measures aimed at suppressing the pandemic and ensuring economic growth.

The methods developed in this paper are quite convenient from a practical point of view and can be easily adapted to current realities. The discrete system (12) performs the function of simulation and generates various scenarios by varying the parameters. Model (12) with control (13) makes it possible to analyze the possibilities that contribute to bringing the socio-economic dynamics to a sustainable trajectory. The resulting model makes it possible to preliminarily estimate the size of managerial decisions aimed at preventing crisis tendencies. Upon using Equation (13), it is possible to calculate the exact size and time for the necessary impulse-like adjustments and model (12) and (13) can be useful in developing and implementing measures aimed at overcoming the pandemic and stabilizing economic development.

The proposed system of models in conceptual terms (1) takes into account the close relationship between the dynamics of the pandemic and economic development; (2) contains control features, thanks to which it not only generates scenario forecasts but forms specific management measures aimed at suppressing the pandemic and restoring economic growth as well; and (3) is a sixth-order discrete non-linear system, which more accurately reflects the nature of real statistical data that are formed at discrete times.

From a management point of view, the model allows the definition of a sequence of steps that includes (1) a preliminary assessment of the financial volumes necessary to prevent crisis tendencies; (2) calculation of the time and size of financial injections (impulse adjustments); and (3) assessment of opportunities that contribute to bringing the socio-economic dynamics to a sustainable trajectory.

As for the time horizons of the proposed model, it seems that its use is most effective in the initial phase of the spread of an epidemic, but it is not suitable for forecasting the epidemic itself. Subsequent studies will be aimed at improving the design scheme of the model.

The organization and implementation of management measures can only be carried out by government structures. Therefore, it is obvious that in a pandemic, the state should play a leading role in overcoming the crisis. Market mechanisms do not work during a pandemic. The market is not able to organize itself in such a way as to carry out preventive measures aimed at anticipating crisis phenomena. Using models (12) and (13), government agencies can analyze and estimate in advance the amount of financial and organizational costs required both to suppress the possible spread of an epidemic and ensure economic recovery. In accordance with the simulation results presented in this work, in the medium term, the priority tasks for governments are the implementation of vaccination and support of the economy in the required volumes and within a certain time frame.

Author Contributions: Conceptualization, A.A., A.I.Z. and A.S.; methodology, T.D.; software, A.T. and Y.I.; validation, A.I.Z.; formal analysis, T.D., A.T. and Y.I.; investigation, A.A. and A.S.; resources, A.T.; data curation, Y.I.; writing—original draft preparation, A.A.; writing—review and editing, T.D. and A.T.; visualization, A.I.Z.; supervision, A.S.; project administration, A.T., funding acquisition, A.T. All authors have read and agreed to the published version of the manuscript.

Funding: The research was supported by the Ministry of Science and Higher Education of the Russian Federation (Grant Agreement No. 075-15-2022-1136 dated 01.07.2022). The APC was partly funded by Óbuda University.

Institutional Review Board Statement: Not applicable.

Informed Consent Statement: Not applicable.

Data Availability Statement: There are no publicly archived datasets generated during the study.

Conflicts of Interest: The authors declare no conflict of interest.

References

1. Wang, P.; Zheng, X.; Li, J.; Zhu, B. Prediction of epidemic trends in COVID-19 with logistic model and machine learning technics. *Chaos Solitons Fractals* **2020**, *139*, 110058. [CrossRef] [PubMed]
2. Barlow, N.S.; Weinstein, S.J. Accurate closed-form solution of the SIR epidemic model. *Phys. Nonlinear Phenom.* **2020**, *408*, 132540. [CrossRef] [PubMed]
3. Marinov, T.T.; Marinova, R.S. Dynamics of COVID-19 using inverse problem for coefficient identification in SIR epidemic models. *Chaos Solitons Fractals X* **2020**, *5*, 100041. [CrossRef]
4. Nkwayep, C.H.; Contreras, S.; Tewa, J.J.; Kurths, J. Short-term forecasts of the COVID-19 pandemic: A study case of Cameroon. *Chaos Solitons Fractals* **2020**, *140*, 110106. [CrossRef] [PubMed]
5. Contreras, S.; Villavicencio, H.A.; Medina-Ortiz, D.; Biron-Lattes, J.P.; Olivera-Nappa, Á. A multi-group SEIRA model for the spread of COVID-19 among heterogeneous populations. *Chaos Solitons Fractals* **2020**, *136*, 109925. [CrossRef] [PubMed]
6. World Bank Group. *Global Economic Prospects. January 2022*; The World Bank Group: Washington, DC, USA, 2022. Available online: https://openknowledge.worldbank.org/bitstream/handle/10986/36519/9781464817601.pdf (accessed on 15 May 2022).
7. International Labour Organization. ILO Monitor: COVID-19 and the World of Work. Seventh Edition. International Labour Organization, 2021. Available online: https://www.ilo.org/wcmsp5/groups/public/---dgreports/---dcomm/documents/briefingnote/wcms_767028.pdf (accessed on 5 May 2022).
8. International Labour Office. *World Social Protection Report 2020-22 Social Protection at the Crossroads—In Pursuit of a Better Future*; International Labour Organisation (ILO): Genève, Switzerland, 2021.
9. Baldwin, R. Keeping the Lights on: Economic Medicine for a Medical Shock. *CEPR*. Available online: https://cepr.org/voxeu/columns/keeping-lights-economic-medicine-medical-shock (accessed on 22 September 2021).
10. Baldwin, R.E.; di Mauro, B.W. *Economics in the Time of COVID-19*; VoxEU.org eBook; CEPR Press: London, UK, 2020. Available online: https://www.sensiblepolicy.com/download/2020/2020_CEPR_McKibbin_Fernando_COVD-19.pdf (accessed on 5 May 2022).
11. Furceri, D.; Loungani, P.; Ostry, J.D.; Pizzuto, P. COVID-19 Will Raise Inequality if Past Pandemics Are a Guide. *CEPR*. Available online: https://cepr.org/voxeu/columns/covid-19-will-raise-inequality-if-past-pandemics-are-guide (accessed on 11 September 2021).
12. Balleer, A.; Gehrke, B.; Hochmuth, B.; Merkl, C. Guidelines for Cost-Effective Use of SURE: Rule-Based Short-Time Work with Workers' Consent and Aligned Replacement Rates. *CEPR*. Available online: https://cepr.org/voxeu/columns/guidelines-cost-effective-use-sure-rule-based-short-time-work-workers-consent-and (accessed on 1 September 2021).

13. Belhadi, A.; Kamble, S.; Jabbour, C.J.C.; Gunasekaran, A.; Ndubisi, N.O.; Venkatesh, M. Manufacturing and service supply chain resilience to the COVID-19 outbreak: Lessons learned from the automobile and airline industries. *Technol. Forecast. Soc. Change* **2021**, *163*, 120447. [CrossRef]
14. Škare, M.; Soriano, D.R.; Porada-Rochoń, M. Impact of COVID-19 on the travel and tourism industry. *Technol. Forecast. Soc. Change* **2021**, *163*, 120469. [CrossRef]
15. Devezas, T. The struggle SARS-CoV-2 vs. homo sapiens–Why the earth stood still, and how will it keep moving on? *Technol. Forecast. Soc. Change* **2020**, *160*, 120264. [CrossRef]
16. Devezas, T. Aeronautics and COVID-19: A Reciprocal Cause-and-Effect Phenomenon. *J. Aerosp. Technol. Manag.* **2020**, *12*, e3420. [CrossRef]
17. Stiglitz, J.E. The proper role of government in the market economy: The case of the post-COVID recovery. *J. Gov. Econ.* **2021**, *1*, 100004. [CrossRef]
18. Haleem, A.; Javaid, M.; Vaishya, R. Effects of COVID-19 pandemic in daily life. *Curr. Med. Res. Pract.* **2020**, *10*, 78–79. [CrossRef] [PubMed]
19. Li, M.Y.; Muldowney, J.S. Global stability for the SEIR model in epidemiology. *Math. Biosci.* **1995**, *125*, 155–164. [CrossRef]
20. Sintunavarat, W.; Turab, A. Mathematical analysis of an extended SEIR model of COVID-19 using the ABC-fractional operator. *Math. Comput. Simul.* **2022**, *198*, 65–84. [CrossRef] [PubMed]
21. Kermack, W.O.; McKendrick, A.G. A contribution to the mathematical theory of epidemics. *Proc. R. Soc. Lond. Ser. Contain. Pap. Math. Phys. Character* **1927**, *115*, 700–721. [CrossRef]
22. Sanderson, W.C. Simulation Models of Demographic, Economic, and Environmental Interactions. In *Population—Development—Environment*; Lutz, W., Ed.; Springer: Berlin/Heidelberg, Germany, 1994; pp. 33–71. [CrossRef]
23. Lutz, W. (Ed.) *Population—Development—Environment: Understanding their Interactions in Mauritius*; Springer: Berlin/Heidelberg, Germany, 1994. [CrossRef]
24. Anderson, J.; Begamini, E.; Brekelmans, S.; Cameron, A.; Darvas, Z.; Domínguez Jiménez, M. The Fiscal Response to the Economic Fallout from the Coronavirus. *Bruegel | The Brussels-Based Economic Think Tank*. Available online: https://www.bruegel.org/publications/datasets/covid-national-dataset/ (accessed on 1 May 2022).
25. RBK. The Accounts Chamber Estimated the Cost of Fighting the Pandemic, Счетная палата оценила величину расходов на борьбу с пандемией (In Russian). Available online: https://www.rbc.ru/economics/24/02/2021/6034d7659a7947b5e4403bdd (accessed on 1 May 2022).
26. Asian Development Bank (ADB). Homepage | ADB COVID-19 Policy Database. 2020. Available online: https://covid19policy.adb.org/ (accessed on 12 May 2022).
27. KPMG. Brazil—Government and Institution Measures in Response to COVID-19. 2020. Available online: https://home.kpmg/xx/en/home/insights/2020/04/brazil-government-and-institution-measures-in-response-to-covid.html (accessed on 1 May 2021).
28. Comunian, A.; Gaburro, R.; Giudici, M. Inversion of a SIR-based model: A critical analysis about the application to COVID-19 epidemic. *Phys. Nonlinear Phenom.* **2020**, *413*, 132674. [CrossRef] [PubMed]
29. Zvyagintsev, A.I. On a Nonlinear Differential System Modeling the Dynamics of the COVID-19 Pandemic, О НЕЛИНЕЙНОЙ ДИффЕРЕНЦИАЛЬНОЙ СИСТЕМЕ, МОДЕЛИРУЮЩЕЙ ДИНАМИКУ ПАНДЕМИИ COVID-19 (In Russian). *Int. Res. J. Международный Научно-Исследовательский Журнал* **2022**, *7*, 116–121. [CrossRef]
30. Sadovnichiy, V.A.; Akaev, A.A.; Zvyagintsev, A.I.; Sarygulov, A.I. Mathematical Modeling of Overcoming the COVID-19 Pandemic and Restoring Economic Growth. *Dokl. Math.* **2022**, *106*, 30–235. [CrossRef]
31. Leonov, G.A.; Zvyagintseva, K.A. Piragas stabilization of discrete delayed feedback systems with a periodic pulse gain, Стабилизация по Пирагасу дискретных систем запаздывающей обратной связью с периодическим импульсным коэффициентом усиления (In Russian). *Vestn. St. Petersburg Univ. Math.* **2015**, *48*, 147–156. [CrossRef]
32. Leonov, G.A.; Zvyagintseva, K.A.; Kuznetsova, O.A. Pyragas stabilization of discrete systems via delayed feedback with periodic control gain. *IFAC-Pap.* **2016**, *49*, 56–61. [CrossRef]
33. Pyragas, K. Continuous control of chaos by self-controlling feedback. *Phys. Lett. A* **1992**, *170*, 421–428. [CrossRef]
34. Ritchie, H.; Mathieu, E.; Rodés-Guirao, L.; Appel, C.; Giattino, C.; Ortiz-Ospina, E.; Hasell, J.; Macdonald, B.; Beltekian, D.; Roser, M. Coronavirus Pandemic (COVID-19). *Our World Data*. 2020. Available online: https://ourworldindata.org/coronavirus (accessed on 27 August 2022).
35. Our World in Data. COVID Live—Coronavirus Statistics—Worldometer. *Coronavirus Pandemic (COVID-19)*. Available online: https://www.worldometers.info/coronavirus/ (accessed on 5 May 2022).
36. International Monetary Fund. *World Economic Outlook, April 2022: War Sets Back the Global Recovery*; IMF: Washington, DC, USA, 2022. Available online: https://www.imf.org/en/Publications/WEO/Issues/2022/04/19/world-economic-outlook-april-2022 (accessed on 5 May 2022).
37. World Bank. GDP (Constant 2015 US$)—India, Indonesia, Brazil, South Africa, Kazakhstan | Data. Available online: https://data.worldbank.org/indicator/NY.GDP.MKTP.KD?end=2020&locations=IN-ID-BR-ZA-KZ&start=2014 (accessed on 1 May 2022).
38. World Bank. Population, Total—India, Indonesia, Brazil, South Africa, Kazakhstan | Data. *The Worldbank*. Available online: https://data.worldbank.org/indicator/SP.POP.TOTL?end=2020&locations=IN-ID-BR-ZA-KZ&start=2014 (accessed on 1 May 2022).
39. Organization for Economic Cooperation and Development. Quarterly National Accounts. Available online: https://stats.oecd.org/Index.aspx?DataSetCode=QNA (accessed on 12 May 2022).

40. Dime, R.; Ginting, E.; Zhuang, J. Estimating Fiscal Multipliers in Selected Asian Economies. *ADB Econ. Work. Pap. Ser.* **2021**, *638*, 1–32. [CrossRef]
41. Raga, S. Fiscal Multipliers: A Review of Fiscal Stimulus Options and Impact on Developing Countries. IDRC. 2022. Available online: https://set.odi.org/fiscal-multipliers-a-review-of-fiscal-stimulus-options-and-impact-on-developing-countries/ (accessed on 5 May 2022).

Article

Forecasting a New Type of Virus Spread: A Case Study of COVID-19 with Stochastic Parameters

Victor Zakharov [1], Yulia Balykina [1], Igor Ilin [2] and Andrea Tick [3,*]

[1] Faculty of Applied Mathematics and Control Processes, Saint Petersburg State University, Universitetskaya Naberezhnaya 7-9, 199034 St. Petersburg, Russia
[2] Graduate School of Business Engineering, Peter the Great St. Petersburg Polytechnic University, 195251 St. Petersburg, Russia
[3] Keleti Károly Faculty of Business and Management, Óbuda University, 1034 Budapest, Hungary
* Correspondence: tick.andrea@kgk.uni-obuda.hu

Abstract: The consideration of infectious diseases from a mathematical point of view can reveal possible options for epidemic control and fighting the spread of infection. However, predicting and modeling the spread of a new, previously unexplored virus is still difficult. The present paper examines the possibility of using a new approach to predicting the statistical indicators of the epidemic of a new type of virus based on the example of COVID-19. The important result of the study is the description of the principle of dynamic balance of epidemiological processes, which has not been previously used by other researchers for epidemic modeling. The new approach is also based on solving the problem of predicting the future dynamics of precisely random values of model parameters, which is used for defining the future values of the total number of: cases (C); recovered and dead (R); and active cases (I). Intelligent heuristic algorithms are proposed for calculating the future trajectories of stochastic parameters, which are called the percentage increase in the total number of confirmed cases of the disease and the dynamic characteristics of epidemiological processes. Examples are given of the application of the proposed approach for making forecasts of the considered indicators of the COVID-19 epidemic, in Russia and European countries, during the first wave of the epidemic.

Keywords: artificial intelligence; balance model; CIR model; COVID-19; forecasting; modeling

MSC: 92D30; 93A30; 68T05; 62-07

1. Introduction

An outbreak of the coronavirus infection, COVID-19, caused by the new SARS-CoV-2 virus, quickly spread around the world at the end of 2019, affecting more than 200 countries. The consideration of infectious diseases caused by new, previously unexplored viruses from a mathematical point of view can reveal not only the important patterns of a pandemic, but also possible options for epidemic control and fighting the spread of infection. Mathematical models of disease transmission, as most experts note, can help to gain insight into the dynamics of the spread of infectious diseases and the potential role of different types of public health intervention strategies. The main problem in the absence of the availability of historical data for modeling the spread of a new virus arises when estimating the values of the input parameters of a particular model. Given the maximum uncertainty in the parameters of the spread of a new virus, difficulties naturally arise in assessing the potential dangers and scale of the epidemic. Traditionally used parameters in the classic Susceptible—Infected—Recovered (SIR)) model cannot be quantified with sufficient accuracy, which does not allow for eliminating the uncertainty of the future dynamics of coronavirus spread. The aim of this work is to study the possibilities of using a case-based rate reasoning (CBRR) model for predicting the rates of COVID-19 spread in Russia and extending it to other countries. The proposed approach considers the percentage increase as one of the most important parameters of the forecasting model, which has a stochastic nature. As part of the approach being developed, it is proposed to evaluate the trends in the key parameters, such as the percentage increase in the number of confirmed cases of the disease and the percentage increase in the number of deaths, and use the resulting trends in the subsequent forecast of the main statistical epidemic spread indicators' dynamics. Another important parameter is the so-called dynamic balance characteristic. It is based on the proposed principle of dynamic balance of epidemiological processes, which has not been previously used in models of epidemic spread. The new approach is based, first of all, on solving the problem of predicting the future dynamics of precisely random values of model parameters, instead of the values of real-time data. Predictive trajectories of the future values of epidemic data (the total number of confirmed cases (C), the number of active cases of the disease (I), and the dynamics of the total number of recovered (R) and dead (D)) were calculated by substituting into the model the values of these random parameters obtained during the analytical study. In the future, this algorithm could be used in modeling complex network systems with emergent intelligence.

2. Literature Overview

A systematic review of the models for predicting epidemics of the novel SARS-CoV-2 (COVID-19) coronavirus was conducted by the authors. The search was carried out in the Web of Science, Scopus and RCI databases. Additionally, the search results were analyzed in the Elsevier database, one of the largest European publishers. The following keywords were used: "forecasting", "prediction", "model", "emerging infection", "coronavirus" and "COVID". There are also several web portals aggregating research related to coronavirus such as, for example, according to the World Health Organization [1]. During the search, special attention was paid to studies related to forecasting models for the dynamics of the epidemic spread in the first wave, when there was a lack of information about the new coronavirus and the population did not develop immunity to the new infection. In general, there are four main approaches to modeling the spread of infectious diseases [2]: time series-based models, compartment models, agent-based (network) models, and models built using machine learning methods and heuristic approaches.

Regarding time-series based models, it should be noted that regression analysis and time series analysis are among the most well-known methods for predicting the incidence of a disease. Several researchers have used the Autoregressive Integrated Moving Average (ARIMA) model to predict the spread of the novel coronavirus pandemic. Examples of such studies include [3–6]. In general, although time series models are a popular forecasting tool, the use of this approach to estimate the spread of new infections has its limitations. In particular, the lack of statistics for previous periods and, as a consequence, unknown parameter values, do not allow building models of a sufficient degree of accuracy.

Most of the research is based on the Susceptible—Infected—Recovered (SIR) paradigm and its variations, which is described as a system of ordinary differential equations [7]. In the basic compartment model, a population of N people is considered. At each moment in time, each person belongs to one of the three groups (compartments): the Susceptible group (S), including people who have not yet encountered an infection; then, as the virus spreads among the population, they move to the Infected group (I); and then, into the Removed (R) group (recovered or deceased). The possibility of reinfection in this model is not taken into consideration. It is also assumed that the size of the population remains unchanged. The so-called SIR epidemic model, described as a system of three ordinary differential equations for the variables S, I and R, was presented for the first time in a paper by Kermack and McKendrick in 1927 [8,9].

Many researchers prefer to use the SIR model because of the small number of input parameters required. However, sometimes this advantage arises from the oversimplification of the model due to relatively unrealistic assumptions. For example, the model assumes the homogeneous mixing of the population, which means that all individuals in the population are equally likely to come into contact with each other. This does not reflect the human social structures in which most contact occurs in localized communities. The SIR model also assumes a closed population with no migration, births or deaths from causes other than epidemics. In addition to the SIR model, various modifications of it are often used such as, for example, the SEIR model, which adds an Exposed (E) parameter for people who have already been infected but are not yet infectious, as well as the SIRD model, which provides separate parameters for recovered people (those who survived the disease and are now immune) and deceased people. In general, up to the present, only several dozen SIR models and their varieties have been published on COVID-19. In particular, these have been the SIR [10–13], SEIR [14–22], SIRD [23,24], SIR-X [25], SEIQR [26] and SEIHR [27] models.

Special mention should be made of research devoted to models that take into account the delay of input parameters [28–33]. Mostly, these research models are focused on evaluating the parameters of compartment models based on the assumptions about the delay between the diagnosis and the time of infection.

Network models can be considered as a discrete variant of compartment models. In modern mathematical epidemiology, network models represent one of the latest methods for analyzing and modeling complex epidemiological systems. Compared to compartment models, network models are more detailed and allow the consideration of each participant separately, while the interaction between people is presented in the form of a complex graph of social connections. Compartment models make it possible to assess the overall dynamics of the rapid spread of the epidemic, whereas network models make it possible to simulate the effectiveness of certain measures to contain the spread of infection [34]. It is also possible to consider the epidemic as a complex network model with emergent intelligence, followed by the use of machine learning and data mining methods to analyze its dynamics [35].

Models using machine learning methods are gaining popularity. Artificial intelligence and machine learning have long been used in epidemiology. Machine leaning is a powerful tool for finding the relationships between inputs and outputs when analytical research is difficult. In some cases, the use of such heuristic approaches for the early detection of epidemiological risks can improve the quality of forecasting. One study examined the possibility of using artificial neural networks (ANNs) to predict the spread of COVID-19 [36]. The results of the network of the developed architecture for some of the regions reached an accuracy of 87%. At the same time, the authors emphasized that a large amount of historical data was required for the correct training of the ANNs. Other examples of machine learning models include the research in [37–42]. It should be noted that the forecasting horizon in this research was several days. There have been attempts to develop models with a longer forecasting horizon (i.e., two weeks and more). An example of such a study is [43]. The studies [44,45] can also be attributed to heuristic models with elements of machine learning. The tSIR model in [44] suggests using a combination of the classical ARIMA model together with neural networks for outbreak spread forecasting.

Another approach worth mentioning is the case-based reasoning approach, which is based on the idea of finding possible solutions to the problem based on existing solutions for similar situations [46]. The research in [47,48] describes a new case-based rate reasoning (CBRR) model based on this approach, for predicting the future values of the main parameters of the coronavirus epidemic in Russia. This model makes it possible to build short-term forecasts based on the similarities in the dynamics of percentage growth in other countries. A new heuristic method is also described for estimating the duration of the transition process of the percentage increase between the given levels, taking into account the information on the dynamics of epidemiological processes in the countries of the distribution chain. A detailed review of the possibilities of using machine learning methods to predict the spread of COVID-19 can be found in [49].

Mention should also be made of a number of studies devoted to predicting the dynamics of mortality of the new coronavirus. In this case, either the number of fatalities was analyzed, or the case fatality ratio (CFR). Examples of studies focused on predicting the mortality of a new coronavirus are [39,50–55].

3. Materials and Methods

The analysis of the available recent literature that provides forecasts of the spread of epidemics shows that the use of traditional classical models rarely provides an acceptable forecasting accuracy for a period of more than 7–10 days in advance [6,39,40]. For example, in [10], Deal and Maken used the SIR model to predict the confirmed cases of COVID-19 in the Eastern Mediterranean region (Iran, Iraq, Saudi Arabia, UAE, Lebanon, Egypt and Pakistan). The authors estimated that by 20 June 2020, 2.12 million cases in Iran, 0.58 million in Saudi Arabia and 0.51 million in Pakistan were expected. In fact, the number of recorded cases of COVID-19 infection as of 20 June 2020 was officially 202,584 cases in Iran, 176,617 cases in Pakistan and 154,233 cases in Saudi Arabia (according to the CSSE Center at Johns Hopkins University) [56]. In [57], a simple moving average method was used to predict the confirmed cases of COVID-19 in Pakistan. The authors predicted over 35,000 cases by the end of May 2020. Real-world data showed that at the end of May 2020, 72,460 cases of new coronavirus infection were recorded (2 times higher than the forecasted values).

In connection with this, a new approach to modeling the spread of the COVID-19 epidemic using the percentage increase in the total number of confirmed cases of the disease is proposed. In the English abbreviation, the new model is called Confirmed Cases C(t)—Infected I(t)—Removed (R(t)) (CIR). To predict the dynamics of the number of confirmed cases in Russia, a case-based rate reasoning (CBRR) model was developed, based on the precedent method and using data on the spread of infection in European countries where the pandemic began earlier than in Russia—namely, in Italy, Spain, France and Great Britain—to construct the future values of the parameters used in the model [48]. These countries in the model are called predecessor countries, whereas Russia is called a follower country.

Let $t \in \overline{1,T}$ be a positive integer, corresponding to day t, and $r(t)$ be the value of the percentage increase in detected cases during the epidemic from the beginning to day T. Let the considered time horizon be divided into M intervals $(T_{m-1}, T_m]$, $0 \leq T_{m-1} < T_m \leq T$, $m = \overline{1,M}$. Then, for any interval $(T_{m-1}, T_m]$ and any $k = 1, 2, \ldots, (T_m - T_{m-1})$, the dynamics of changes in the number of confirmed cases can be described as follows:

$$C(T_{m-1}+k) = \left(1 + \frac{r(T_{m-1}+k)}{100}\right) C(T_{m-1}+k-1), \quad k = 1, 2, \ldots, (T_m - T_{m-1}).$$

In the process of constructing the predicted trajectory of the confirmed cases of dynamics of changes over the interval $(T_{m-1}, T_m]$, a sequence of predicted values of the percentage increase $\widetilde{r}(T_{m-1}+k), k = 1,2,\ldots,(T_m - T_{m-1})$ is generated. For example, these estimates can be obtained by piecewise linear approximations on each of the M intervals. At the same time, the lengths of the intervals in the considered country are estimated by taking into account information received from the predecessor countries. Then, the predicted value of the number of confirmed cases can be calculated as follows:

$$\widetilde{C}(T_m) = \left(1 + \frac{\widetilde{r}(T_{m-1}+1)}{100}\right) \times \cdots \times \left(1 + \frac{\widetilde{r}(T_m)}{100}\right) \widetilde{C}(T_{m-1}).$$

The effectiveness of the CBRR model was confirmed by examples of fairly accurate predictions of future values of the total number of confirmed cases of the disease, built in the course of the experiments, for several prediction intervals in April to June 2020. The results of using the CIR model to predict the dynamics of the number of active cases in Russia in the second half of 2021 were presented in the paper, [58], and in the reports by the Izvestia newspaper [59].

The main idea of the principle of dynamic balance of epidemiological processes lies in the fact that the past values of the total number of cases are close enough to the values of the total number of recovered and deceased patients at the current time. Concurrently, the time intervals through which such a balance is observed change rather predictably over time, although, like the percentage increase, they take random values.

It should be noted that the new approach is based primarily on the problem of modeling the dynamics of random values of model parameters. The trajectories of the model describing the dynamics of the total number of active cases, and the number of confirmed and dead cases are calculated by substituting the obtained values of these parameters into the model. The results of using the proposed approach can be found in the notes and scientific papers posted on the page by the Center for Intelligent Logistics of St. Petersburg State University [60].

3.1. CIRD Balance Model

The CIR model proposed in [58] can be generalized to the case of isolating the total number of deaths as a separate variable. Let us denote by $C(t)$, the total number of confirmed cases of infection (cumulative cases and confirmed cases) from the beginning of the epidemic to day t, and by $\Delta(t)$, the number of new cases registered per day. The number of infected for $t = 0$ is set equal to one. Both functions are random variables with non-negative values. Taking into account the introduced notation for $t = 1, 2, 3, \ldots$, we have:

$$C(t) = C(t-1) + \Delta(t).$$

Let

$$r(t) = 100 \frac{\Delta(t)}{C(t-1)} \quad (1)$$

where $\Delta(t) = C(t) - C(t-1)$.

Then, we have the following formula:

$$C(t) = \left(1 + \frac{r(t)}{100}\right) \times C(t-1). \quad (2)$$

Here, the parameter $r(t)$ is interpreted as the ratio of the percentage of the absolute increase in the total number of detected infections per day t to the total number of detected infections on the previous day.

We will further call this parameter $r(t)$, for the percentage increase in the total number of detected cases per day t. Considering that $\Delta(t)$ is a random variable that takes non-negative values, the percentage increase is also a non-negative random variable.

The value of $C(t)$ on any day T is calculated by the following formula:

$$C(T) = \prod_{t=1}^{T}\left(1 + \frac{r(t)}{100}\right). \quad (3)$$

Let us fix some values $t_0 \geq 0$ and $T > t_0$, such that $R(T) > C(t_0) > 1$, where $R(t)$ is the number of recoveries and deaths, and $C(t)$ is the number of confirmed cases. Taking into account the non-decreasing functions $C(t)$ and $R(t)$, as well as the fact that $C(t) \geq R(t)$ for any $t \geq t_0$, such a value T exists. In fact, the existence of such values t_0 and T means that patients who become ill by a point in time t_0 will recover or die in a finite time.

Let us consider the integer programming problem:

$$\min_{t_0 \leq t \leq T} t \quad (4)$$

$$C(t) \geq R(T). \quad (5)$$

Considering the properties of the functions $C(t)$ and $R(t)$, the set of feasible solutions to such a problem is not empty. Let us denote the solution to problem (4)–(5) by $\tau(T)$.

Note that condition (5) is not fulfilled for $t = \tau(T) - 1$. Then, taking into account that the function $C(t)$ does not decrease, the inequality $R(T) \geq C(\tau(T) - 1)$ holds. Thus, the following theorem is true:

Theorem 1. *(Principle of dynamic balance) Let the values $t_0 \geq 0$ and $T > t_0$ be given, such that $R(T) > C(t_0) > 1$. Then, for the solution $\tau(T)$ of the problem (4)–(5), the value of $R(T)$ per day T satisfies the inequalities:*

$$C(\tau(T)) \geq R(T) \geq C(\tau(T) - 1). \tag{6}$$

Condition (6) means that the number of people who recovered and died on a certain day depends on the total number of cases recorded in the past, namely, $t - \tau(t)$ days ago. Thus, using condition (6), it is possible to establish a dynamic balance between the values of the functions $R(t)$, $C(\tau(t) - 1)$ and $C(\tau(t))$.

Note that for any T the following equality holds:

$$C(T) = I(T) + R(T) \tag{7}$$

where $C(t)$ is the number of confirmed cases, $R(t)$ is the number of recoveries and deaths, and $I(t)$ is the number of infected cases.

This balance ratio means that the group of detected cases on any day can be divided into those who are still ill and those who have recovered or died by that day. Let us use Formula (2), inequality (6) and balance relation (7), and write down the system of discrete equations:

$$C(t) = \left(1 + \frac{r(t)}{100}\right) \times C(t-1) \tag{8}$$

$$I(t) = \left(1 + \frac{r(t)}{100}\right) \times C(t-1) - R(t) \tag{9}$$

$$R(t) = \lambda_t C(\tau(t) - 1) + (1 - \lambda_t) C(\tau(t)). \tag{10}$$

The choice of the parameter $\lambda_t \in [0, 1]$ for any t guarantees that the $R(t)$ lies in the interval $[C(\tau(t) - 1), C(\tau(t))]$.

The model, the dynamics of which are described as the system of discrete Equations (8)–(10), will be abbreviated as the CIR model in what follows.

3.1.1. CIR Model

The CIR model uses the concept of the percentage increase in the total number of $C(t)$ and takes into account the balance ratio between the total number of cases at $\tau(t)$ and $\tau(t) - 1$, and the total number of $R(t)$ of those who recovered and died at the moment t. For this reason, we may also refer to this model as the balanced epidemic model based on percentage growth.

Definition 1. *The function $\vartheta(t) = t - \tau(t)$ will be called the dynamic balance characteristic of the epidemic.*

Definition 2. *Let us call the epidemiological process stationary over a period of time $[t_1, t_2]$, if $\vartheta(t) = const$ for all $t \in [t_1, t_2]$.*

Let $\sigma(t)$ denote the segment of the total number of people who died during the epidemic, $D(t)$, out of the total number $R(t)$. The value $\sigma(t)$ is the coefficient (indicator) of the current fatality of the epidemiological process. Then, the total number of deceased people by the time t can be written as the equation:

$$D(t) = \sigma(t) R(t).$$

Taking into account the representation of $R(t)$ in Equation (10), we obtain the following expression:

$$D(t) = \sigma(t)(\lambda_t C(\tau(t) - 1) + (1 - \lambda_t) C(\tau(t))) \tag{11}$$

When the system of Equations (8)–(10) is supplemented with Equation (11), we obtain a new system of discrete equations, which we will call the CIRD model.

Note that the number $d(t)$ of people who died on day t can be calculated by the formula:

$$d(t) = \sigma(t)(\lambda_t C(\tau(t) - 1) + (1 - \lambda_t) C(\tau(t))) - \sigma(t-1)(\lambda_{t-1} C(\tau(t-1) - 1) + (1 - \lambda_{t-1}) C(\tau(t-1)))$$

If we approximate the trends for the values of $r(t)$, $\sigma(t)$ and λ_t for the forecast interval $[t, t + \nu(t)]$ and use the obtained values, then we can obtain a forecast of future values of daily mortality for this interval.

4. Results

4.1. Modeling Experiments

Studying the statistics of the COVID-19 pandemic in Russia and other countries, it can be noted that at certain intervals, the dynamic balance characteristic $\vartheta(t)$ is constant; however, in general, it changes over time, although its volatility is limited. Examples of the behaviour of the dynamic balance characteristic, introduced by the authors, are shown in Figure 1. Thus, in the period from May to December 2020 in Russia, the values for $\vartheta(t)$ were in the range from 19 to 35 days. For Italy, the values of the characteristic of the dynamic balance were higher, from 25 to 55 days. In Germany, the characteristic value varied from 12 to 20 days.

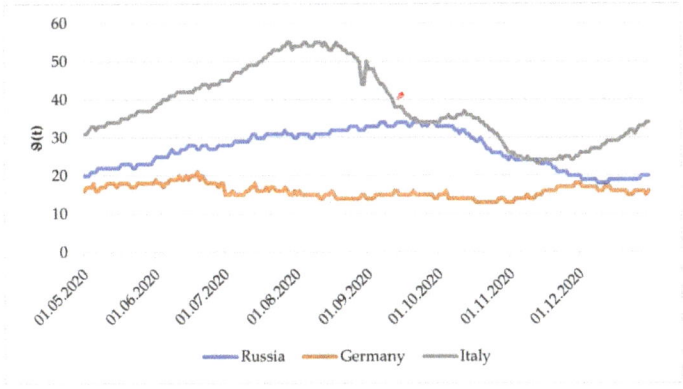

Figure 1. $\vartheta(t)$ dynamics for Russia, Germany and Italy during the time period of May to December 2020.

4.1.1. Case of Russia

Let us consider in more detail the data of the epidemiological process in Russia. As a demonstration of the proposed algorithm, according to the available data on $C(t)$ and $R(t)$, we determined the moments of time corresponding to admissible solutions of the problem (8)–(9) in the period from 12 May to 19 May 2020, where information was available on the previous 22 days, that is, for $T = 23, 24, \ldots, 30$, respectively. We estimated the values of the functions $\tau(t)$ and $\vartheta(t)$, the intervals of the possible values of the function $R(t)$ in the interval [23–30], as well as the deviations $R(t)$ from these intervals. Table 1 shows the solutions obtained, as well as the calculated values of the characteristic $\vartheta(t)$, the estimate of the intervals of future possible values $R(T)$ with a constant characteristic of the dynamic balance ($\vartheta(t) = 22$), and the deviation of the actual value from the estimated interval. In the period from 12 May to 19 May, the function $\vartheta(t)$ was not equal to a constant; that is, the epidemiological process under consideration during this time period was not stationary.

Note also that an increase in the value of the function $\vartheta(t) = t - \tau(t)$ by one day on 18 and 19 May in relation to the period from 12 May to 17 May 2020, when it was constant (22 days), led to a deviation of the actual values R(18) and R(19) by 1657 and 1982, respectively.

Table 1. Values of the dynamic balance characteristic and its assessment (period from 12 May to 19 May 2020).

t	Date	Number of $C(t)$	Number of $R(t)$	$\tau(t)$	$\vartheta(t)$	Intervals of Possible Values of $R(t)$	Deviation $R(t)$ from the Interval
23	12 May 2020	232,243	45,628	1	22	≤47, 121	0
24	13 May 2020	242,271	50,215	2	22	[47,121; 52,763]	0
25	14 May 2020	252,245	55,835	3	22	[52,763; 57,999]	0
26	15 May 2020	262,843	60,644	4	22	[57,999; 62,773]	0
27	16 May 2020	272,043	65,739	5	22	[62,773; 68,622]	0
28	17 May 2020	281,752	70,004	6	22	[68,622; 74,588]	0
29	18 May 2020	290,678	72,931	6	23	[74,588; 80,949]	1657
30	19 May 2020	299,941	78,967	7	23	[80,949; 87,147]	1982

Let us consider the results of applying the developed approach to forecasting the indicators of the epidemic in Russia, for the period from 6 June to 30 June 2020, based on the information available on 5 June 2020.

Figure 2 shows the predicted trajectories built over the period from 6 June to 30 June 2020, which were compared with the actual trajectories. The average deviation of the trajectory of the total number of cases forecasted compared to the actual one was 0.37%. On 30 June 2020, the deviation was found to be minus 1.17%.

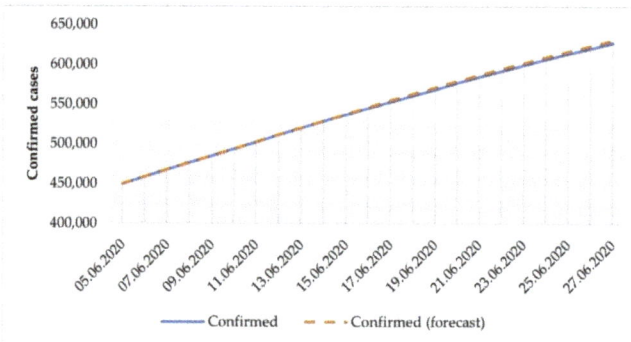

Figure 2. Dynamics of the total number of $C(t)$, the actual and predicted trajectories.

The dynamics of the function $\vartheta(t)$ for the period under consideration are shown in Figure 3.

Assuming the stationarity of the epidemiological process from 6 June to 30 June 2020, a constant value of the dynamic balance characteristic corresponding to the actual value at the beginning of the forecasting horizon (25 days) was taken for the calculation. The graphs of the calculated and actual values of the total number of deaths $D(t)$ and active cases $I(t)$ are shown in Figures 4 and 5.

The deviation of the modeled trajectory from the actual trajectory of the number of active patients was, on average, 3.63%. The mean absolute percentage error (MAPE) value for this forecast was 4.14%. For the dynamics of the number of deaths, the forecasted MAPE of the considered interval was 3.17%. The maximum deviation was 6.5%.

By using the actual values of $\vartheta(t)$, the graphs of the actual and predicted values of $D(t)$ practically coincide. Similarly, the graphs of the total number of active patients in the case of using the actual $\vartheta(t)$ values practically coincide. In the interval from 6 June to 30 June 2020, the deviation from the actual trajectory of the number of active patients averaged to 1.13% and the number of deaths was 0.4%.

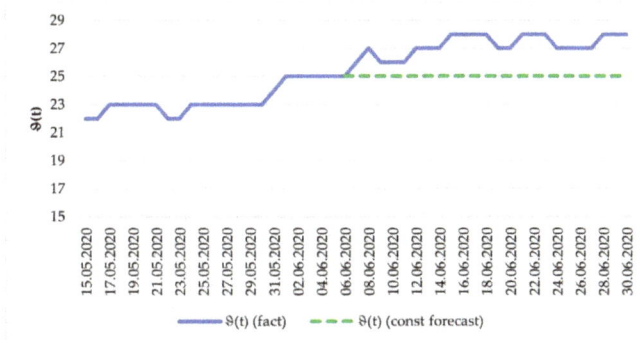

Figure 3. Actual and modeled values of the dynamic balance characteristic $\vartheta(t)$ from 15 May to 30 June 2022.

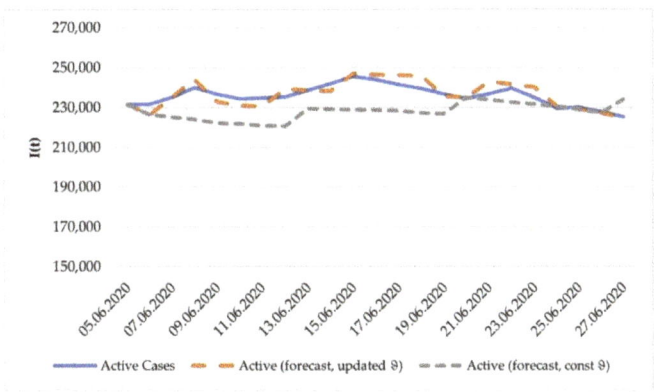

Figure 4. Dynamics of $I(t)$, the actual and modeled trajectories in the interval from 6 June to 30 June 2020.

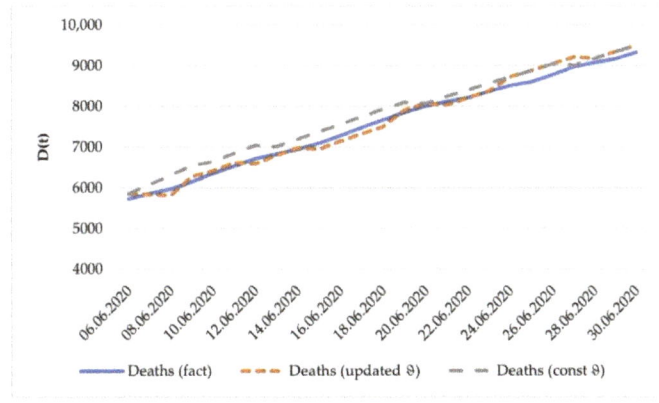

Figure 5. Dynamics of $D(t)$, the actual and calculated trajectories in the interval from 6 June to 30 June 2020 in a stationary case and in the case of updated values of $\vartheta(t)$.

Let us consider the results of applying the developed approach to forecasting the indicators of the epidemic in Russia in the period from 29 November to 22 December 2020, based on the information available on 28 November. Assuming the stationarity of the epidemiological process, the theta value taken was equal to 20. The change of the function $\vartheta(t)$ over the considered time interval is shown in Figure 6.

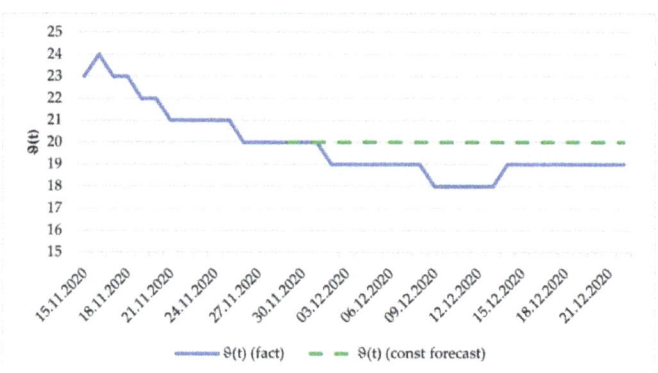

Figure 6. Actual and modeled values of the dynamic balance characteristic $\vartheta(t)$ from 15 November to 30 December 2020.

The dynamics of the modeled and actual values of the total number of confirmed cases $C(t)$, deaths $D(t)$ and active cases $I(t)$ are shown in Figures 7–9.

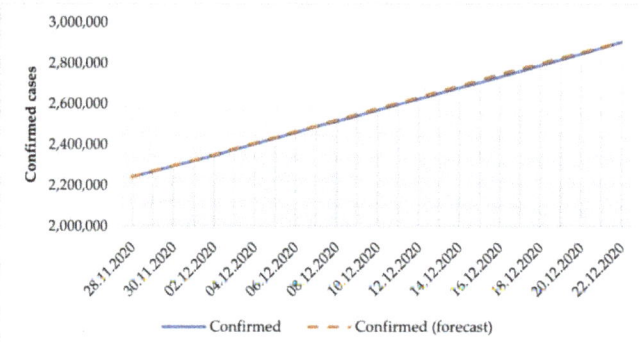

Figure 7. Dynamics of $C(t)$, the actual and modeled trajectories in the interval from 1 December to 22 December 2020.

The deviation of the modeled trajectory from the actual trajectory of the number of active patients in terms of MAPE was 1.1%. For the dynamics of the predicted trajectory of the deceased, the forecasted MAPE of the considered interval was 3.1%. However, by taking into account the update in information about the dynamic balance characteristic, the deviations of the actual graphs from the predicted values of both $I(t)$ and $D(t)$ decreased. Thus, in the considered interval, the deviation from the actual trajectory of the number of active patients in terms of MAPE was 0.9% and the number of deaths was 2.9%.

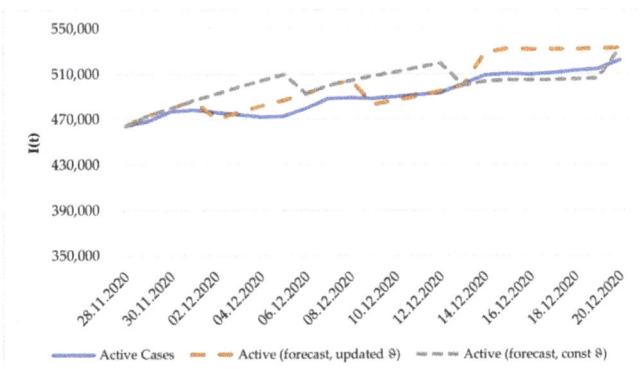

Figure 8. Dynamics of $I(t)$, the actual and modeled trajectories in the interval from 1 December to 22 December 2020, in a stationary case and at updated values of $\vartheta(t)$.

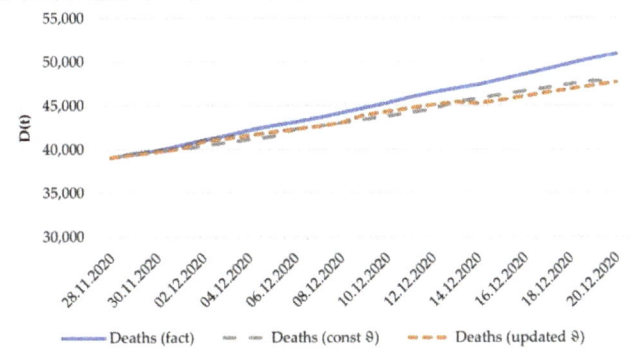

Figure 9. Dynamics of $D(t)$, the actual and modeled trajectories in the interval from 1 December to 22 December 2020, in a stationary case and at updated values of $\vartheta(t)$.

4.1.2. Case of Europe

Let us consider the possibilities of applying this approach in other countries. For consistency, similar forecasting time intervals were analyzed. The daily updated statistical data published on the Johns Hopkins University portal were used as the initial information for modeling [56]. The dynamics of $\vartheta(t)$ are shown in Figure 10.

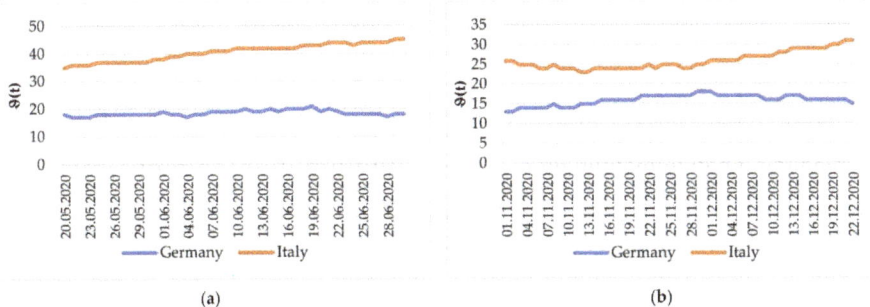

Figure 10. Dynamics of changes in the values of $\vartheta(t)$ in Germany and Italy (**a**) from May to June 2020 and (**b**) from November to December 2020.

Under the assumption of the piecewise stationarity of the epidemiological process, a constant value of the dynamic balance characteristic corresponding to the actual value at the beginning of the forecasting horizon was used. The case of the dynamic updating of information when the value of the characteristic was updated was also considered. The modeling was carried out at the same time as the intervals that were used to analyze the situation in Russia. The graphs of the modeled and actual values of the total number of confirmed cases, the number of deaths $D(t)$ and active cases $I(t)$ are shown in Figures 11–16.

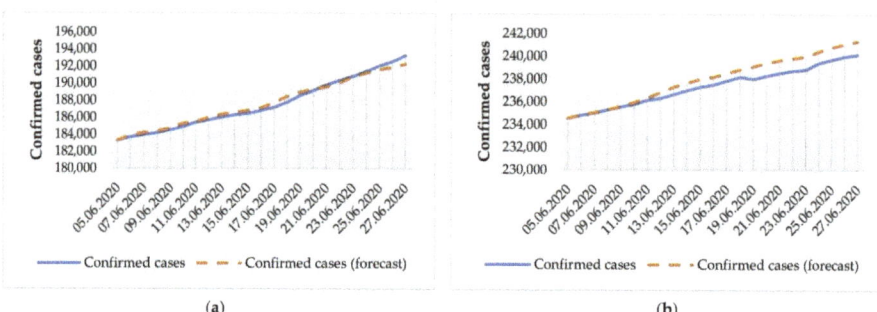

Figure 11. Dynamics of $C(t)$, the actual and modeled trajectories in the interval, from 6 June to 30 June 2020, in (**a**) Germany and (**b**) Italy.

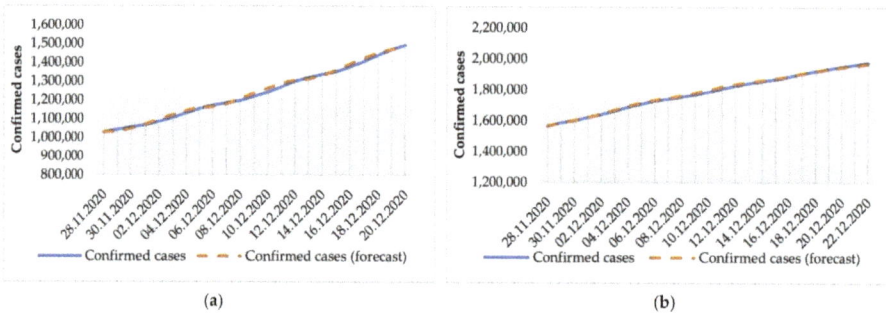

Figure 12. Dynamics of $C(t)$, the actual and modeled trajectories in December 2020, in (**a**) Germany and (**b**) Italy.

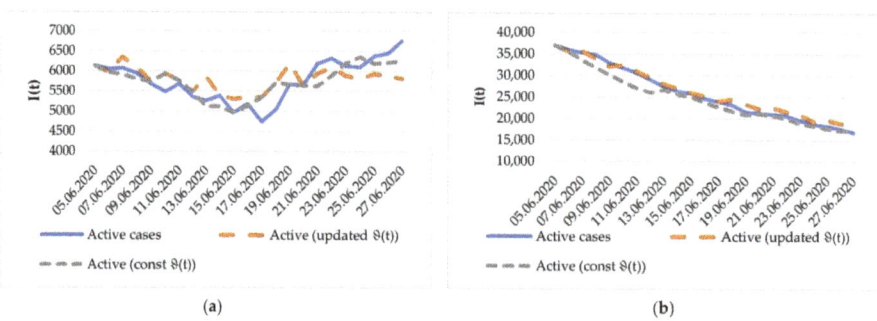

Figure 13. Dynamics of $I(t)$, the actual and modeled trajectories in the interval, from 6 June to 30 June 2020, in (**a**) Germany and (**b**) Italy.

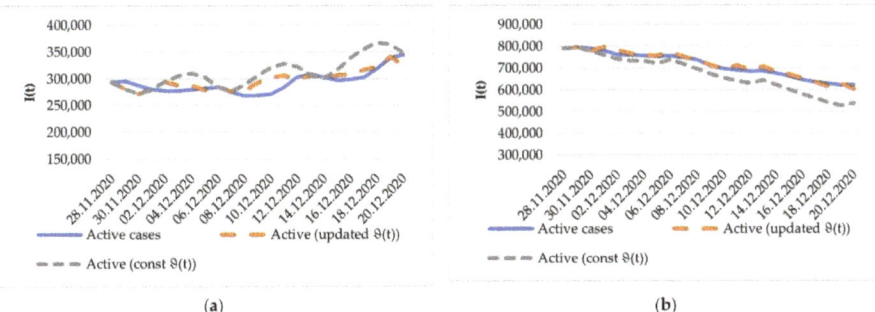

Figure 14. Dynamics of $I(t)$, the actual and calculated trajectories for the interval of 1 to 20 December 2020, in a stationary case and with the updated values of $\vartheta(t)$, in (**a**) Germany and (**b**) Italy.

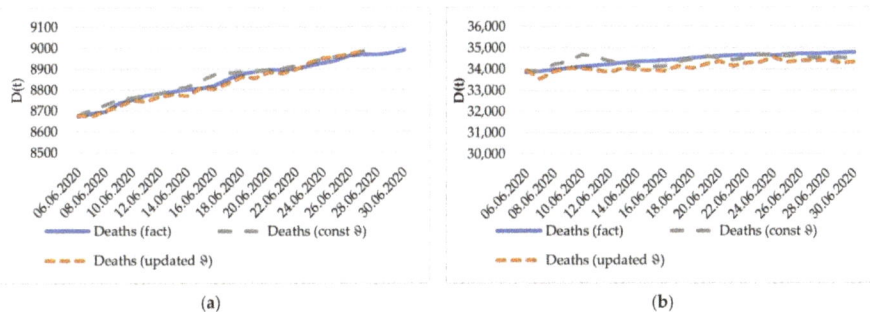

Figure 15. Dynamics of $D(t)$, the actual and calculated trajectories in the interval from 6 June to 30 June 2020, in a stationary case and with the updated values of $\vartheta(t)$, in (**a**) Germany and (**b**) Italy.

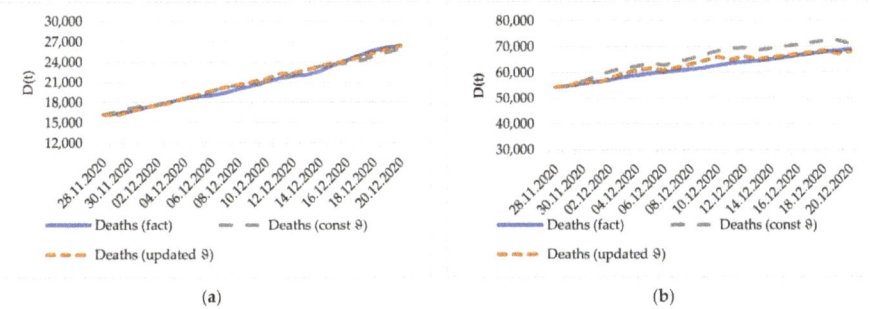

Figure 16. Dynamics of $D(t)$, the actual and calculated trajectories for the interval of 1 to 20 December 2020, in a stationary case and with the updated values of $\vartheta(t)$, in (**a**) Germany and (**b**) Italy.

For Germany, the value of $\vartheta(t) = 18$ days was used in the summer 2020 forecast. In the winter 2020 case, the $\vartheta(t)$ value was 17 days. The deviation of the predicted trajectory of the number of active cases in summer was 3.22% in terms of the MAPE indicator. When we used the dynamically updated $\vartheta(t)$, the MAPE value dropped to 2.8%. When assessing the dynamics of epidemic indicators in December 2020, the MAPE for the trajectory of active infections was 2.7%. When the information was dynamically updated and actual theta values were used, the variance was reduced to 1.25%. The deviation in the total number of cases at both prediction intervals did not exceed 0.25%. For the dynamics of the total number of deaths, the forecast MAPE of the intervals under consideration was 0.7%.

In the case of Italy, the value of $\vartheta(t) = 40$ days was used when constructing the forecast in summer. In winter, the value of $\vartheta(t)$ was 25 days. The deviation of the number of active cases in the interval from 6 June to 30 June 2020 was 3.85% (3.3% in the case of a dynamic update of information about the $\vartheta(t)$ value). When assessing the dynamics of epidemic indicators in December 2020, the MAPE for the estimated trajectory of active infections was 9%. When the information was dynamically updated and actual theta values were used, the variance was reduced to 0.5%. The deviation of the total number of cases at both prediction intervals did not exceed 0.1%. For the dynamics of the total number of deaths, the forecast MAPE of the intervals under consideration did not exceed 5.7%. Summary results of the assessment of forecast accuracy in terms of MAPE are presented in Table 2.

Table 2. Forecast results for the constant and dynamically updated $\vartheta(t)$.

Country	$\vartheta(t)$	$C(t)$ Forecast (MAPE)	Constant $\vartheta(t)$		Dynamically Updated $\vartheta(t)$	
			$I(t)$ Forecast (MAPE)	$D(t)$ Forecast (MAPE)	$I(t)$ Forecast (MAPE)	$D(t)$ Forecast (MAPE)
			6–30 June 2020			
Russia	25	0.37%	4.14%	3.17%	1.13%	0.4%
Germany	18	0.25%	3.22%	0.7%	2.8%	0.7%
Italy	40	0.1%	3.85%	5.7%	3.3%	5.7%
			1–20 December 2020			
Russia	20	0.30%	1.10%	3.10%	0.9%	2.9%
Germany	17	0.25%	2.70%	0.7%	1.25%	0.7%
Italy	25	0.10%	9%	5.7%	0.5%	5.7%

Based on [58], we described an example of using the SIR model for forecasting the dynamics of active cases in Russia during the period from 20 April 2020 to 19 May 2020. The results of the $I(t)$ forecasts using the SIR and CIRD models are presented in Table 3. It contains information on real data of the $I(t)$ dynamics, the forecasted trajectories based on the SIR and CIRD models, as well as the deviations of the predicted values from the real ones in the form of differences in values. The forecast MAPE for the trajectory of the active cases using the SIR model was 11%, while the application of the proposed technique based on the principle of dynamic balance gave a value of 1.85% MAPE for the same time period.

Table 3. The $I(t)$ forecast results for the SIR and CIRD models, for the period from 20 April 2020 to 19 May 2020, in Russia.

Data	Actual $I(t)$	$I(t)$ Forecast, SIR	$I(t)$ Forecast, CIRD	SIR Difference	CIRD Difference
20 April 20	43,270	43,270		0	
21 April 20	48,434	46,332		−2102	
22 April 20	53,066	49,612	53,268	−3454	−202
23 April 20	57,327	53,122	58,366	−4205	−1039
24 April 20	62,421	56,881	63,526	−5540	−1105
25 April 20	67,657	60,906	68,840	−6751	−1183
26 April 20	73,435	65,215	75,652	−8220	−2217
27 April 20	79,007	69,828	81,619	−9179	−2612
28 April 20	84,235	74,767	87,616	−9468	−3381
29 April 20	88,138	80,055	93,575	−8083	−5437
30 April 20	93,806	85,716	101,582	−8090	−7776
01 May 20	100,042	91,777	108,151	−8265	−8109
02 May 20	107,819	98,265	114,394	−9554	−6575
03 May 20	116,768	105,211	120,414	−11,557	−3646
04 May 20	125,817	112647	129,095	−13,170	−3278
05 May 20	134,054	120,607	135,054	−13,447	−1000

Table 3. Cont.

Data	Actual $I(t)$	$I(t)$ Forecast, SIR	$I(t)$ Forecast, CIRD	SIR Difference	CIRD Difference
06 May 20	143,065	129,128	140,395	−13,937	2670
07 May 20	151,732	138,249	149,063	−13,483	2669
08 May 20	159,528	148,012	154,160	−11,516	5368
09 May 20	164,933	158,462	158,861	−6471	6072
10 May 20	173,467	169,647	163,065	−3820	10,402
11 May 20	179,534	181,619	166,203	2085	13,331
12 May 20	186,615	194,432	171,332	7817	15,283
13 May 20	192,056	208,144	175,275	16,088	16,781
14 May 20	196,410	222,819	179,801	26,409	16,609
15 May 20	202,199	238,524	184,952	36,325	17,247
16 May 20	206,304	255,329	189,177	49,025	17,127
17 May 20	211,748	273,311	193,418	61,563	18,330
18 May 20	217,747	292,551	197,381	74,804	20,366
19 May 20	220,974	313,136	201,606	92,162	19,368

5. Discussion and Conclusions

The present study focused on the possibilities of analyzing the dynamics of changes in the indicators of the spread of the COVID-19 during the first wave of epidemic. A distinctive feature of such periods is the lack of sufficient statistics. On the other hand, vaccination had not yet been initiated, so everyone could be considered as having been a potential patient. Percentage growth was used as the most important parameter of the model, which has a stochastic nature. Another important result of the study was the description of the principle of dynamic balance of epidemiological processes previously not used by other researchers to model epidemics. The principle lies in the fact that the past values of the total number of cases are close enough to the values of the total number of recovered and deceased patients at the current time. At the same time, the time intervals through which such a balance is observed change in time rather predictably, although like the percentage increase, they take random values. This study considered the case when the dynamic balance characteristic was constant during the forecasting horizon.

Following the proposed approach for choosing the parameters of epidemic modeling, this paper presented a new discrete CIRD model of the epidemic spread, which uses random values of the percentage increase and characteristics of the dynamic balance to describe the dynamics of the total number of cases, the total number of recovered and deaths, and the number of active cases of the disease. To predict the values of the total number of cases, the CBRR model was used, which was previously proposed and tested by the authors. It first predicts the percentage increase in the total number of cases in the medium term using the precedent method and intelligent algorithms for extracting the necessary data using a special class of the predictive dynamic regression model (PDRM) model [58], and then uses the obtained values to predict the total number of confirmed cases. The forecasting horizon of the CIRD model, as a rule, is limited by a value equal to the value of the dynamic balance characteristic of the epidemic, calculated on the basis of the principle of dynamic balance of the epidemiological processes formulated in Theorem 1. Since the characteristic of the dynamic balance is a random variable that depends on time and takes integer values, the possibilities for using the CIRD model to predict the dynamics of non-stationary epidemiological processes are still limited. However, as the calculations show, the forecasting accuracy when using the proposed model, even in the case of the non-stationarity of epidemiological processes in the forecasting interval, is quite high (see Figures 11–16). For example, when predicting the dynamics of active cases of the disease, MAPE deviations in Germany ranged from 2.7% to 3.2%, in Italy, from 3.8% to 9%, and in Russia, from 1.1% to 4.1%. At the same time, the MAPE indicator for predicting the dynamics of deaths did not exceed 5.7% for all countries and time intervals considered.

Author Contributions: Conceptualization, V.Z. and Y.B.; methodology, V.Z. and Y.B.; validation, V.Z., Y.B., I.I. and A.T.; writing—original draft, V.Z., Y.B., I.I. and A.T.; writing—review and editing, V.Z., Y.B., I.I. and A.T. All authors have read and agreed to the published version of the manuscript.

Funding: This research was partially funded by the Ministry of Science and Higher Education of the Russian Federation as part of the World-Class Research Center program: Advanced Digital Technologies (contract No. 075-15-2020-934 dated by 17 November 2020).

Conflicts of Interest: The authors declare no conflict of interest. The funders had no role in the design of the study; in the collection, analyses or interpretation of data; in the writing of the manuscript; or in the decision to publish the results.

References

1. World Health Organization. Available online: https://search.bvsalud.org/global-literature-on-novel-coronavirus-2019-ncov/ (accessed on 15 December 2021).
2. Shinde, G.R.; Kalamkar, A.B.; Mahalle, P.N.; Dey, N.; Chaki, J.; Hassanien, A.E. Forecasting Models for Coronavirus (COVID-19): A Survey of the State-of-the-Art. *SN Comput. Sci.* **2020**, *1*, 197. [CrossRef]
3. Moftakhar, L.; Seif, M.; Safe, M.S. Exponentially increasing trend of infected patients with COVID-19 in Iran: A comparison of neural network and ARIMA forecasting models. *Iran. J. Public Health* **2020**, *49*, 92–100. [CrossRef]
4. Singh, S.; Chowdhury, C.; Panja, A.K.; Neogy, S. Time Series Analysis of COVID-19 Data to Study the Effect of Lockdown and Unlock in India. *J. Inst. Eng. India Ser. B* **2021**, *102*, 1275–1281. [CrossRef]
5. Harvey, A.; Kattuman, P. A farewell to R: Time-series models for tracking and forecasting epidemics. *J. R. Soc. Interface* **2021**, *18*, 20210179. [CrossRef]
6. Aditya Satrio, C.B.; Darmawan, W.; Nadia, B.U.; Hanafiah, N. Time series analysis and forecasting of coronavirus disease in Indonesia using ARIMA model and PROPHET. *Procedia Comput. Sci.* **2021**, *179*, 524–532. [CrossRef]
7. Chen, S.; Robinson, P.; Janies, D.; Dulin, M. Four Challenges Associated with Current Mathematical Modeling Paradigm of Infectious Diseases and Call for a Shift. *Open Forum Infect. Dis.* **2020**, *7*, ofaa333. [CrossRef] [PubMed]
8. Kermack, W.O.; McKendrick, A.G. A contribution to the mathematical theory of epidemics. *Proc. R. Soc. Lond. Ser. A* **1927**, *115*, 700–721. [CrossRef]
9. Anderson, R.M.; May, R.M. *Infectious Diseases of Humans: Dynamics and Control*; Oxford University Press: Oxford, UK, 1991.
10. Dil, S.; Dil, N.; Maken, Z.H. COVID-19 Trends and Forecast in the Eastern Mediterranean Region with a particular focus on Pakistan. *Cureus* **2020**, *12*, e8582. [CrossRef] [PubMed]
11. Rodrigues, H.S. Application of SIR epidemiological model: New trends. *Int. J. Appl. Math. Inform.* **2016**, *10*, 92–97. [CrossRef]
12. Iwami, S.; Takeuchi, Y.; Liu, X. Avian–human influenza epidemic model. *Math. Biosci.* **2007**, *207*, 1–25. [CrossRef]
13. Teles, P. Predicting the evolution of SARS-COVID-2 in Portugal using an adapted SIR model previously used in South Korea for the MERS outbreak. *MedRxiv* **2020**. [CrossRef]
14. Rădulescu, A.; Williams, C.; Cavanagh, K. Management strategies in a SEIR-type model of COVID 19 community spread. *Sci. Rep.* **2020**, *10*, 21256. [CrossRef] [PubMed]
15. Fanelli, D.; Piazza, F. Analysis and forecast of COVID-19 spreading in China, Italy and France. *Chaos Solitons Fractals* **2020**, *134*, 109761. [CrossRef]
16. Zhao, C.; Tepekule, B.; Criscuolo, N.G.; Wendel, G.P.; Hilty, M.P.; RISC-ICU Consortium Investigators in Switzerland; Fumeaux, T.; Van Boeckel, T. Icumonitoring.ch: A platform for short-term forecasting of intensive care unit occupancy during the COVID-19 epidemic in Switzerland. *Swiss Med. Wkly.* **2020**, *150*, w20277. [CrossRef]
17. Chinazzi, M.; Davis, J.T.; Ajelli, M.; Gioannini, C.; Litvinova, M.; Merler, S.; Pastore, Y.; Piontti, A.; Mu, K.; Rossi, L.; et al. The effect of travel restrictions on the spread of the 2019 novel coronavirus (COVID-19) outbreak. *Science* **2020**, *368*, 395–400. [CrossRef]
18. Tang, B.; Wang, X.; Li, Q.; Bragazzi, N.L.; Tang, S.; Xiao, Y.; Wu, J. Estimation of the transmission risk of 2019-nCov and its implication for public health interventions. *J. Clin. Med.* **2020**, *9*, 462. [CrossRef] [PubMed]
19. Tian, H.; Liu, Y.; Li, Y. An investigation of transmission control measures during the first 50 days of the COVID-19 epidemic in China. *Science* **2020**, *368*, 638–642. [CrossRef] [PubMed]
20. López, L.; Rodó, X. A modified SEIR model to predict the COVID-19 outbreak in Spain and Italy: Simulating control scenarios and multi-scale epidemics. *Results Phys.* **2021**, *21*, 103746. [CrossRef] [PubMed]
21. Feng, S.; Feng, Z.; Ling, C.; Chang, C.; Feng, Z. Prediction of the COVID-19 epidemic trends based on SEIR and AI models. *PLoS ONE* **2021**, *16*, e0245101. [CrossRef]
22. Matveev, A. The mathematical modeling of the effective measures against the COVID-19 spread. *Natl. Secur. Strateg. Plan.* **2020**, *1*, 23–39. [CrossRef]
23. Anastassopoulou, C.; Russo, L.; Tsakris, A.; Siettos, C. Data-based analysis, modelling and forecasting of the COVID-19 outbreak. *PLoS ONE* **2020**, *15*, e0230405. [CrossRef]
24. Calafiore, G.C.; Novara, C.; Possieri, C. A time-varying SIRD model for the COVID-19 contagion in Italy. *Annu. Rev. Control* **2020**, *50*, 361–372. [CrossRef]

25. Maier, B.F.; Brockmann, D. Effective containment explains subexponential growth in recent confirmed COVID-19 cases in China. *Science* **2020**, *368*, 742–746. [CrossRef] [PubMed]
26. Mandal, S.; Bhatnagar, T.; Arinaminpathy, N. Prudent public health intervention strategies to control the coronavirus disease 2019 transmission in India. A mathematical model-based approach. *Indian J. Med. Res.* **2020**, *151*, 190–199. [CrossRef] [PubMed]
27. Choi, S.; Ki, M. Estimating the reproductive number and the outbreak size of COVID-19 in Korea. *Epidemiol. Health* **2020**, *42*, e2020011. [CrossRef] [PubMed]
28. Guglielmi, N.; Iacomini, E.; Viguerie, A. Delay differential equations for the spatially re-solved simulation of epidemics with specific application to COVID-19. *Math. Methods Appl. Sci.* **2022**, *45*, 4752–4771. [CrossRef] [PubMed]
29. Guglielmi, N.; Elisa Iacomini, E.; Viguerie, A. Identification of Time Delays in COVID-19 Data. *arXiv* **2021**, arXiv:2111.13368.
30. Pugliese, A.; Sottile, S. Inferring the COVID-19 infection curve in Italy. *arXiv* **2020**, arXiv:2004.09404.
31. Paul, S.; Lorin, E. Estimation of COVID-19 recovery and decease periods in Canada using delay model. *Sci. Rep.* **2021**, *11*, 23763. [CrossRef] [PubMed]
32. Dell'Anna, L. Solvable delay model for epidemic spreading: The case of Covid-19 in Italy. *Sci. Rep.* **2020**, *10*, 15763. [CrossRef] [PubMed]
33. Yang, C.; Yang, Y.; Li, Z.; Zhang, L. Modeling and analysis of COVID-19 based on a time delay dynamic model. *Math. Biosci. Eng.* **2021**, *18*, 154–165. [CrossRef]
34. Adam, D. Special report: The simulations driving the world's response to COVID-19. *Nature* **2020**, *580*, 316–318. [CrossRef]
35. Amelin, K.; Granichin, O.; Sergeenko, A.; Volkovich, Z.V. Emergent Intelligence via Self-Organization in a Group of Robotic Devices. *Mathematics* **2021**, *9*, 1314. [CrossRef]
36. Wieczorek, M.; Siłka, J.; Woźniak, M. Neural network powered COVID-19 spread forecasting model. *Chaos Solitons Fractals* **2020**, *140*, 110203. [CrossRef]
37. Liu, D.; Clemente, L.; Poirier, C.; Ding, X.; Chinazzi, M.; Davis, J.; Vespignani, A.; Santillana, M. Real-Time Forecasting of the COVID-19 Outbreak in Chinese Provinces: Machine Learning Approach Using Novel Digital Data and Estimates from Mechanistic Models. *J. Med. Internet Res.* **2020**, *22*, e20285. [CrossRef]
38. Kumar, R.L.; Khan, F.; Din, S.; Band, S.S.; Mosavi, A.; Ibeke, E. Recurrent Neural Network and Reinforcement Learning Model for COVID-19 Prediction. *Front. Public Health* **2021**, *9*, 744100. [CrossRef]
39. Ayoobi, N.; Sharifrazi, D.; Alizadehsani, R.; Shoeibi, A.; Gorriz, J.M.; Moosaei, H.; Khosravi, A.; Nahavandi, S.; Gholamzadeh Chofreh, A.; Goni, F.A.; et al. Time series forecasting of new cases and new deaths rate for COVID-19 using deep learning methods. *Results Phys.* **2021**, *27*, 104495. [CrossRef]
40. Alali, Y.; Harrou, F.; Sun, Y. A proficient approach to forecast COVID-19 spread via optimized dynamic machine learning models. *Sci. Rep.* **2022**, *12*, 2467. [CrossRef]
41. Zhao, H.; Merchant, N.N.; McNulty, A.; Radcliff, T.A.; Cote, M.J.; Fischer, R.S.B.; Sang, H.; Ory, M.G. COVID-19: Short term prediction model using daily incidence data. *PLoS ONE* **2021**, *16*, e0250110. [CrossRef]
42. Amaral, F.; Casaca, W.; Oishi, C.M.; Cuminato, J.A. Towards Providing Effective Data-Driven Responses to Predict the COVID-19 in São Paulo and Brazil. *Sensors* **2021**, *21*, 540. [CrossRef]
43. Pavlyutin, M.; Samoyavcheva, M.; Kochkarov, R.; Pleshakova, E.; Korchagin, S.; Gataullin, T.; Nikitin, P.; Hidirova, M. COVID-19 Spread Forecasting, Mathematical Methods vs. Machine Learning, Moscow Case. *Mathematics* **2022**, *10*, 195. [CrossRef]
44. Katris, C. A time series-based statistical approach for outbreak spread forecasting: Application of COVID-19 in Greece. *Expert Syst. Appl.* **2021**, *166*, 114077. [CrossRef]
45. Finkenstädt, B.F.; Grenfell, B.T. Time series modelling of childhood diseases: A dynamical systems approach. *J. R. Stat. Soc. Ser. C (Appl. Stat.)* **2000**, *49*, 187–205. [CrossRef]
46. Kondratyev, M. Forecasting methods and models of disease spread. *Comput. Res. Modeling* **2013**, *5*, 863–882. [CrossRef]
47. Zakharov, V.; Balykina, Y.; Petrosian, O.; Gao, H. CBRR Model for Predicting the Dynamics of the COVID-19 Epidemic in Real Time. *Mathematics* **2020**, *8*, 1727. [CrossRef]
48. Zakharov, V.; Balykina, Y. Predicting the dynamics of the coronavirus (COVID-19) epidemic based on the case-based reasoning approach. *Vestn. St.-Peterb. Univ. Appl. Math. Comput. Sci. Control Process* **2020**, *16*, 249–259. [CrossRef]
49. Dairi, A.; Harrou, F.; Zeroual, A.; Hittawe, M.M.; Sun, Y. Comparative study of machine learning methods for COVID-19 transmission forecasting. *J. Biomed. Inform.* **2021**, *18*, 103791. [CrossRef]
50. CDC. Reported and Forecasted New and Total COVID-19 Deaths in USA. Available online: https://www.cdc.gov/coronavirus/2019-ncov/science/forecasting/forecasting-us.html (accessed on 7 December 2021).
51. Congdon, P. Mid-Epidemic Forecasts of COVID-19 Cases and Deaths: A Bivariate Model Applied to the UK. *Interdiscip. Perspect. Infect. Dis.* **2021**, *2021*, 8847116. [CrossRef]
52. Banoei, M.M.; Dinparastisaleh, R.; Zadeh, A.V.; Mirsaeidi, M. Machine-learning-based COVID-19 mortality prediction model and identification of patients at low and high risk of dying. *Crit. Care* **2021**, *25*, 328. [CrossRef]
53. Ahmadini, A.A.H.; Naeem, M.; Aamir, M.; Dewan, R.; Alshqaq, S.S.A.; Mashwani, W.K. Analysis and Forecast of the Number of Deaths, Recovered Cases, and Confirmed Cases from COVID-19 for the Top Four Affected Countries Using Kalman Filter. *Front. Phys.* **2021**, *9*, 629320. [CrossRef]
54. Ioannidis, J.; Cripps, S.; Tanner, M.A. Forecasting for COVID-19 has failed. *Int. J. Forecast.* **2020**. advance online publication. [CrossRef] [PubMed]

55. Hasan, M.N.; Haider, N.; Stigler, F.L.; Khan, R.A.; McCoy, D.; Zumla, A.; Kock, R.A.; Uddin, M.J. The Global Case-Fatality Rate of COVID-19 Has Been Declining Since May 2020. *Am. J. Trop. Med. Hyg.* **2021**, *104*, 2176–2184. [CrossRef] [PubMed]
56. Johns Hopkins Coronavirus Resource Center. Available online: https://coronavirus.jhu.edu/map.html (accessed on 15 December 2021).
57. Chaudhry, R.M.; Hanif, A.; Chaudhary, M.; Minhas, S., II; Mirza, K.; Asif, T.; Gilani, S.A.; Kashif, M. Coronavirus disease 2019 (COVID-19): Forecast of an emerging urgency in Pakistan. *Cureus* **2020**, *12*, e8346. [CrossRef] [PubMed]
58. Zakharov, V.; Balykina, Y. Balance Model of COVID-19 Epidemic Based on Percentage Growth Rate. *Inform. Autom.* **2021**, *20*, 1034–1064. [CrossRef]
59. Izvestia Newspaper. Available online: https://iz.ru/1233744/olga-kolentcova/podem-s-povorotom-matematiki-dali-novyi-prognoz-po-zabolevaemosti-covid (accessed on 15 December 2021).
60. Center for Intelligent Logistics of St. Petersburg State University. Available online: http://old.apmath.spbu.ru/cil/index_en.html (accessed on 12 December 2021).

Article

A New COVID-19 Pandemic Model Including the Compartment of Vaccinated Individuals: Global Stability of the Disease-Free Fixed Point

Isra Al-Shbeil [1], Noureddine Djenina [2,*], Ali Jaradat [3], Abdallah Al-Husban [4], Adel Ouannas [2] and Giuseppe Grassi [5]

1. Department of Mathematics, Faculty of Sciences, University of Jordan, Amman 11942, Jordan
2. Laboratory of Dynamical Systems and Control, University of Larbi Ben M'hidi, Oum El Bouaghi 04000, Algeria
3. Department of Mathematics, Faculty of Arts and Science, Amman Arab University, Amman 11953, Jordan
4. Department of Mathematics, Faculty of Science and Technology, Irbid National University, Irbid 2600, Jordan
5. Dipartimento Ingegneria Innovazione, Universita del Salento, 73100 Lecce, Italy
* Correspondence: noureddine.djenina@univ-oeb.dz

Abstract: Owing to the COVID-19 pandemic, which broke out in December 2019 and is still disrupting human life across the world, attention has been recently focused on the study of epidemic mathematical models able to describe the spread of the disease. The number of people who have received vaccinations is a new state variable in the COVID-19 model that this paper introduces to further the discussion of the subject. The study demonstrates that the proposed compartment model, which is described by differential equations of integer order, has two fixed points, a disease-free fixed point and an endemic fixed point. The global stability of the disease-free fixed point is guaranteed by a new theorem that is proven. This implies the disappearance of the pandemic, provided that an inequality involving the vaccination rate is satisfied. Finally, simulation results are carried out, with the aim of highlighting the usefulness of the conceived COVID-19 compartment model.

Keywords: dynamical systems; epidemics; stability; disease; bio mathematics modeling; COVID 19 model; basic reproduction number

MSC: 92B05; 37N25; 34D20; 34D23

1. Introduction

The first study of epidemics can be found in an Egyptian medical papyrus of about 3500 years ago [1]. However, mathematical models for precisely analyzing the spread of epidemics were introduced only at the beginning of the twentieth century [2]. In [3], a numerical investigation of discrete oscillating epidemic models was investigated. Recently, attention has been focused on the COVID-19 pandemic, which broke out in December 2019, but is still affecting social and economic life across the world. Because of this, several epidemic models have been proposed, described by integer-order differential/difference equations [4], as well as fractional differential/difference equations [5–7]. For example, some works have presented dynamic compartment models (described by integer-order operators) that analyze the evolution of the disease over time by dividing the communities into some classes (i.e., susceptible, exposed, infected, etc). In particular, ref. [8] propose a model that considers eight stages of infection for the COVID-19 disease: susceptible (S), infected (I), diagnosed (D), ailing (A), recognized (R), threatened (T), healed (H) and extinct (E). The dynamic model, called SIDARTHE, introduces the distinction between diagnosed and non-diagnosed individuals, because the former are typically isolated and hence less likely to spread the infection. In [9], the dynamic behavior of the classical susceptible-infectious-removed (SIR) model when applied to the transmission of COVID-19 disease has been studied. The model includes both a nonlinear removal rate (related to the hospital bed population ratio) and the effects of media on public awareness. In [10], the inverse problem in epidemiology is exploited via a SIR model. At first, the method is used for estimating the infectivity and recovery rates from real COVID-19 data. Then, the estimated rates have been used to compute the evolution of the COVID-19 disease. In [11], a novel SIR model for simulating the COVID-19 epidemic in Wuhan is presented. The transmission dynamic model in [11] is updated with real-time input data and enriched with additional data sources, in order to infer a preliminary set of clinical parameters that could guide public health decision-making. In [12], a nonlinear SIR epidemic dynamic system is introduced to model the spread of COVID-19 under the effect of social distancing imposed by the government measures. In [13], a novel deterministic mathematical model of the COVID-19 pandemic is developed. In particular, by using time-series plots and phase portraits, references [13,14] show that the conceived COVID-19 model exhibits chaotic behaviors. Referring to systems described by fractional operators, in [15], a non-integer order COVID-19 dynamic model is proposed, which incorporates the reinfection rate in the individuals recovered from the disease. In [16], a fractional-order COVID-19 model involving the Caputo derivative is introduced. Moreover, the stability of the steady-states and the existence of non-negative solutions are investigated. In [17], a fractional compartmental model for predicting the spread of the COVID-19 pandemic is developed. In particular, reference [17] discusses the conditions under which the disease-free and the endemic equilibrium points become asymptotically stable. In [18], the spread of the COVID-19 pandemic is modeled via the Caputo–Fabrizio derivative, as well as via the Atangana–Baleanu–Caputo derivative. Then, the local stability of the equilibrium points is investigated for both modeling approaches. In [19], a fractional dynamic model for evaluating the consequences of adaptive immune responses to the COVID-19 viral mutation is presented. In [14], the dynamics of a novel COVID-19 pandemic model described by fractional derivatives are investigated. Several chaotic behaviours, obtained by varying the order of the derivative, are proven to exist in the conceived model.

Some papers, starting in the beginning of the year 2021, began to take into account, in different ways, the role of vaccination in reducing the spread of the COVID-19 pandemic. A dynamic model based on eight state variables is proposed in [20], for instance, where vaccination is viewed as a preventive action (rather than as a system variable). In [21], a new discrete susceptible-exposed-infectious-recovered (SEIR) epidemic model is illustrated. In particular, a feedback vaccination control law on the susceptible population is incorporated to stabilize the system dynamics. In [22], novel SEIR dynamic models (both integer-order and fractional-order) that include vaccine rate are proposed. In [23], a new SEIR epidemic model (the so-called SE(Is)(Ih)AR epidemic model) is presented. The conceived system incorporates two control laws, a feedback vaccination law and an antiviral treatment control law. In [24], some basic properties of a SEIR COVID-19 epidemic model subject to vaccination and treatment controls are studied. In particular, stability, boundedness, and non-negativity of the state trajectories are investigated in detail. In [25], the impact of multiple vaccination strategies on the dynamics of a fractional COVID-19 model is analyzed. The existence and uniqueness of the system solution is proven using Banach's fixed point theorem. In [26], a set of ordinary differential equations that use vaccinations as the control input signals describe how COVID-19 behaves in Iranian and Russian societies. A Lyapunov-based methodology is used in [27] to examine how vaccination affects COVID-19 inhibition in Canada.

However, many of the proposed systems describe the virus accurately but use compartments such that it is difficult to find the initial conditions and some parameters. For example, the presence of a cabin for people who carry the disease but do not show symptoms increases the accuracy of the system, but after applying it, we find it is impossible to find the initial condition for this cabin, as well as the rate of settlement in it or the rate of exit from it. However, some systems are easy to apply, but they neglect many compartments. Therefore, in this work we formulated a system that took into account all possible compartments and whose initial conditions can be found to make the system applicable, as well as without neglecting any possible compartment that may reduce the accuracy of the system. The result was a system characterized by the maximum possible accuracy and applicability.

This paper introduces a new COVID-19 model, which incorporates the number of immunized individuals as an additional state variable describing the system dynamics, in an effort to further the discussion of the mathematical modeling of epidemics. The study demonstrates the existence of two fixed points, a disease-free fixed point and an endemic fixed point. The proposed compartment model is described by integer-order differential equations. When an inequality involving the vaccination rate is satisfied, the pandemic vanishes, according to the results of a stability analysis of the disease-free fixed point. This finding of the proposed approach is noteworthy because it is rigorous (proven by a theorem), which may aid decision-makers in better understanding the epidemiological behavior of the disease over time. The arrangement of the manuscript is as follows. A brand new compartment model for explaining the spread of the COVID-19 pandemic is presented in Section 2. Five state variables, namely the Susceptible class S, the Recovered class R, the Infection class I, the Infection dangerous class Id and the Vaccinated class V, which denotes the number of vaccinated people, are used to describe the dynamics of the system. The existence of the solution for the analyzed COVID-19 model is demonstrated in Section 3. It is demonstrated in Section 4 that the system has two fixed points, namely the epidemic fixed point and the disease-free fixed point. Additionally, the fundamental reproduction number is calculated. The stability of the disease-free fixed point is thoroughly examined in Section 5. The pandemic vanishes provided that an inequality involving the vaccinated rate and some system parameters are satisfied, according to a new theorem that ensures the global stability of the disease-free fixed point. Finally, simulation results are presented in Section 6 with the goal of demonstrating the value of the novel COVID-19 compartment model.

2. A New Compartment Model Including Vaccinated Individuals

We divided the study population (N) into three main classes in order to better understand the behavior of the epidemic's spread: a class for those who were exposed to the infection, a class for those who became infected, a class for those who died as a result of the disease. Each of these classes are also divided into secondary classes, so that the class of people who are exposed to infection is divided into three sub-classes: people who are exposed to infection but were not previously infected and did not receive a vaccine S, people who were previously infected but recovered from the disease and are at risk of being infected again R, and people who were vaccinated against the epidemic V. As for the class of infected persons, it is divided into two secondary classes: for persons with good immunity and for whom infection does not pose a great risk, suppose that their ratio in society is λ, ($\lambda \leq 1$), such that the class of infected people from this group are assigned to sub-class I. The infected persons for whom the infection is dangerous consists of the elderly, pregnant women and people with chronic diseases (whose ratios in society are $(1-\lambda)$) I_d. And finally, the class of deaths due to the epidemic D. Now we will explain the immigration in each class and compare it to the others.

Susceptible class S: This class acquires Ω persons in the unit of time. In the event that the studied area is isolated, then Ω represents the birth number. This class loses persons who are exposed to infection at a rate $r_1(I+I_d)$, where $r_1 = \frac{p_1 k}{N}$, k is the average numbers of contacts per person (per unit of time), p_1 is the probability of contagion and N is the total population (it can be considered as the maximum value of the population). Additionally, this class loses patients who have received a vaccination against the disease at a rate of v and at a natural death rate of μ.

Recovered class R: This class acquires newly recovered individuals from classes I and I_d at a rate of ρ, and loses persons who are exposed to infection at a rate of $r_2(I+I_d)$, where $r_2 = \frac{p_2 k}{N}$, which is less than r_1. Similar to S, this class loses patients who have received a vaccination against the disease at a rate of v and at a natural death rate of μ.

Vaccinated class V: This class represents people who have been vaccinated and may have been infected and cured of the virus (they are people from the class R) or people who have never been infected (people from the class S). This category is not immune to the virus, but the infection rate is lower than in other classes. This class acquires persons at the vaccination rate v, representing newly vaccinated persons from classes S and R. This class loses persons who are infected at a rate of $r_3(I + I_d)$, where $r_3 = \frac{p_3 k}{N}$, which is less than r_1 and r_2 (the probability of infection p_3 in this class is less due to vaccination). This class also loses deceased persons (natural deaths) at a rate of μ.

Infection class I: This class acquires newly infected persons from classes S, R and V at a rate of $(\lambda(r_1 + r_2) + r_3)(I + I_d)$. This class loses recovered persons at a rate of ρ, as well as due to natural deaths at a rate of μ (we assume that in this class there are no deaths due to the epidemic).

The class of dangerous infection I_d: This class acquires newly infected persons from classes S and R (the class V does not go into this class because vaccination protects against dangerous infection) with a rate of $(1 - \lambda)(r_1 + r_2)(I + I_d)$. This class loses recovered persons at a rate of ρ, a rate of μ (natural deaths) and due to death related to infection at a rate of δ.

Death class D: This class acquires persons who have died due to the epidemic at rate of δ.

From these, we can now make the following model, which gives the mathematical explanation for all of the above:

$$\begin{cases} \frac{dS}{dt} = \Omega - r_1(I(t) + I_d(t))S(t) - (\mu + v)S(t), \\ \frac{dR}{dt} = \rho(I(t) + I_d(t)) - r_2(I(t) + I_d(t))R(t) - (v + \mu)R(t), \\ \frac{dV}{dt} = v(S(t) + R(t)) + \rho' I(t) - r_3(I(t) + I_d(t))V(t) - \mu V(t), \\ \frac{dI}{dt} = (\lambda(r_1 S(t) + r_2 R(t)) + r_3 V(t))(I(t) + I_d(t)) - (\mu + \rho + \rho')I(t), \\ \frac{dI_d}{dt} = (1 - \lambda)(r_1 S(t) + r_2 R(t))(I(t) + I_d(t)) - (\mu + \delta + \rho)I_d(t), \\ \frac{dD}{dt} = \delta I_d(t). \end{cases} \quad t \in \mathbb{R}^+. \quad (1)$$

We note that the last equation in the system is isolated and can be neglected, so we will neglect it in future studies to simplify the system slightly. Thus, we find the following system:

$$\begin{cases} \frac{dS}{dt} = \Omega - r_1(I(t) + I_d(t))S(t) - (\mu + v)S(t), \\ \frac{dR}{dt} = \rho(I(t) + I_d(t)) - r_2(I(t) + I_d(t))R(t) - (v + \mu)R(t), \\ \frac{dV}{dt} = v(S(t) + R(t)) - r_3(I + I_d)V(t) - \mu V(t), \\ \frac{dI}{dt} = (\lambda(r_1 S(t) + r_2 R(t)) + r_3 V(t))(I(t) + I_d(t)) - (\mu + \rho)I(t), \\ \frac{dI_d}{dt} = (1 - \lambda)(r_1 S(t) + r_2 R(t))(I(t) + I_d(t)) - (\mu + \delta + \rho)I_d(t). \end{cases} \quad t \in \mathbb{R}^+. \quad (2)$$

With the following initial conditions:

$$S(0), R(0), V(0), I(0), I_d(0) \geq 0. \quad (3)$$

The sum of all the compounds represents the total number of living persons in the studied population (N), thus:

$$N(t) = S(t) + R(t) + V(t) + I(t) + I_d(t). \quad (4)$$

This system must be subject to the following assumptions:

Assumption 1. *The number of new infections is directly proportional to infection rates r_1, r_2 and r_3.*

Assumption 2. *The number of new infections is inversely proportional to the cure rate ρ and the vaccination rate v.*

3. Existence, Positivity and Invariant Region

3.1. Existence and Uniqueness

To prove the existence and uniqueness of the solution, we use the Cauchy–Lipshitz theorem. First, let the problem

$$\frac{dy}{dt} = f(t, y(t)), \tag{5}$$

where $f : U \to \mathbb{R}^n$, is continuous and U is an open set of $\mathbb{R} \times \mathbb{R}^n$.

Theorem 1 (Cauchy–Lipschitz [28]). *Suppose that f is locally Lipschitz in y, then problem (5) with the initial conditions $(t_0, y_0) \in \mathbb{R} \times \mathbb{R}^n$ implies a unique maximal solution.*

Based on the above theorem, we find the following result regarding the existence and uniqueness of the solution to problem (2).

Theorem 2. *System (2) with the initial conditions (3) has a unique maximal solution.*

Proof. The second part of system (2) is continuous and also belongs to class C^∞, with respect to (S, R, V, I, I_d), hence it is a locally Lipschitzien. From it, according to the **Cauchy–Lipschitz** theorem system (2) has a unique maximal solution. □

3.2. Positivity and Invariant Region

We are interested here in studying whether the solution is positive. Because it must be, we will also find an invariant region such that when we take the initial conditions from it, the solution remains bonded and belongs to this region. First, we start with the positivity of the solution through the following Lemma.

Lemma 1. *The solution of (2) is positive when the initial conditions are positives.*

Proof. First, we note that the initial condition is positive and the solution is continuous. Therefore, in order for one of the components of the solution to become negative, it must first be null and the derivative at the zero point be negative. But we note from:

$$\begin{cases} \left.\frac{dS}{dt}\right|_{S=0} = \Omega, \\ \left.\frac{dR}{dt}\right|_{R=0} = \rho(I + I_d), \\ \left.\frac{dV}{dt}\right|_{V=0} = v(S + R), \\ \left.\frac{dI}{dt}\right|_{I=0} = (\lambda(r_1 S + r_2 R) + r_3 V)(I_d), \\ \left.\frac{dI_d}{dt}\right|_{I_d=0} = (1 - \lambda)(r_1 S + r_2 R) I. \end{cases} \tag{6}$$

Hence, the first element that vanishes from (S, R, V, I, I_d), its derivative at the point of vanishing is non-negative and therefore increases again and remains positive. Therefore, all elements of (S, R, V, I, I_d) remain positive and never become negative as long as the initial condition is positive, i.e., $(S, R, V, I, I_d) \in \mathbb{R}^5_+$, where $\mathbb{R}^5_+ = \{(x_1, x_2, x_3, x_4, x_5) \in \mathbb{R}^5$ and $x_i \geq 0$, for $i = 1.5\}$. □

Based on the following theorem:

Theorem 3 ([29] **Comparison Theorem**). *Let $f;g : \mathbb{R} \to \mathbb{R}$ two Lipschitz function. We consider the solutions $x(t)$ and $y(t)$ of the Cauchy problems:*

$$\begin{cases} x'(t) = f(t, x(t)), \\ x(0) = x_0, \end{cases} \qquad \begin{cases} y'(t) = g(t, y(t)), \\ y(0) = y_0. \end{cases}$$

Suppose that $f(t;x) \leq g(t;x)$ for all $(t;x) \in \mathbb{R} \times \mathbb{R}$ and that $x_0 \leq y_0$. Then $x(t) \leq y(t)$ for all t.

This immediately follows for the invariant region, and we give the following result.

Theorem 4. *System (2) has*

$$\Psi = \left\{ (S, R, V, I, I_d) \in \mathbb{R}_+^5 \text{ and } S + R + V + I + I_d \leq \frac{\Omega}{\mu} \right\}, \tag{7}$$

as invariant region.

Proof. Adding the equations of the system (2), we get:

$$\frac{dS(t)}{dt} + \frac{dR(t)}{dt} + \frac{dV(t)}{dt} + \frac{dI(t)}{dt} + \frac{dI_d(t)}{dt} = \Omega - \mu N(t) - \delta I_d(t),$$

which means

$$\frac{dN(t)}{dt} = \Omega - \mu N(t) - \delta I_d(t),$$

Since I_d is positive, we get:

$$\frac{dN(t)}{dt} \leq \Omega - \mu N(t),$$

where, according to Equation (4)

$$N(0) = S(0) + R(0) + V(0) + I(0) + I_d(0).$$

Applying the comparison theorem by replacing $x = N$, $x(0) = y(0) = N(0)$, $f = \frac{dN}{dt}$ and $g = \Omega - \mu N(t)$, we find that

$$N(t) \leq \frac{\Omega}{\mu} - \left(\frac{\Omega}{\mu} - N(0) \right) e^{-\mu t}.$$

So the following is achieved

$$0 \leq N(t) \leq \frac{\Omega}{\mu},$$

where $N(0) \leq \frac{\Omega}{\mu}$.

Therefore, the solution to the equation belongs to Ψ. □

4. Fixed Points and Basic Reproduction Number

In this section, we will study the fixed points and the basic reproduction number, which are important for studying the stability of epidemic systems.

4.1. Fixed Points

Finding the fixed points is necessary before studying the dynamics of the model (2), and finding the fixed points requires solving the equation:

$$\begin{cases} \Omega - r_1(I^* + I_d^*)S - (\mu + v)S^* = 0, \\ \rho(I^* + I_d^*) - r_2(I^* + I_d^*)R^* - (v + \mu)R^* = 0, \\ v(S^* + R^*) - r_3(I^* + I_d^*)V^* - \mu V^* = 0, \\ (\lambda(r_1 S^* + r_2 R^*) + r_3 V^*)(I^* + I_d^*) - (\mu + \rho)I^* = 0 \\ (1 - \lambda)(r_1 S^* + r_2 R^*)(I^* + I_d^*) - (\mu + \delta + \rho)I_d^* = 0. \end{cases} \tag{8}$$

The previous equation has the point $E_0 = \left(\frac{\Omega}{(\mu+v)}, 0, \frac{v\Omega}{\mu(\mu+v)}, 0, 0 \right)$ as a solution. As can be seen, there is no disease at this point, so it is referred to as the disease-free fixed point, and we will study its stability later.

If we suppose that $(I^* + I_d^*) \neq 0$, we will get:

$$\begin{aligned}
\frac{\Omega}{r_1(I^*+I_d^*)+(\mu+v)} &= S^*, \\
\frac{\rho(I^*+I_d^*)}{r_2(I^*+I_d^*)+(v+\mu)} &= R^*, \\
\frac{v(S^*+R^*)}{r_3(I^*+I_d^*)+\mu} &= V^*, \\
\frac{(\lambda(r_1 S^*+r_3 R^*)+r_2 V^*)}{(\mu+\rho)} &= \frac{I^*}{(I^*+I_d^*)}, \\
\frac{(1-\lambda)(r_1 S^*+r_2 R^*)}{(\mu+\delta+\rho)} &= \frac{I_d^*}{(I^*+I_d^*)}.
\end{aligned} \qquad (9)$$

This system is a classical non-linear system, which can be solved numerically and then studied for its stability. Overall, this point is called the endemic equilibrium point $E^* = (S^*, R^*, V^*, I^*, I_d^*)$.

4.2. Basic Reproduction Number

We will now calculate a key number called the basic reproduction number R_0. In the study of stability for the disease-free fixed point, this is crucial. We will follow the steps described in [30] to calculate this number, which represents the rate of new people being infected by one sick person until their recovery. We determine the class expressing the new infection, which corresponds to the final two equations in system (2), and hence it can be calculated using:

$$\begin{aligned}
\frac{dI}{dt} &= (\lambda(r_1 S(t) + r_2 R(t)) + r_3 V(t))(I(t) + I_d(t)) - (\mu+\rho)I(t), \\
\frac{dI_d}{dt} &= (1-\lambda)(r_1 S(t) + r_2 R(t))(I(t) + I_d(t)) - (\mu+\delta+\rho)I_d(t).
\end{aligned} \qquad (10)$$

We reformulate this system as follows

$$\frac{d}{dt}\begin{pmatrix} I \\ I_d \end{pmatrix} = \mathcal{F} - \mathcal{V},$$

where $\mathcal{F}$ is the rate of the appearance of new infections,

$$\mathcal{F} = \begin{pmatrix} (\lambda(r_1 S(t) + r_2 R(t)) + r_3 V(t))(I(t) + I_d(t)) \\ (1-\lambda)(r_1 S(t) + r_2 R(t))(I(t) + I_d(t)) \end{pmatrix},$$

and $\mathcal{V}$ is the rate of the transfer of individuals into other compartments,

$$\mathcal{V} = \begin{pmatrix} (\mu+\rho)I(t) \\ (\mu+\delta+\rho)I_d(t) \end{pmatrix}.$$

Calculating the Jacobian matrix F and V for $\mathcal{F}$ and $\mathcal{V}$ respectively at the disease-free fixed point, i.e., at $\left(S = \frac{\Omega}{(\mu+v)}, R = 0, V = \frac{v\Omega}{\mu(\mu+v)}, I = 0, I_d = 0\right)$, we get:

$$F = \begin{pmatrix} \frac{\Omega}{(\mu+v)}\left(\lambda r_1 + \frac{vr_3}{\mu}\right) & \frac{\Omega}{(\mu+v)}\left(\lambda r_1 + \frac{vr_3}{\mu}\right) \\ (1-\lambda)\left(\frac{\Omega r_1}{(\mu+v)}\right) & (1-\lambda)\left(\frac{\Omega r_1}{(\mu+v)}\right) \end{pmatrix},$$

and

$$V = \begin{pmatrix} (\mu+\rho) & 0 \\ 0 & (\mu+\delta+\rho) \end{pmatrix}.$$

The next generation matrix is:

$$FV^{-1} = \begin{pmatrix} \frac{\Omega}{\mu}\frac{vr_3+\lambda\mu r_1}{(\mu+v)(\mu+\rho)} & \frac{\Omega}{\mu}\frac{vr_3+\lambda\mu r_1}{(\mu+v)(\mu+\delta+\rho)} \\ -\Omega r_1 \frac{\lambda-1}{(\mu+v)(\mu+\rho)} & -\Omega r_1 \frac{\lambda-1}{(\mu+v)(\mu+\delta+\rho)} \end{pmatrix}. \qquad (11)$$

The basic reproductive number is given as the spectral radius (the greater length of the eigenvalues) of FV^{-1}. When calculating the characteristic polynomial $P(FV^{-1})$, for FV^{-1}, we find

$$P(FV^{-1}) = X\left(X - \frac{\Omega}{(\mu+v)}\left(\frac{(1-\lambda)r_1}{(\mu+\delta+\rho)} + \frac{vr_3+\lambda\mu r_1}{\mu(\mu+\rho)}\right)\right)$$

We notice that FV^{-1} has 0 and $\frac{\Omega}{(\mu+v)}\left(\frac{(1-\lambda)r_1}{(\mu+\delta+\rho)} + \frac{vr_3+\lambda\mu r_1}{\mu(\mu+\rho)}\right)$ as eigenvalues, thus

$$R_0 = \frac{\Omega}{(\mu+v)}\left(\frac{(1-\lambda)r_1}{(\mu+\delta+\rho)} + \frac{vr_3+\lambda\mu r_1}{\mu(\mu+\rho)}\right). \tag{12}$$

5. Stability Analysis of the Disease-Free Fixed Point

What matters to us is the disappearance of the disease. In this section, we will formulate the conditions to ensure the stability of the disease-free fixed point, that is, formulate the conditions to ensure the disappearance of the disease. When applying the model (2) in a specific region, we need the initial conditions and the parameters of that region. The parameters are divided into two parts. Firstly, the fixed part, which cannot be modified, is represented by: the birth rate Ω, natural death rate μ, recovered rate ρ and death due to infection rate δ. Although it is possible to modify ρ and δ by improving the health conditions, we assume that the authorities are doing everything possible, meaning that these rates are the best they can be. The second kind of parameters can be modified, which are: infection rates r_1, r_2 and r_3 and the vaccinated rate v. Authorities can impose measures such as closing some facilities and imposing quarantine to reduce the rate of infection that reduces rates r_1, r_2 and r_3. By increasing the vaccination rate, we can decrease classes S and R and increase class V, in which the probability of infection is less than the previous two classes.

5.1. Local Stability

The stability of the disease-free fixed point will be examined in this subsection using an imposed condition on R_0.

Theorem 5. *Assume that $R_0 < 1$. Therefore, the disease-free fixed point E_0 of system (2) is locally asymptotically stable.*

Proof. The system at E_0 has the Jacobian matrix:

$$J = \begin{pmatrix} -(\mu+v) & 0 & 0 & -\frac{\Omega r_1}{(\mu+v)} & -\frac{\Omega r_1}{(\mu+v)} \\ 0 & -(v+\mu) & 0 & \rho & \rho \\ v & v & -\mu & -\frac{\Omega v r_3}{\mu(\mu+v)} & -\frac{\Omega v r_3}{\mu(\mu+v)} \\ 0 & 0 & 0 & \left(\frac{\lambda r_1 \Omega}{(\mu+v)} + \frac{v r_3 \Omega}{\mu(\mu+v)}\right) - (\mu+\rho) & \left(\frac{\lambda r_1 \Omega}{(\mu+v)} + \frac{v r_3 \Omega}{\mu(\mu+v)}\right) \\ 0 & 0 & 0 & \frac{(1-\lambda)r_1 \Omega}{(\mu+v)} & \frac{(1-\lambda)r_1 \Omega}{(\mu+v)} - (\mu+\delta+\rho) \end{pmatrix} \tag{13}$$

Then, the characteristic polynomial:

$$P(J) = (X+\mu)(X+\mu+v)^2\left(X^2 + AX + B\right)$$

where

$$A = (\mu+\rho) + (\mu+\delta+\rho) - \frac{\Omega(\mu r_1 + v r_3)}{\mu(\mu+v)},$$
$$B = (\mu+\rho)(\mu+\delta+\rho) - \frac{\Omega(\mu r_1(\mu+\rho+\lambda\delta) + v r_3(\mu+\delta+\rho))}{\mu(\mu+v)},$$

As a result, the matrix J has two eigenvalues that are both less than zero: $-\mu$ as a normal eigenvalue and $-(\mu + v)$ as a double eigenvalue. The rest of the roots of the polynomial:

$$X^2 + AX + B, \tag{14}$$

A and B can be written as follows:

$$A = (\mu + \rho) + (\mu + \delta + \rho)(1 - R_0) + \frac{\Omega \delta (vr_3 + \lambda \mu r_1)}{\mu(\mu + \rho)(\mu + v)},$$
$$B = (\mu + \rho)(\mu + \delta + \rho)(1 - R_0).$$

If $R_0 < 1$, then according to the Routh–Hurwitz criterion, the roots of the polynomial (14) have a negative real part. Therefore, all eigenvalues of matrix J have a negative real part, so the disease-free fixed point is locally asymptotically stable. □

This theorem can be applied easily, but it remains only local, that is, when the initial conditions are far from the fixed point. Hence, we need to study its global stability.

5.2. Global Stability

The last two equations of the system added together describe infection, and yield the following equation:

$$\frac{d}{dt}(I + I_d) = (r_1 S(t) + r_2 R(t) + r_3 V(t) - (\mu + \rho))(I(t) + I_d(t)) - \delta I_d(t).$$

Since I is positive:

$$\frac{d}{dt}(I + I_d) \leq (r_1 S(t) + r_2 R(t) + r_3 V(t) - (\mu + \rho))(I(t) + I_d(t)).$$

Since $r_i = \frac{p_i k}{N}, i = 1, 2, 3$:

$$\begin{aligned}
\frac{d}{dt}(I + I_d) &\leq \left(\frac{p_1 k}{N} S(t) + \frac{p_2 k}{N} R(t) + \frac{p_3 k}{N} V(t) - (\mu + \rho)\right)(I(t) + I_d(t)), \\
&\leq \left(\frac{p_1 k}{N}(S(t) + R(t) + V(t)) - (\mu + \rho)\right)(I(t) + I_d(t)), \\
&\leq (p_1 k - (\mu + \rho))(I(t) + I_d(t)).
\end{aligned}$$

Using the comparison theorem, we find that

$$(I + I_d)(t) \leq (I + I_d)(0) e^{(p_1 k - (\mu + \rho))t}.$$

We note that if $(p_1 k - (\mu + \rho)) < 0$, then $(I + I_d)(t) \to 0$ when $t \to \infty$. Thus, we get the following result:

Theorem 6. *If*

$$k < \frac{\mu + \rho}{p_1}, \tag{15}$$

then the disease will disappear.

Remark 1. *We note that this condition is very logical. Factors contributing to the disappearance of the disease are when the rate of infection is small, or the recovery rate is increased, or the probability of injury is small.*

This condition is sufficient for the disappearance of the disease, but it may not be achievable and may be very expensive. Sometimes it is not possible to reduce the rate of infection more than a certain limit. In the following, we will study the disappearance of the disease by studying the stability of the disease-free fixed point.

To prove global stability we use the method of Carlos Castillo-Chavez described in [31]. First, system (2) must be written in the form:

$$\frac{dX}{dt} = F(X,Y),$$
$$\frac{dY}{dt} = G(X,Y), \qquad (16)$$

where $X = (S, R, V)$ denotes the number of uninfected individuals and $Y = (I, I_d)$ denotes the number of infected, infectious. $E_0 = (X_0, 0)$ denotes the disease-free equilibrium of this system.

(H_1) In $\frac{dX}{dt} = F(X,0)$, X_0 is globally asymptotically stable.

(H_2) $G(X,Y) = AY - \hat{G}(X,Y), \hat{G}(X,Y) \geq 0$ for $(X,Y) \in \Psi$, where A is an M-matrix (the off diagonal elements of A are non negative) and Ψ is the region where the model makes biological sense.

If system (2) satisfies the above two conditions, then the following theorem holds:

Theorem 7 ([31]). *$E_0 = (X_0, 0)$, is a globally asymptotically stable equilibrium of (2) provided that $R_0 < 1$ and that assumptions (H_1) and (H_2) are satisfied.*

First, we will prove that the two conditions (H_1) and (H_2) are satisfied for system (2): the system $\frac{dX}{dt} = F(X,0)$ is written as follows:

$$\frac{dS}{dt} = \Omega - (\mu + v)S(t),$$
$$\frac{dR}{dt} = -(\mu + v)R(t),$$
$$\frac{dV}{dt} = v(S(t) + R(t)) - \mu V(t),$$

and $X_0 = \left(\frac{\Omega}{(\mu+v)}, 0, \frac{v\Omega}{\mu(\mu+v)} \right)$, and from it

$$S(t) = \frac{\Omega}{(\mu+v)} - \left(\frac{\Omega}{(\mu+v)} - S(0) \right) e^{-(\mu+v)t},$$

and

$$R(t) = R(0)e^{-(\mu+v)t}.$$

Additionally, from

$$\frac{d(S+R+V)}{dt} = \Omega - \mu(S+R+V)(t),$$

we get

$$(S+R+V)(t) = \frac{\Omega}{\mu} - \left(\frac{\Omega}{\mu} - (S+R+V)(0) \right) e^{-vt},$$

when $t \to \infty : S(t) \to \frac{\Omega}{(\mu+v)}, R(t) \to 0, V(t) \to \frac{v\Omega}{\mu(\mu+v)}$. Hence, the global stability of X_0.

We have, on the other hand:

$$G(X,Y) = \begin{pmatrix} (\lambda(r_1 S(t) + r_2 R(t)) + r_3 V(t))(I(t) + I_d(t)) - (\mu + \rho)I(t) \\ (1-\lambda)(r_1 S(t) + r_2 R(t))(I(t) + I_d(t)) - (\mu + \delta + \rho)I_d(t) \end{pmatrix},$$

thus

$$G(X,Y) = \hat{A}Y - \hat{G}(X,Y),$$

where

$$\hat{A} = \begin{pmatrix} \left(\frac{\lambda r_1 \Omega}{(\mu+v)} + \frac{v r_3 \Omega}{\mu(\mu+v)} \right) - (\mu + \rho) & \left(\frac{\lambda r_1 \Omega}{(\mu+v)} + \frac{v r_3 \Omega}{\mu(\mu+v)} \right) \\ \frac{(1-\lambda)r_1 \Omega}{(\mu+v)} & \frac{(1-\lambda)r_1 \Omega}{(\mu+v)} - (\mu + \delta + \rho) \end{pmatrix}$$

and

$$\hat{G}(X,Y) = \begin{pmatrix} \left(\left(\frac{\lambda r_1 \Omega}{(\mu+v)} + \frac{v r_3 \Omega}{\mu(\mu+v)}\right) - \lambda((r_1 S(t) + r_2 R(t)) + r_3 V(t))\right)(I + I_d) \\ (1-\lambda)\left(\frac{r_1 \Omega}{(\mu+v)} - (r_1 S(t) + r_2 R(t))\right)(I + I_d) \end{pmatrix}$$

$\hat{A}$ is an M-matrix, thus we get the following result:

Theorem 8. *Suppose that $R_0 < 1$. If*

$$\rho - \frac{r_2 \Omega}{(\mu+v)} \leq 0, \tag{17}$$

then E_0 is globally asymptotically stable.

Proof. Suppose the initial conditions are as follows: $S(0) + R(0) \leq \frac{\Omega}{(\mu+v)}$, $V(0) \leq \frac{v\Omega}{\mu(\mu+v)}$, such that

$$\left(\frac{r_1 \Omega}{(\mu+v)} - (r_1 S(t) + r_2 R(t))\right) \geq r_1 \left(\frac{\Omega}{(\mu+v)} - (S(t) + R(t))\right).$$

On the other hand:

$$\frac{d(S(t)+R(t))}{dt} = \Omega + \rho(I(t) + I_d(t)) - r_1(I(t) + I_d(t))S(t) - r_2(I(t) + I_d(t))R(t) \\ - (v+\mu)(R(t) + S(t)),$$

which gives us

$$\begin{aligned} \frac{d(S(t)+R(t))}{dt}\bigg|_{S+R=\frac{\Omega}{(\mu+v)}} &= \Omega + \rho(I(t) + I_d(t)) - r_1(I(t) + I_d(t))S(t) - r_2(I(t) + I_d(t))R(t) \\ &\quad - (v+\mu)(R(t) + S(t)) \\ &\leq \Omega + \rho(I(t) + I_d(t)) - ((v+\mu) + r_2(I(t) + I_d(t)))(R(t) + S(t)) \\ &\leq \left(\rho - \frac{r_2 \Omega}{(\mu+v)}\right)(I(t) + I_d(t)), \end{aligned}$$

and according to (15):

$$\frac{d(S(t)+R(t))}{dt}\bigg|_{S+R=\frac{\Omega}{(\mu+v)}} \leq 0,$$

thus

$$(S(t) + R(t)) \leq \frac{\Omega}{(\mu+v)}, \tag{18}$$

and

$$(1-\lambda)\left(\frac{r_1 \Omega}{(\mu+v)} - (r_1 S(t) + r_2 R(t))\right)(I + I_d) \geq 0.$$

On the other hand:

$$\frac{dV}{dt}\bigg|_{V=\frac{v\Omega}{\mu(\mu+v)}} = v(S(t) + R(t)) - r_3(I + I_d)\frac{v\Omega}{\mu(\mu+v)} - \mu\frac{v\Omega}{\mu(\mu+v)},$$

and from (18)

$$\frac{dV}{dt}\bigg|_{V=\frac{v\Omega}{\mu(\mu+v)}} \leq -r_3(I + I_d)\frac{v\Omega}{\mu(\mu+v)} \leq 0,$$

then $V(t) \leq \frac{v\Omega}{\mu(\mu+v)}$, and from it

$$\left(\left(\frac{\lambda r_1 \Omega}{(\mu+v)} + \frac{v r_3 \Omega}{\mu(\mu+v)}\right) - \lambda((r_1 S(t) + r_2 R(t)) + r_3 V(t))\right)(I + I_d) \geq 0.$$

Finally, $\hat{G}(X, Y) \geq 0$. According to Carlos Castillo-Chavez [31], E_0 is globally asymptotically stable. □

6. Numerical Simulations

In this section, we'll apply the system under study to the country of Brazil and contrast the results with real data to determine how effective it is. Based on [32], we can divide the initial population as follows:

$$S(0) = 17300532, \quad R(0) = 30921318, \quad V(0) = 169017000, \quad I(0) = 616214, \quad I_d(0) = 410809. \tag{19}$$

The values of the system's parameters can be calculated according to the same source [32], however, we discover the following:

$$\begin{array}{lll} \Omega = 287010; & \mu = 0.00066; & \lambda = 0.6; \\ r_1 = 4.4 \times 10^{-11}; & r_2 = 3.215 \times 10^{-11}; & r_3 = 1.4925 \times 10^{-11}; \\ \rho = 0.0352; & v = 0.0166; & \delta = 5.6259 \times 10^{-4}. \end{array} \tag{20}$$

We also take the real data of active cases in Brazil in the period from 23 July to 15 August 2022, shown in Figure 1.

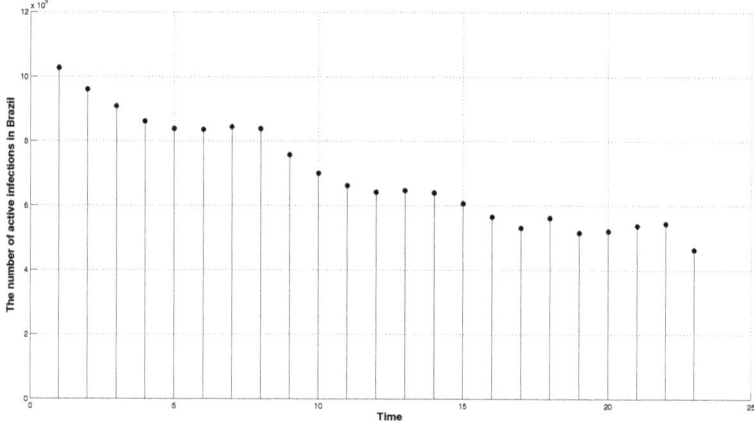

Figure 1. The number of active infections in Brazil in the period from 23 July to 15 August 2022 [32].

We apply system (2) using the previous data, and we find the numerical simulation as shown in Figure 2.

Note that:

$$\rho - \frac{r_2 \Omega}{(\mu + v)} = 3.466510^{-2} > 0, \text{ and } R_0 = 0.19435. \tag{21}$$

Thus, the condition of Theorem 7 is not fulfilled, we can then say that E_0 is locally asymptotically stable (according to Theorem 7) but not necessarily globally asymptotically stable. We notice from the simulations in Figure 3 that the model gives a good result in terms of its prediction.

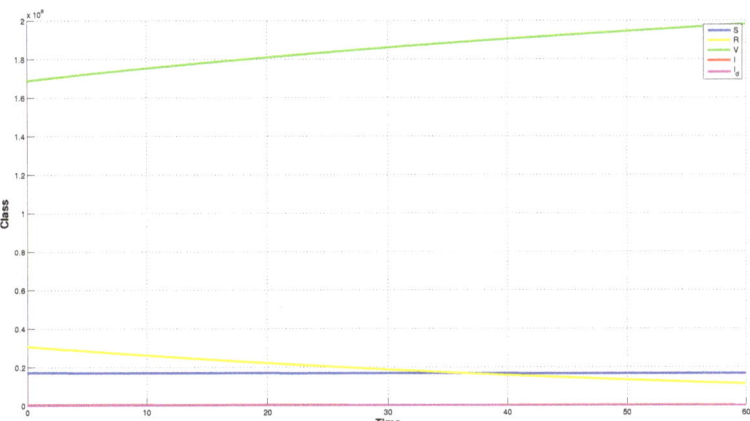

Figure 2. Numerical simulation for system (2) using (19) and (20).

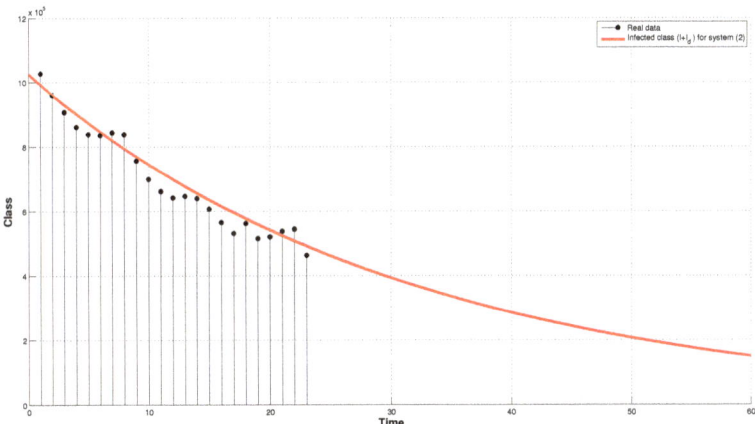

Figure 3. Numerical simulation of the infected class and comparison with real-world data.

7. Conclusions

The number of immunized people is a new state variable in the COVID-19 compartment model that this paper has used to illustrate the dynamics of the system. The proposed model, which is described by integer-order differential equations and is demonstrated in this paper, has a disease-free fixed point and an endemic fixed point. We also proved a new theorem, which has assured the global stability of the disease-free fixed point. The theorem has highlighted that the pandemic can disappear, provided that an inequality involving the vaccination rate is satisfied. This is a noteworthy discovery of the suggested approach, as it may aid in the comprehension of decision-makers when it comes to the disease's evolving epidemiological behavior. Last but not least, numerical simulations have been conducted to demonstrate the value of the COVID-19 epidemic model.

Author Contributions: Conceptualization, I.A.-S. and A.J.; methodology, A.O.; software, A.O.; validation, A.A.-H., N.D. and A.O.; formal analysis, N.D.; investigation, G.G.; resources, N.D.; data curation, A.O.; writing—original draft preparation, N.D.; writing—review and editing, N.D.; visualization, N.D.; supervision, A.O.; project administration, A.O.; funding acquisition, G.G. All authors have read and agreed to the published version of the manuscript.

Funding: This research received no external funding.

Institutional Review Board Statement: Not applicable.

Informed Consent Statement: Not applicable.

Data Availability Statement: The data that support the findings of this study are available upon request.

Conflicts of Interest: The authors declare that they have no conflict of interest.

References

1. Thanin, S.; Anwar, Z.; Saowaluck, C.; Zohreh, E.; Mouhcine, T.; Salih, D.; Analysis of a discrete mathematical COVID-19 model. *Results Phys.* **2021**, *28*, 104668.
2. Kermack, W.; McKendrick, A. Contributions to the mathematical theory of epidemics. *Proc. R. Soc. A* **1927**, *115*, 700–721.
3. DInnocenzo, A.; Paladini, F.; Renna, L. A numerical investigation of discrete oscillating epidemic models. *Physica A* **2006**, *364*, 497–512. [CrossRef]
4. Albadarneh, R.B.; Batiha, I.M.; Ouannas, A.; Momani, S. Modeling COVID-19 pandemic outbreak using fractional-order systems. *Int. J. Math. Comput. Sci.* **2021**, *16*, 1405–1421.
5. Djenina, N.; Ouannas, A.; Batiha, I.M.; Grassi, G.; Oussaeif, T.E.; Momani, S. A novel fractional-order discrete SIR model for predicting COVID-19 behavior. *Mathematics* **2022**, *10*, 2224. [CrossRef]
6. Batiha, I.M.; Momani, S.; Ouannas, A.; Momani, Z.; Hadid, S.B. Fractional-order COVID-19 pandemic outbreak: Modeling and stability analysis IM Batiha. *Int. J. Biomath.* **2022**, *15*, 2150090. [CrossRef]
7. Batiha, I.M.; Al-Nana, A.A.; Albadarneh, R.B.; Ouannas, A.; Al-Khasawneh, A.; Momani, S. Fractional-order coronavirus models with vaccination strategies impacted on Saudi Arabia's infections. *AIMS Math.* **2022**, *7*, 12842–12858. [CrossRef]
8. Giordano, G.; Blanchini, F.; Bruno, R.; Colaneri, P.; Di Filippo, A.; Di Matteo, A.; Colaneri, M. Modelling the COVID-19 epidemic and implementation of population-wide interventions in Italy. *Nat. Med.* **2020**, *26*, 855–860. [CrossRef]
9. Ajbar, A.; Alqahtani, R.T.; Boumaza, M. Dynamics of an SIR-Based COVID-19 Model with Linear Incidence Rate, Nonlinear Removal Rate, and Public Awareness. *Front. Phys.* **2021**, *9*, 634251. [CrossRef]
10. Marinov, T.T.; Marinova, R.S. Dynamics of COVID-19 using inverse problem for coefficient identification in SIR epidemic models. *Chaos Solitons Fractals X* **2020**, *5*, 100041. [CrossRef]
11. JWu, J.T.; Leung, K.; Bushman, M.; Kishore, N.; Niehus, R.; de Salazar, P.M.; Cowling, B.J.; Lipsitch, M.; Leung, G.M. Estimating clinical severity of COVID-19 from the transmission dynamics in Wuhan, China. *Nat. Med.* **2020**, *26*, 506–510.
12. Gounane, S.; Barkouch, Y.; Atlas, A.; Bendahmane, M.; Karami, F.; Meskine, D. An adaptive social distancing SIR model for COVID-19 disease spreading and forecasting. *Epidemiol. Methods* **2021**, *10*, 20200044. [CrossRef]
13. Mangiarotti, S.; Peyre, M.; Zhang, Y.; Huc, M.; Roger, F.; Kerr, Y. Chaos theory applied to the outbreak of COVID-19: An ancillary approach to decision making in pandemic context. *Epidemiol. Infect.* **2020**, *148*, 1–29. [CrossRef]
14. Debbouche, N.; Ouannas, A.; Batiha, I.M.; Grassi, G. Chaotic dynamics in a novel COVID-19 pandemic model described by commensurate and incommensurate fractional-order derivatives. *Nonlinear Dyn.* **2022**, *109*, 33–45. [CrossRef]
15. de Carvalho, J.P.M.; Moreira-Pinto, B. A fractional-order model for CoViD-19 dynamics with reinfection and the importance of quarantine. *Chaos Solitons Fractals* **2021**, *151*, 111275. [CrossRef]
16. Mohammadi, H.; Rezapour, S.; Jajarmi, A. On the fractional SIRD mathematical model and control for the transmission of COVID-19: The first and the second waves of the disease in Iran and Japan. *ISA Trans.* **2021**, *124*, 103–114. [CrossRef]
17. Biala, T.A.; Khaliq, A.Q.M. A fractional-order compartmental model for the spread of the COVID-19 pandemic. *Commun. Nonlinear Sci. Numer. Simul.* **2021**, *98*, 105764. [CrossRef]
18. Panwar, V.S.; Uduman, P.S.; Gómez-Aguilar, J.F. Mathematical modeling of coronavirus disease COVID-19 dynamics using CF and ABC non-singular fractional derivatives. *Chaos Solitons Fractals* **2021**, *145*, 110757. [CrossRef]
19. Chatterjee, A.N.; Ahmad, B. A fractional-order differential equation model of COVID-19 infection of epithelial cells. *Chaos Solitons Fractals* **2021**, *147*, 110952. [CrossRef]
20. Gozalpour, N.; Badfar, E.; Nikofard, A. Transmission dynamics of novel coronavirus sars-cov-2 among healthcare workers, a case study in Iran. *Nonlinear Dyn.* **2021**, *105*, 3749–3761. [CrossRef]
21. De la Sen, M.; Alonso-Quesada, S.; Ibeas, A. On a discrete SEIR epidemic model with exposed infectivity, feedback vaccination and partial delayed re-susceptibility. *Mathematics* **2021**, *9*, 520. [CrossRef]
22. Kumar, P.; Erturk, V.S.; Murillo-Arcila, M. A new fractional mathematical modelling of COVID-19 with the availability of vaccine. *Results Phys.* **2021**, *24*, 104213. [CrossRef] [PubMed]
23. De la Sen, M.; Ibeas, A. On an SE(Is)(Ih)AR epidemic model with combined vaccination and antiviral controls for COVID-19 pandemic. *Adv. Differ. Equ.* **2021**, *2021*, 92. [CrossRef] [PubMed]
24. De la Sen, M.; Ibeas, A.; Nistal, R. About partial reachability issues in an SEIR epidemic model and related infectious disease tracking in finite time under vaccination and treatment controls. *Disc. Dyn. Nat. Soc.* **2021**, *2021*, 1026–1226. [CrossRef]
25. Omame, A.; Okuonghae, D.; Nwajeri, U.K.; Onyenegecha, C.P. A fractional-order multi-vaccination model for COVID-19 with non-singular kernel. *Alex. Eng. J.* **2022**, *61*, 6089–6104. [CrossRef]
26. Badfar, E.; Zaferani, E.J.; Nikofard, A. Design a robust sliding mode controller based on the state and parameter estimation for the nonlinear epidemiological model of COVID-19. *Nonlinear Dyn.* **2021**, *109*, 5–18. [CrossRef]
27. Rajaei, A.; Raeiszadeh, M.; Azimi, V.; Sharifi, M. State estimation-based control of covid-19 epidemic before and after vaccine development. *J. Process Control.* **2021**, *102*, 1–14. [CrossRef]

28. Feng, Z.; Li, F.; Lv, Y.; Zhang, S. A note on Cauchy-Lipschitz-Picard theorem. *J. Inequal Appl.* **2016**, *271*. [CrossRef]
29. Andrica, D.; Rassias, T.M. *Differential and Integral Inequalities; Springer Optimization and Its Applications*; Springer: Berlin, Germany, 2019; Volume 151.
30. Van den Driessche, P.; Watmough, J. Reproduction numbers and sub-threshold endemic equilibria for compartmental models of disease transmission. *Math. Biosci.* **2002**, *180*, 29–48. [CrossRef]
31. Castillo-Chavez, C.; Feng, Z.; Huang, W. Mathematical approaches for emerging and reemerging infectious diseases: An introduction. In *Proceedings of the IMA*; Springer: Berlin/Heidelberg, Germany; New York, NY, USA, 2002; pp. 229–250.
32. Available online: https://www.worldometers.info (accessed on 19 August 2022).

Disclaimer/Publisher's Note: The statements, opinions and data contained in all publications are solely those of the individual author(s) and contributor(s) and not of MDPI and/or the editor(s). MDPI and/or the editor(s) disclaim responsibility for any injury to people or property resulting from any ideas, methods, instructions or products referred to in the content.

Article

On the Impact of Quarantine Policies and Recurrence Rate in Epidemic Spreading Using a Spatial Agent-Based Model

Alexandru Topîrceanu

Department of Computer and Information Technology, Politehnica University Timișoara, 300006 Timisoara, Romania; alext@cs.upt.ro

Abstract: Pandemic outbreaks often determine swift global reaction, proven by for example the more recent COVID-19, H1N1, Ebola, or SARS outbreaks. Therefore, policy makers now rely more than ever on computational tools to establish various protection policies, including contact tracing, quarantine, regional or national lockdowns, and vaccination strategies. In support of this, we introduce a novel agent-based simulation framework based on: (i) unique mobility patterns for agents between their home location and a point of interest, and (ii) the augmented SICARQD epidemic model. Our numerical simulation results provide a qualitative assessment of how quarantine policies and the patient recurrence rate impact the society in terms of the infected population ratio. We investigate three possible quarantine policies (proactive, reactive, and no quarantine), a variable quarantine restrictiveness (0–100%), respectively, and three recurrence scenarios (short, long, and no recurrence). Overall, our results show that the proactive quarantine in correlation to a higher quarantine ratio (i.e., stricter quarantine policy) triggers a phase transition reducing the total infected population by over 90% compared to the reactive quarantine. The timing of imposing quarantine is also paramount, as a proactive quarantine policy can reduce the peak infected ratio by over ×2 times compared to a reactive quarantine, and by over ×3 times compared to no quarantine. Our framework can also reproduce the impactful subsequent epidemic waves, as observed during the COVID-19 pandemic, according to the adopted recurrence scenario. The suggested solution against residual infection hotspots is mobility reduction and proactive quarantine policies. In the end, we propose several nonpharmaceutical guidelines with direct applicability by global policy makers.

Keywords: agent-based model; epidemic model; computational epidemics; epidemic control policies; nonpharmaceutical interventions

MSC: 93A16; 91D10; 92D30; 68T09

Citation: Topirceanu, A. On the Impact of Quarantine Policies and Recurrence Rate in Epidemic Spreading Using a Spatial Agent-Based Model. *Mathematics* 2023, 11, 1336. https://doi.org/10.3390/math11061336

Academic Editors: Cristiano Maria Verrelli and Fabio Della Rossa

Received: 15 January 2023
Revised: 23 February 2023
Accepted: 8 March 2023
Published: 9 March 2023

Copyright: © 2023 by the author. Licensee MDPI, Basel, Switzerland. This article is an open access article distributed under the terms and conditions of the Creative Commons Attribution (CC BY) license (https://creativecommons.org/licenses/by/4.0/).

1. Introduction

Understanding the dynamics of pandemics (e.g., COVID-19, H1N1, Ebola, SARS, etc.) is an ongoing multidisciplinary scientific challenge and a public health priority for many governments around the world [1–3]. The conventional epidemiological approach, or the analytic approach, offers solutions by employing compartmental models that divide the population into groups based on economics, demographics, and other characteristics. It has been shown that even though compartmental models lack the complexity of individuals behavior, they are exceedingly successful in informing and developing public health policies [4–6].

The analytic approach involves gradually changing one variable at a time to deduce general laws that allow for predictions about the system's properties under varying conditions, which hold true in homogeneous systems consisting of similar elements with weak interactions between them [7]. In addition, systems consisting of a diversity of elements linked together by strong interactions employ the newer paradigm of complex systems. These complex systems methods aim to examine the system as a whole, taking into account its complexity and inherent dynamics. Through discrete event computer simulation of the system, one can observe the effects of various interactions between its elements in real time. Through the study of behavior, rules that can alter the system can eventually be determined [7,8].

Over the last decade, following the paradigm of complex systems, agent-based modeling (ABM) approaches have been developed to simulate epidemic outbreaks that enable embedding the behavior of individuals and their inherent stochasticity by characterizing every individual as a (simplified) social agent in an agent population. Each social agent is characterized by several parameters considered relevant for the viral spreading process, e.g., mobility patterns, physical social networking, social, minority, and economical status, or age. For example, during the debut of the COVID-19 pandemic, ABM was successfully used in exchange for simpler compartmental models, like SIR, SIS [9,10]. Later approaches also focused on developing epidemic models which consider more specific patient states like quarantined, infected and aware, vaccinated, asymptomatic, etc. One of these early compartmental models is SICARS, which was introduced by the authors in 2020 [11], and which we aim to further develop in this study.

In the face of existing challenges, we enumerate several state-of-the-art impactful studies that mainly improve mass action models into tools applicable to analyzing large epidemics [4,5,12–14]. Nevertheless, the underlying epidemic models used by the majority of papers are limiting (e.g., SI, SIS, SIR, SEIR, SIRS), as they are unable to implement the effects gradual quarantining, variable recurrence rates, and vaccination policies. Furthermore, their simplification of spatial and social organization lacks the complexity of population organization [15]. Therefore, the numerical results of such models can lead to over-/underestimations of the impact of an epidemic [6,14].

To address these limitations, our main motivation is to investigate the impact of two important nonpharmaceutical interventions, i.e., quarantine timing and strength of isolation policies, corroborated with variable patient recurrence rates based on ABM combined with an augmented epidemic model. Thus, our research aims to answer the following questions:

1. How does the *timing of quarantine* affect the maximum and total estimated ratios of the infected population?
2. How does the *recurrence rate* (i.e., duration of natural immunity) affect the ratio of infected population in time?
3. How does the *ratio of quarantined* infected population (i.e., isolation policy strength) affect the dynamics of an epidemic?
4. How does every control measure compare to the baseline (i.e., no action whatsoever) in terms of the epidemic outcome?

Answering these questions can lead us toward developing a set of qualitative "guidelines" for effective epidemic control with direct social impact, as well as impact in interdisciplinary applications, like epidemiology, modeling and simulation, and healthcare.

Closely related work on ABM for epidemic control is limited in the sense that most studies are either confined to specific geographical regions, or make use of epidemic models which are not aimed at studying control policies. For example, authors Hoertel et al. define an ABM with embedded data for France [16]; similarly, Datta et al. embed their ABM with data specific to the state of New York [17]. Hinch et al. provide a more robust ABM, called OpenABM-Covid19, tuned for the UK [10], but one that can be adjusted for other countries; however, their model is specifically a "flattening the curve" solution that does not consider reinfection and possible subsequent infection waves. In contrast, our simulation framework uses an epidemic model that permits, under the right parameters, epidemic waves to recur. In another study, authors Frias-Martinez et al. [18] propose an ABM aimed at H1N1 that does not consider quarantine policies. The work of Alzu'bi et al. is similar in terms of methodology, but aims towards a different research goal, without discussing the implications of quarantine policies [19].

Taken together, our main contribution is studying the effect of control procedures—through quarantine timing and strength, in the wake of different recurrence rates—to contain the epidemiological effect over closed populations. Our ABM offers an original contribution in regard to agent mobility patterns based on points-of-interest (POIs), which have been proven to represent infection hotspots in densely populated areas [20–22].

The rest of the paper is structured in the following order: the Section 2 section presents the epidemic model, the COVID-19 parameter settings, and the agent-based spatial model used in our simulations; the Section 3 section summarizes the simulation output and details the analysis over each experiment; the Section 4 section outlines a meta-analysis to understand the impact of each ABM parameter and presents conclusions from the experiments; the Section 5 outline the main results and enumerate the contributions of this work, including future directions for the presented research.

2. Materials and Methods

The study presents a multi-compartment model that tracks the spread of an epidemic at a macro-level within a community with a variable micro-structure, which is based on the local movements of agents. By considering the infectious status of agents in different compartments and controlling their movements, the model is able to accurately simulate the real-world mobility of the population and the progression of the outbreak.

We further detail the reasoning and methodology for developing the SICARQD epidemic model and the agent-based model used to simulate the target population. Given the multitude of possible parameters in an epidemic model as well as an ABM, we consider incorporating an intuitive set of rules for the individuals' mobility patterns and their spatial distribution. Specifically, to achieve our research goals, we augment the previously developed SICARS epidemic model [11] into the newly proposed SICARQD model to allow the evaluation of quarantine policies.

As a note, SICARS is a generalization of the popular SIR model [23,24], which we explicitly aimed at the analysis of isolation strategies that were relevant to the COVID-19 outbreak. To this end, SICARS defines five possible states for an individual: *susceptible S, incubating I, contagious C, aware A,* and *removed R*, with the note that R merges both *recovered* and *dead* compartments. Nevertheless, SICARS was developed as an *edge-removal model* to be used in a complex network context, rather than as an ABM. Further, our SICARQD epidemic model does not consider asymptomatic cases, which can occur during SARS-CoV-2 infection, so the estimated number of infected agents and the impact of the discussed quarantine policies can vary based on the actual real-world setting.

In addition, we make use of discrete event simulation which is a recognized option in the computational epidemiology literature for modeling high complexity and detail, where complexity is specifically the result of multiple random processes and the inherent structure of the system [25].

2.1. The SICARQD Epidemic Model

The proposed epidemic model is summarized in Figure 1 which defines the seven possible states and the particular transition rates. An agent can be in one of the following states: *susceptible S, incubating I, contagious C, aware A, quarantined Q, recovered R,* or *dead D*. The transition $S \rightarrow I \rightarrow C \rightarrow A \rightarrow R$ is determined by the following infection and recovery rates:

- λ_{inf}: rate for *susceptible* agents to become infected in the vicinity of an infectious agent (which is in the C or A states). An agent in the I state does not infect other agents.
- λ_{ctg}: rate of becoming contagious C after a specific period (depending on the modeled virus). In this state, an agent does not know that it is infected, yet it infects others.
- λ_{awr}: rate of becoming aware A after a specific period. In this state, an agent knows that it is infected, and it infects other agents in its vicinity.
- λ_{rec}: recovery rate after a specific infectious period. The transition determines whether an agent has fully recovered, becoming temporarily immune (R), or whether the agent has died (D) based on the death ratio r_{death}. Recovered agents R may not be infected.
- λ_{rcr}: recurrence rate to *susceptible* after a specific period of recovery from infection.

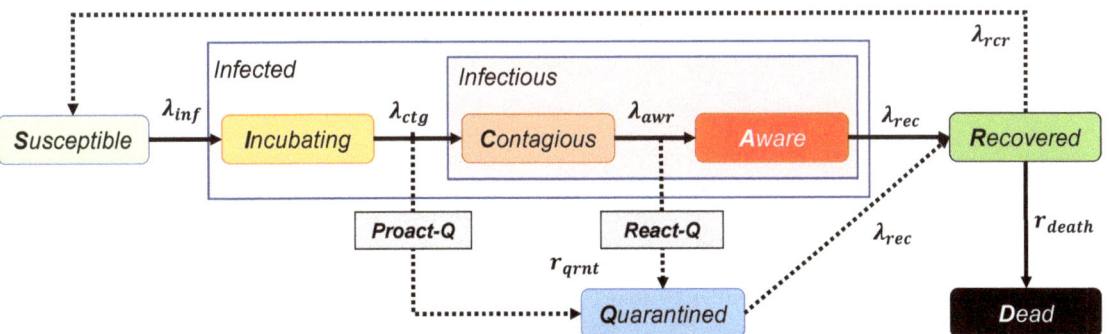

Figure 1. The states and parameters that define the SICARQD epidemic model. By splitting the infectious stage into two states—*contagious* and *aware*—our model offers the capability of implementing so-called proactive or reactive quarantine policies. The rates specific to transitions between infectious states are denoted with λ, and the ratios of quarantined and dead patients are denoted with r.

Furthermore, the transition to the quarantined state Q may occur when an agent moves to the contagious state C or to the aware state A. The number of quarantined agents corresponds to r_{qrnt}, which represents the ratio of infected patients that can be isolated from others, further removing their infectious capacity. As suggested by the dashed arrows in Figure 1, we implement three possible quarantine policies:

1. Proactive quarantine (**proact-Q**), which means that a proportion of agents will be quarantined immediately as they transit to become contagious C. This policy is the most expensive in the real world and involves contract tracing and intensive population-wide testing; also, this policy can prove to be the most effective way to mitigate infectious spreading.
2. Reactive quarantine (**react-Q**), which means that a proportion of agents will be quarantined right when they transit to become aware A. This policy is less costly, but requires population-wide testing and population awareness such that symptomatic individuals will auto-quarantine.
3. No quarantine (**no-Q**), which means that state Q is never used and that all infected agents remain active in the population. This is the baseline policy used for comparison.

Based on the recurrence rate λ_{rcr}, we determine whether the recovered state R is a final state or not. Consequently, we implement three recurrence scenarios:

1. Short recurrence scenario (**short-R**), corresponding to an average, normally distributed, 3 month immunity period.
2. Long recurrence scenario (**long-R**), corresponding to an average, normally distributed, 12 month immunity period.
3. Never-recurrence scenario (**never-R**), corresponding to a $\lambda_{rcr} = 0$, thus granting permanent immunity to all agents that reach the recovered state. This scenario is less realistic in the case of most infectious diseases, but serves as a baseline for comparison.

2.2. Literature Review for COVID-19 Infectious Parameters

Although our SICARQD model may be parameterized for other viruses, including future viruses, in this paper, we focus on customizing the experimental setup for the SARS-CoV-2 virus responsible for the COVID-19 pandemic.

Consequently, Table 1 details the settings for each model parameter, as supported by recent COVID-19 studies. All parameters correspond to those illustrated in Figure 1 and are based on an extensive review of the literature. In general, where we found several estimations, we picked either the worst case (see superscript [1]), or a rounded average value (without superscript). The last two parameters in Table 1 (see superscript [2]) are being investigated in our paper, so they take values in a wider range.

More precisely, we set the recurrence rates for recovered individuals to be normally distributed around 3 months for short-R and 12 months for long-R, respectively, as supported by recent estimates of COVID-19 immunity found in the recent literature: 3 months [26], 4–5 months [27], 6 months [28], and up to 1 year [29]. We also assign the quarantine ratio with values from 0 to 1, with a step of 0.1, that is, $r_{qrnt} = \{0, 0.1, \ldots 0.9, 1\}$.

With this study we aim towards a qualitative, general-purpose oriented model, rather than a model tuned to a specific dataset, region, or time frame, such that we use averaged fixed infectious parameters in our experiments, similar to other ABM approaches [30–33]. However, to prove that our model can fit real data, we present several fitting results in the Supplementary Materials Sections S1 and S2. Here, we use daily infectious and deaths data on COVID-19 from Romania and Hungary during October 2020–June 2021.

In contrast to our complex system approach on modeling and analyzing epidemic spreading, complementary studies employing an analytic approach in epidemics [5,34,35] include a time-dependent transmission rate to model changes in the infection rate caused by viral strain evolution, seasonality, social interactions, or governmental policies. Complementary to the differential equations of the compartmental models applied on the recent COVID-19 pandemic [5,34,35], ABMs study models of global mobility mechanisms based on emergent transmission dynamics. In this study, instead of a variable transmission rate, we use distributed simulation on the ABM resulting in a complex emergent population mixing instead of the compartmental models based on random uniform contact networks that are typically used to study epidemics spreading. Thus, modeling and quantifying human mobility is critical for studying large-scale spatial transmission of infectious diseases [31,32,36].

We note that the epidemic model considered in this study does not take into account active virus mutations, which means that it cannot simulate multiple outbreaks that arise from new strains. Instead, the model focuses on examining the dynamics related to quarantine for a single virus strain, i.e., the original SARS-CoV-2 strain. Nevertheless, in order to adjust the SICARQD model to another virus or strain, it suffices to redefine the seven parameters in Table 1 according to the available epidemiological data.

Table 1. Parameter setting for the SICARQD epidemic model, highlighting the values found in the literature, the values chosen in our model, and the supporting references. [1] Assumed worst case scenario based on literature review. [2] Parameters are being investigated in our study.

Parameter Name	Model Parameter	Value in Literature	Value in Model	References
Infection rate	λ_{inf}	0–5%	5% [1]	[24,37]
Incubation period	λ_{ctg}	5–7 days	5 days	[38,39]
Delay—contagion to onset	λ_{awr}	4–7 days	5 days	[37,38]
Delay—onset to recovery	λ_{rec}	10–14…56 days	14 days	[40–42]
Death ratio	r_{death}	1–3.4%	3.4% [1]	[41,43]
Recurrence rate	λ_{rcr}	3–12 months	3/12/∞ months [2]	[26,29]
Quarantine ratio	r_{qrnt}	–	0–100% [2]	

2.3. The Spatial Agent-Based Model

We start from the premise that social agents will move freely within a delimited area, such as a human settlement, or more specifically, an airport, a shopping mall, a hotel, a university campus, a conference venue, etc. In this sense, we define a two-dimensional rectangular space $S = 1600 \times 1200$ inside which we define a number of N agents to serve as the simulation population. All agents are considered equal and uniformly distributed inside the space S. By keeping the area size fixed, we influence the population density $\rho = N/S$ through the population parameter N.

Next, we assign each agent a_i a $home_i$ location, given by randomly generated coordinates (hx_i, hy_i), and a point of interest (POI) poi_i location given by a second set of coordinates (px_i, py_i), generated based on the home location. More precisely, the POI is assigned by generating a power-law distributed distance d_i from the home location, then a random angle is chosen to define the direction of the POI location. The distance is calculated as follows:

$$d_i = d_{max} * ((\phi_m - 1) * d_{min} * rand(0,1))^{-1*\phi_m}$$
$$d_i = max\{min\{d_i, d_{max}\}, d_{min}\} \quad (1)$$

Two constants are defined in Equation (1), namely a minimum distance to the POI of $d_{min} = 100$ and a maximum distance to the POI of $d_{max} = 1000$. Both values are based on the chosen space S and on empirical analysis through simulation. The two constants are also used to limit all values of generated distances within $[d_{min}, d_{max}]$. Another important parameter of the model, introduced in Equation (1), is the mobility factor ϕ_m. This factor determines how close (contracted) or far away (expanded) the population is when referring to its mobility patterns. A smaller ϕ_m will generate more distant POIs and vice versa. Based on the parameters of the model mentioned above, we found the interval $1.05 \leq \phi_m \leq 1.5$ that allows us to simulate restriction policies in the agent-based model. For a chosen mobility factor or $\phi_m = 1.1$ we obtain a distribution of POI distances with a power-law exponent of -2.3 (measured using the *poweRlaw* package in R). Power-laws are found in many empirical studies to represent the realistic distribution of human activity, interactions, and mobility [44,45].

Each agent a_i continuously moves between its home and POI (e.g., school, workplace, mall, grocery store, restaurant); its position at any time is given by the coordinates (x_i, y_i). The movement towards a POI (and back home) is implemented by adjusting the current coordinates with a constant node speed on the horizontal and vertical axes, resulting in a 45-degree movement in one of four possible directions: NW, NE, SE, SW. Once the same coordinate as the POI is reached on either axis, the agent will move straight until the POI is reached. This intuitive process is explained in Figure 2a.

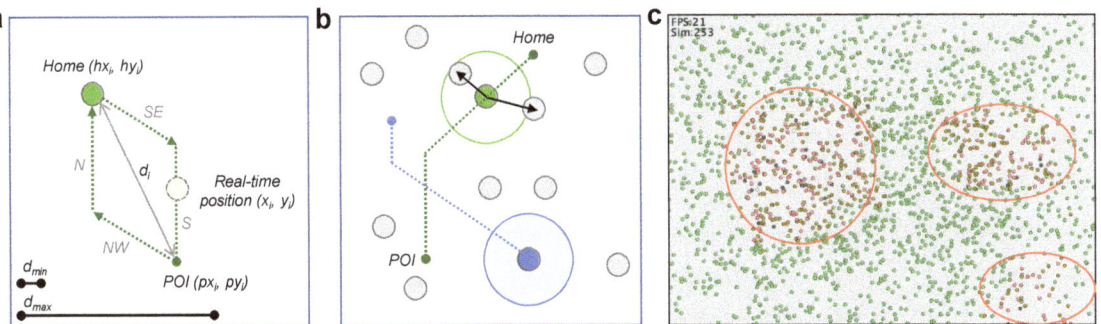

Figure 2. Overview of the agent-based model. (**a**) The relationship between an agent's home, POI and real-time locations. An agent will move to a POI (and back home) based on a combination of 45-degree directions followed by orthogonal movement. In the lower-left corner, the minimum and maximum POI distances are illustrated relative to the exemplified simulation space. (**b**) Multiple agents in the simulation space, of which two are highlighted with green and blue. Each agent moves independently to its POI and back. Each infected agent has an infection radius around which it can infect other susceptible agents based on the infection rate. (**c**) A screenshot from the simulation tool developed in Processing for educational purpose. Here, three initial infection clusters (orange and red dots) emerge and spread out based on each agent's mobility to engulf a larger population.

As the same mobility process is repeated for each agent a_i, a stochastic mixing of the population emerges. Here comes into play the SICARQD epidemic model. That is, almost every agent is initialized as susceptible S, except for several seed agents N_s that are initially infected. In our experiments, we set the total population to $N \in \{1000, 2500, 5000\}$ agents and the number of seed agents $N_s = 25$ (0.5–2.5% of the population).

An infectious agent a_i (i.e., contagious or aware, but not quarantined) will infect any other agent a_j found in its real-time vicinity based on the infection rate λ_{inf}. The infection distance is set to 4× the agent size = 4 × 8 = 24 [30,46]. This process is exemplified in Figure 2b. Since we run a discrete event simulation, we define the discrete simulation time t, where $0 \leq t \leq K$, K being the maximum simulation length. More precisely, we impose two stop conditions for our experiments: either a number of $K = 5000$ iterations are completed, or the number of infected agents drops to two or less, i.e., $I(t) + C(t) + A(t) + Q(t) \leq 2$. We chose these values on the basis of running several calibration experiments.

Based on the terms introduced, we quantify the effects of the epidemic using the following parameters: the epidemic size over time $0 < \phi(t) \leq 1$ (expressed as the ratio of the total infected population), the peak infected ratio $0 < \psi \leq 1$ (expressed as the maximum ratio of the infected population throughout the simulation), and the residual infection ratio at the end of the simulation $0 \leq \theta < 1$. The epidemic size, or the ratio of the total infected population $\phi(t)$ at any time t, is measured as follows:

$$\phi(t) = \frac{1}{N}[I(t) + C(t) + A(t) + Q(t)] \qquad (2)$$

The peak infected ratio ψ, measured at the end of the simulation time $t = K$, is expressed as follows:

$$\psi = max\{\phi(t)\}, \; for \; 0 \leq t \leq K \qquad (3)$$

The residual infection ratio θ, measured as the average epidemic size over the last 50 iterations, is expressed as follows:

$$\theta = \frac{1}{50} \sum_{t=K-50}^{K} (\phi(t)) \qquad (4)$$

To further explain the way our simulation environment is initialized, how the agents' position and infectious status are updated, and what the simulation framework measures, we provide Algorithm 1 with the goal of clarifying the relationship between the ABM and the epidemic model.

The spatial agent-based model presented and the embedded SICARQD epidemic model are implemented both as a Java applet with a user interface (UI) in the language *Processing* (mainly for educational purposes at our university; exemplified in Figure 2c), as well as in classic *Java* language for running the large-scale simulations (from the console) presented in the next section.

Algorithm 1 Infectious status and position updates for any agent a_i during each iteration t

procedure SETUPABM($A, S = [w \times h]$)   ▷ Agent and POI placement
    for $\forall a_i \in A$ **do**
        $home_i(hx_i, hy_i) \leftarrow random([0, w], [0, h])$ ▷ Random within $S = [w \times h]$
        $d_i \leftarrow d_{max} \cdot ((\Phi_m - 1) \cdot d_{min} \cdot random(0, 1))^{-\Phi_m}$
        $\alpha_i = random(2\pi)$
        $poi_i(px_i, py_i) \leftarrow (d_i \cdot sin(\alpha_i), d_i \cdot cos(\alpha_i))$ ▷ Must be inside S, else regenerate
    end for
end procedure
procedure UPDATEAGENTS($A, SICARQD, t$)   ▷ Agent update
    for $\forall a_i \in A$ **do**
        if $a_i.state = D$ **then** continue;   ▷ Ignore dead agents
        end if
        update $a_i.status(t+1) \leftarrow (a_i.status(t), SICARQD)$ ▷ Update infection status
        update $a_i(x_i, y_i) \leftarrow (d_x, d_y)$ ▷ Update position towards home or POI
    end for
    for $\forall (a_i, a_j) \in A$ **do**
        if distance$\{a_i, a_j\} \leq$ infection distance **then**   ▷ Check agent–agent interaction
            if $a_i.state = I$ & $a_j.state = S$, with λ_{inf} **then**
                $a_j.state \leftarrow I$
            end if
            if $a_j.state = I$ l; & $a_i.state = S$, with λ_{inf} **then**
                $a_i.state \leftarrow I$
            end if
        end if
        update(S,I,C,A,R,Q,D)
    end for
    $\phi(t) \leftarrow 1/N \cdot (I + C + A + Q)$
end procedure

3. Results

The experimental setup consists of repeated discrete event computer simulations that alter one of the following model parameters: quarantine policy, patient recurrence rate, ratio of quarantined infected population, and agent population size. Specifically, the three quarantine policies are proactive, reactive, and no quarantine (see Section 2.1); the three epidemic recurrence scenarios are short, long, and never recur (see Section 2.1); the quarantine ratios range within $r_{qrnt} \in \{0, 0.1, \ldots 0.9, 1\}$; population size ranges within $N \in \{1000, 2500, 5000\}$. Furthermore, the duration of the simulation is limited to $K = 5000$ iterations or until the number of infected agents drops back to two (or fewer) agents. We made several considerations to simplify the analysis of all simulation results, such as: (i) treating all social agents as identical, (ii) limiting the agent population to reduce simulation time, (iii) assigned a reasonable amount of infectious spreader agents (i.e., $|N_s| = 25$, which corresponds to $\approx$2% of all agents in the network), (iv) assigned power-law distributed travel distances to all agents in the same manner, (vi) fixed a long enough simulation duration to allow convergence of the epidemic size in time.

To ensure statistical validity, all experimental results represent the average values measured over 100 repeated simulations using the same settings. Overall, we conducted a total of 3 (quarantine policies) $\times$3 (recurrence scenarios) $\times$11 (quarantine ratios r_{qrnt}) $\times$3 (population sizes N) $\times$100 (repetitions) = 29,700 experiments that correspond to 297 unique simulation settings. Therefore, we prefer to provide a graphical representation of the numerical results in various settings rather than providing unusually long tables. To fully recognize the impact of each simulation parameter, we study the results using the graphical representations in Figures 3–8.

We first measure the infected population ratio $\phi(t)$ (i.e., total number of infected relative to the whole population, at every moment in time) by comparing the proactive (proact-Q) and reactive (react-Q) quarantine policies for the never-recurrence scenario (never-R). Figure 3 details the two quarantine policies for three representative quarantine ratios ($r_{qrnt} = \{0.2, 0.6, 0.8\}$). By adopting a proactive quarantine, as illustrated in Figure 3a, a distinctive impact of the quarantine ratios becomes visible. The measured peak infection ratios are $\psi = 0.36$ (for $r_{qrnt} = 0.2$ quarantined infected patients), respectively, $\psi = 0.14$ (for $r_{qrnt} = 0.6$ quarantined) and $\psi = 0.036$ (for $r_{qrnt} = 0.8$ quarantined). The differences measured in ψ result in a decrease of approximately 61% in the size of the epidemic when the strength of the quarantine increases from 20% to 60%, and another 74% when the quarantine is strengthened further to 80%. A noticeable effect of increasing the quarantine ratio past 0.6 is that the duration of the epidemic increases significantly. On the other hand, by adopting the reactive quarantine depicted in Figure 3b the impact of the quarantine ratio is significantly reduced. We measure a drop of 17% from $\psi = 0.41$ (for $r_{qrnt} = 0.2$) to $\psi = 0.33$–0.34 (for $r_{qrnt} = 0.6$–0.8). Additionally, comparing the numerical results in Figure 3a,b, we notice an increase in the epidemic size determined by switching from proact-Q to react-Q. The increases are 14% for $r_{qrnt} = 0.2$, 43% for $r_{qrnt} = 0.6$, and more than 700% for $r_{qrnt} = 0.8$.

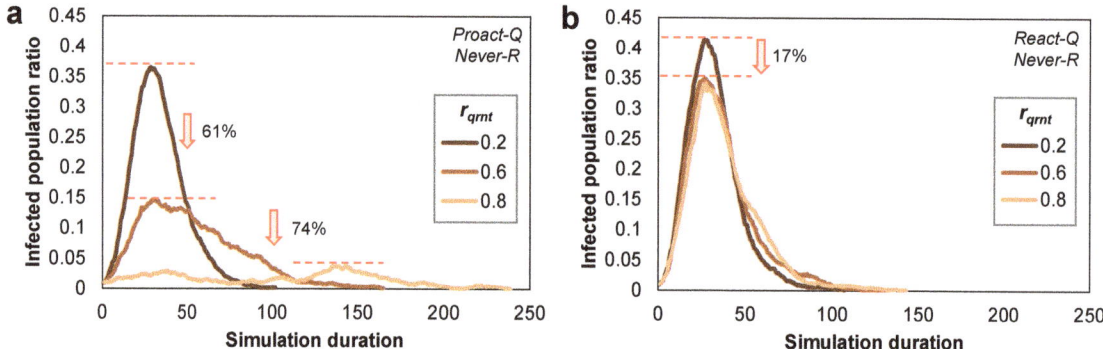

Figure 3. Overview of the evolution of an epidemic outbreak without patient recurrence (never-R), in time, measured by the infected population ratio $\phi(t)$. (**a**) The proactive quarantine policy (proact-Q) is applied. (**b**) The reactive quarantine policy (react-Q) is applied. The relative drops in infected population are depicted in percentages.

Next, in Figure 4 we aim to highlight the impact of quarantine policy on the total infected population ratio and the peak infection ratio. Given a long-recurrence (long-R) scenario, we notice very distinctive evolution patterns of the epidemic outbreaks. In Figure 4a, in terms of the total population infected throughout the simulation, when no agents are quarantined (i.e, $r_{qrnt} = 0$, up to 120% of the population becomes infected; this value results from counting all new infections that occur before the outbreaks dampen. In case of no-Q, the r_{qrnt} parameter does not have an influence, therefore, the infected population remains at 120%. In the case of react-Q, we measure a slight drop of 21% with increasing quarantine ratio. Finally, in the case of proact-Q, we observe an almost complete reduction (>99%) of the epidemic. However, more important is the visible phase transition around the values $r_{qrnt} = 0.6 - 0.8$, when for just a 20% strengthening of quarantine, a 92% relative drop of the total infected population is triggered. In Figure 4b, the decrease in the peak infected ratio ψ is similar to the total infected ratio, for the no-Q and react-Q policies; we measure an average drop of 28% for ψ for react-Q. The evolution of the peak infected ratio for proact-Q is different from the total infected ratio in the left panel; here, we see an almost perfectly linear drop with the quarantine ratio. The amplitude of the highest epidemic wave is 47% given the simulation parameters described.

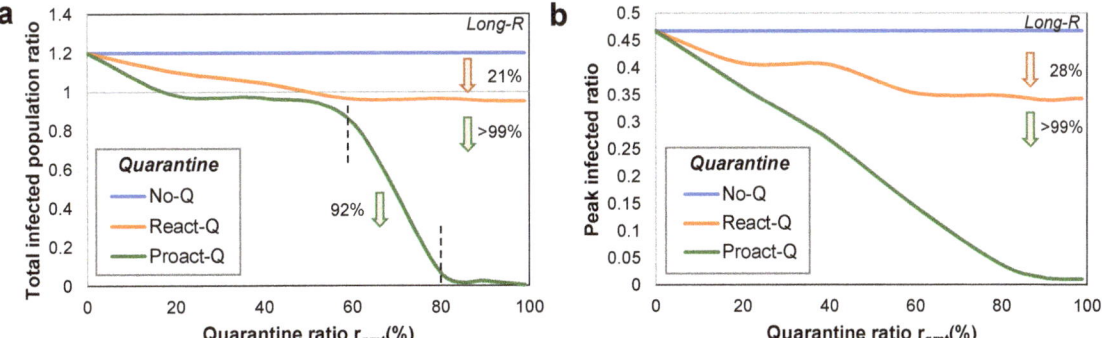

Figure 4. Impact of each quarantine policy in terms of an increasing quarantine ratio r_{qrnt}. (**a**) The total infected population ratio measured as the sum of all new cases throughout the simulation until the epidemic dissipates. The two vertical dotted bars delimit the phase transition area between $r_{qrnt} = 0.6$–0.8. (**b**) The peak infection ratio expressed as the maximum amplitude of the epidemic wave relative to the population of agents. The percentages along the right border of each panel represent relative decreases from the maximum to the minimum infected ratio.

Next, in Figure 5 we emphasize the impact of the patient recurrence rate on the total infected population. In Figure 5a we apply a reactive quarantine policy and notice a significant difference in epidemic size between short-R, on the one hand, and long-R and never-R, on the other. In the case of a rapidly recurring disease (i.e., short-R) as many as ×3 of the total population becomes infected throughout the epidemic; practically, each agent is infected, on average 3 times. However, if the immunity granted is about one year or more, the total infected ratio drops to 1–1.2 ×N, which means that all agents will become infected approximately once. The drops in the total infected ratio are small, of about 19–21% as the quarantine ratio increases. In case of a proactive quarantine policy, as depicted in Figure 5b, the epidemic impact is significantly reduced with increasing r_{qrnt}. Specifically, the same phase transition mentioned previously occurs between $r_{qrnt} = 0.6$–0.8, where we register drops in the total infected ratio of 85% for the short-R scenario, 60% for the long-R scenario, respectively, 92% for never-R.

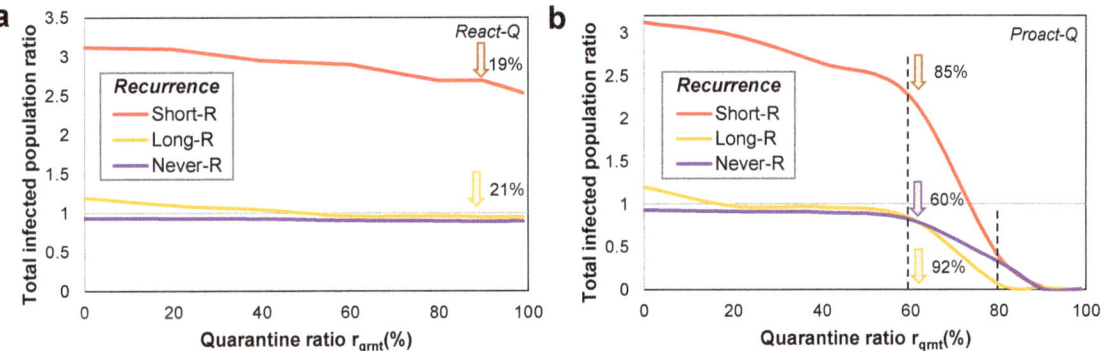

Figure 5. Impact of the recurrence scenario in terms of an increasing quarantine ratio r_{qrnt}. (**a**) The total infected population ratio measured for a reactive quarantine policy. (**b**) The total infected population ratio measured for a proactive quarantine policy. The two vertical dotted bars delimit the phase transition area between $r_{qrnt} = 0.6$–0.8; the respective decreases in the infected population ratios are displayed in relative percentage for each scenario of recurrence.

Our next analysis focuses on the dynamics of the infected population ratio $\phi(t)$ from the perspective of two fixed quarantine ratios. In Figure 6 we chose to depict the two ratios that delimit the phase transition previously observed, namely $r_{qrnt} = 0.6$ and $r_{qrnt} = 0.8$. We keep the recurrence scenario fixed to long-R and observe the qualitative differences between the three quarantine policies. Figure 6a corresponds to the snapshot of the epidemic before the phase transition triggers, and we easily distinguish between quarantine policies. The highest peak infection ratio is determined by a no-Q policy; the react-Q policy reduces the maximum infection ratio by 23%, and the proact-Q policy further reduces the infection ratio by 59%. Figure 6b corresponds to the measurements after the phase transition has been initiated, and, again, we can easily distinguish between quarantine policies. The highest peak infection ratio is determined by the no-Q policy, followed by the react-Q policy, which reduces the infection ratio by 26%. The proact-Q policy further reduces the infection ratio by up to 89% compared to react-Q. Although the duration of the proact-Q epidemic increases, the total infected ratio remains much smaller than for the other two quarantine policies. When comparing the green line infected ratios $\phi(t)$ for proact-Q, in the two panels, we measure a relative decrease of 74%.

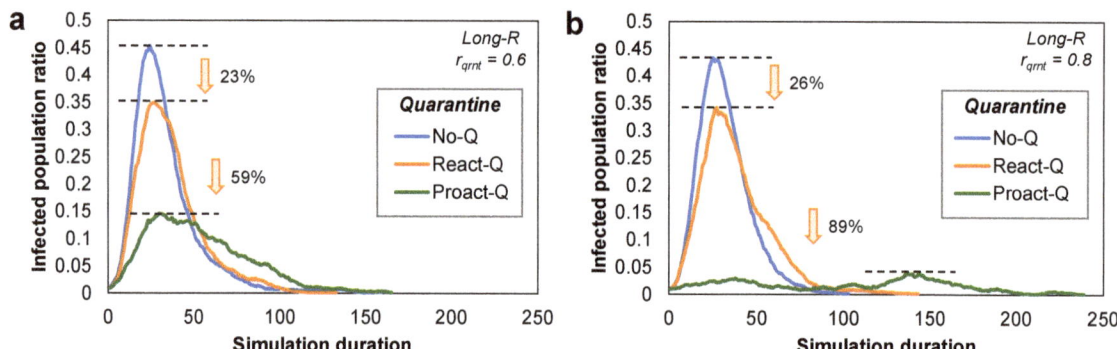

Figure 6. Overview of the evolution of epidemic outbreaks with long patient recurrence (long-R) from the perspective of the three quarantine policies. (**a**) The quarantine ratio is set to $r_{qrnt} = 0.6$, before which the phase transition in the infected population ratio occurs. (**b**) The quarantine ratio is set to $r_{qrnt} = 0.8$, after which the phase transition in the infected population ratio occurs. The decreases in the infected population ratios are shown as relative percentage for each quarantine policy.

In addition to the previous analysis, in Figure 7 we further illustrate the significant effect of the restrictiveness of quarantine, measured by the three quarantine ratios $r_{qrnt} = 0.2$, 0.6, 0.8. In all simulations, we use the proactive quarantine policy (proact-Q). Figure 7a exemplifies the evolution of the infected population ratio $\phi(t)$, in what we call a "less restrictive" quarantine, given by the small $r_{qrnt} = 0.2$. Although the long-R and never-R scenarios are characterized by a single epidemic wave, the short-R scenario leaves a significant residual (ongoing) infection of $\theta \approx 12\%$ in the population, with subsequent smaller waves. The peak infection rate for all recurrence scenarios is $\psi = 0.36$, but only short-R remains active throughout the duration of the simulation. In Figure 7b we observe a drop in the peaks of all epidemics. Notably, for this "moderately restrictive" quarantine ($r_{qrnt} = 0.6$), the peak of the short-R scenario drops to $\psi = 0.2$, by 44% compared to the "less restrictive" quarantine. The other two recurrence scenarios reduce the peaks even further to $\psi = 0.13$–0.14, and their corresponding epidemics dissipate. Again, the short-R scenario leaves a significant residual infection of $\theta \approx 10\%$ in the population, with subsequent smaller waves. Figure 7c presents the infected population ratio by adopting a "highly restrictive" quarantine, given by the higher $r_{qrnt} = 0.8$. Here, all recurrence scenarios are greatly dampened, and we measure infection peaks of only $\psi = 0.002$–0.038. Thus, the short-R peak is reduced by 81% compared to the peak measured for $r_{qrnt} = 0.6$;

nevertheless, larger secondary epidemic waves are produced (around t = 150–200) even in case of the highly restrictive quarantine. In this case, only a negligible amount of residual infection of $\theta \leq 1\%$ remains in the population at the end of the simulation.

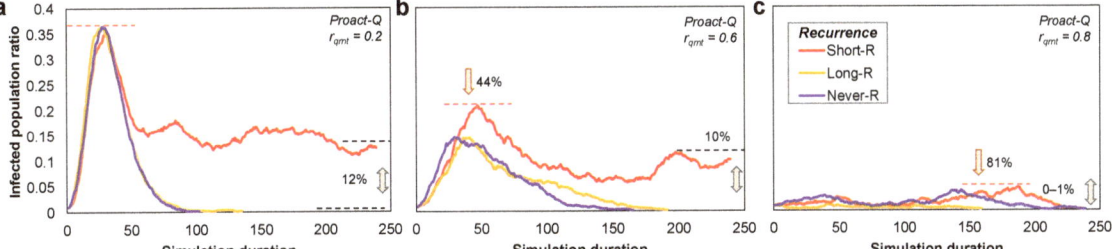

Figure 7. Evolution of epidemic outbreaks in time by employing a proactive quarantine policy (proact-Q) from the perspective of the three recurrence scenarios, using different quarantine ratios in each of the three panels. (**a**) A "less restrictive" quarantine ($r_{qrnt} = 0.2$) which triggers the highest infection peaks ($\psi = 0.36$) for all recurrence scenarios. The short-R scenario leaves a residual (ongoing) infection of 12% at the end of the simulation. (**b**) A "moderately restrictive" quarantine ($r_{qrnt} = 0.6$) which lowers the infection peaks ($\psi = 0.20$ for short-R, in red) for all recurrence scenarios. Again, the short-R scenario leaves a residual infection of around 10%. (**c**) A "highly restrictive" quarantine ($r_{qrnt} = 0.8$) which lowers the infection peaks even further ($\psi = 0.002$–0.038) for all recurrence scenarios. Almost no residual infection remains in the population. The decreases in the peak infection ratio for short-R (red) are displayed for the quarantine ratios $r_{qrnt} = 0.6, 0.8$ relative to the highest peak measured for $r_{qrnt} = 0.2$.

Finally, we discuss the impact of population density in the agent-based model. Figure 8 shows the infected population ratio $\phi(t)$ in two different scenarios that we consider representative of the entire experimental setup. In Figure 8a we apply a reactive quarantine and a short recurrence to distinguish between the evolution of three different populations. The largest agent population, $N = 5000$, results in the highest population density, which is translated into the most impactful outbreak. To this end, we measure a peak of $\psi = 0.68$ followed by a permanent infected ratio of $\phi(t) \approx 25\%$ (caused by the short-R scenario). The smallest population of $N = 1000$ produces a peak of only $\psi = 0.28$ and a lower residual infection rate of $\theta \approx 10\%$. The reduction in the peak infection rate is 38% if the population is reduced to $N = 2500$, and further by 33% for $N = 1000$. By contrast, in Figure 8b we employ a proactive quarantine and a long recurrence to distinguish between the evolution of three different populations. The largest population produces a peak of $\psi = 0.36$ followed by a permanent infected ratio of $\phi(t) \approx 5\%$. The smallest population produces a peak of $\psi = 0.083$ and a lower residual infection rate of $\theta \leq 1\%$. The peak infection rate decreases by 31% if the population is reduced from $N = 5000$ to $N = 2500$, and further decreases by 67% for $N = 1000$.

An important remark is that we did not plot the 95% confidence intervals (SD or any similar statistical reliability measure) for any of the plots in the results section because it would add a lot of complexity to the already detailed plots. However, all results from each experimental set-up present reliable convergence over the 100 repeated simulations. In our numerical analysis, we found that the maximum deviation from the plotted averages is not greater than $\pm 3\%$.

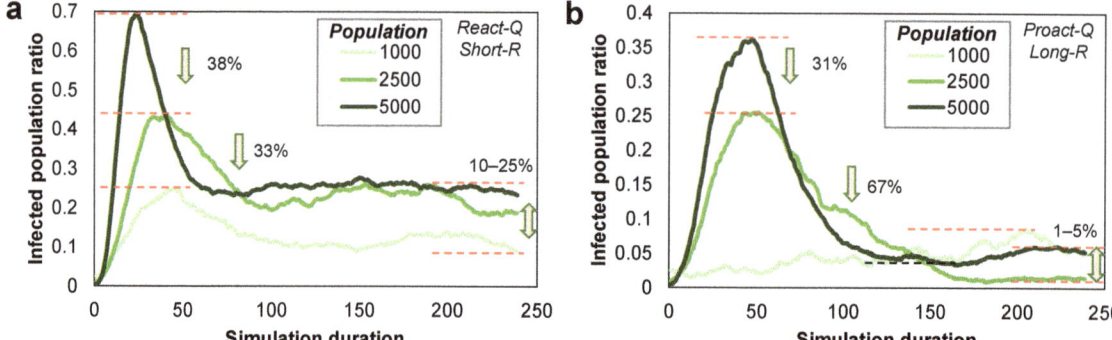

Figure 8. The impact of the population density on the evolution of epidemic outbreaks measured by $\phi(t)$. Note that density is proportional to the population size N, as we keep the simulation space S fixed. (**a**) A reactive quarantine policy and a short recurrence determine high infection ratios in the population (up to $\psi = 0.68$). A final, residual infection rate of 10–25% remains until the end of the simulation time. (**b**) A proactive quarantine policy and a long recurrence determine relatively lower infection ratios (up to $\psi = 0.36$). A final, residual infection rate of 1–5% remains until the end of the simulation time. The decreases in the peak infected ratio are shown as relative percentage for each population size N.

4. Discussion

Agent-based modeling is an effective interdisciplinary means of tracking and controlling epidemics, capable of incorporating the stochasticity of human behavior, which remains unaccounted for with just the classical compartmental model approach [4,5,47,48]. Specifically, by combining a spatial ABM with our proposed SICARQD epidemic model, we can replicate, monitor, and understand the parameters that affect the control of large-scale epidemics, with an overarching impact in epidemiology, mathematical modeling and public health, which are all very important social and scientific challenges [1,2,49,50].

In this study, we devise a spatial ABM represented by a closed-space agent population, in which agents receive two generated locations—a home and a point of interest—between which they continuously travel back and forth. The emergent mobility patterns result in a stochastic mixing of agents that replicates the real-world transmission of airborne diseases between each other (i.e., through physical proximity) to distances greater than their own vicinity. Furthermore, we take our previously introduced COVID-19 specific SICARS epidemic model [11] and augment it to a more elaborate model that: (i) can be used to incorporate quarantine policies (e.g., time and strength of these) and different patient recurrence rates, (ii) and can be parameterized to different viruses based on available epidemiological data.

We are able to reproduce similar epidemic dynamics to that of time-dependent transmission rates [34,35] to account for the social interaction and variable policies, by modeling and controlling the spatio-temporal movement of contagious and healthy agents through discrete event computer simulation [31,32]. Therefore, we defined a stochastic mobility model for agents and integrated an epidemic model to account for the specific SARS-CoV-2 transmission rates to provide a qualitative overview that emphasizes global perception of the complex system.

Compared to the classic infected (I) state of the SIR model, in SICARQD we define three different infected states: incubating I (timeout period from contracting a virus until the patient becomes a spreader), contagious C (first state in which the patient infects its peers, but without displaying symptoms), and aware A (second state in which the patient infects peers, but with symptoms). The details of our SICARQD model are illustrated in Figure 1. Furthermore, we address the issue of quarantine, which represents another possible patient state (Q). By controlling when and how many patients transit to Q, we can simulate different levels of policy and restriction (using the r_{qrnt} parameter). To this end, we analyze a proactive quarantine policy (proact-Q) that is applied before the patient becomes contagious, a reactive policy (react-Q) that is applied before the patient becomes aware, and no quarantine at all (no-Q). In a public health system, these policies can be applied by continuous contact tracing (proact-Q) or by intensive testing (react-Q). Another parameter we aim to study is the impact of the disease recurrence rate; in this sense, we consider a short recurrence (short-R) of 3 months, a long recurrence (long-R) of 1 year, and no recurrence (never-R). The purpose of adding these parameters to SICARQD is to observe whether (i) proactive quarantine is worth the extra costs, (ii) whether more restrictive quarantine has an impact on reducing the outbreak size, and (iii) whether the patient recurrence rate plays an important role in our efforts to mitigate the disease spreading.

We rely on discrete event simulation using a custom-developed computer program (using Processing and Java) to implement the described methodology. A total of 297 unique simulation scenarios were run, each repeated 100 times, and their results were averaged. Our results focus on the qualitative aspects of the described epidemiological policy parameters rather than replicating a specific geographical setting or demographic constraints.

The main observations drawn from our experiments are explained in the following three paragraphs. The proactive quarantine policy in correlation to a higher quarantine ratio (i.e., stricter quarantine policy) triggers a phase transition that reduces the total infected population by more than 90%, compared to reactive quarantine. Therefore, a proactive quarantine policy associated with a strict quarantine ratio can almost entirely inhibit infectious spreading, compared to a reactive quarantine which limits infectious spreading by about 20%, compared to no quarantine at all.

Indeed, proactive quarantine policy plays a significant role in reducing the peak infection rate, but only if combined with a more restrictive isolation of >60% quarantined infected patients. Specifically, when restrictiveness increases from 20% to 60%, the epidemic size decreases by 61% (see Figure 3). By increasing the restrictiveness to 80%, we can drastically reduce the peak infection ratio by another 74% relative to the previous peak. In contrast, if we rely on reactive quarantine, then the impact of increasing quarantine restrictiveness is relatively small, a reduction of just 17% on the infection peak. Furthermore, we investigated the restrictiveness interval 60-80% where we found a phase transition in the epidemic size; namely, the reduction of the epidemic is significant, by up to 92% if we increase the restrictiveness by only 20%. Of course, in real-world settings it is difficult to achieve an 80% isolation of infected patients, but the rewards of achieving such an isolation level are very high (see Figure 4).

The recurrence rate plays a very important role in terms of the total infected population over the course of the epidemic. Although the amplitude of epidemic waves for short-R remains similar to that of long-R, a residual (ongoing) infection remains in the population, with subsequent smaller waves, which add up to the total number of infected. More precisely, approximately ×3 times more agents are infected compared to long-R and never-R. However, all recurrence scenarios are positively influenced by the increase in quarantine restrictiveness, since we measure similar phase transitions in epidemic size, with drops of 85%, 60% and 92% for the short-R, long-R, and never-R recurrence scenarios (see Figure 5). Furthermore, as supported by our analysis of the population size (see Figure 8 for details), we found that the short recurrence scenario (short-R) significantly favors the epidemic to remain active on the long-term. Here, we suggest that short-R may determine a residual infection ratio of 10–25% in the population. When switching to a long-R scenario, the residual infection ratio remains within a more manageable 1–5% of the population.

The timing of quarantine is an important parameter that directly influences the peak infection ratio (i.e., amplitude of the highest wave). For a moderately restrictive quarantine ($r_{qrnt} = 0.6$), a no quarantine policy (no-Q) translates into having a peak infection ratio of 44% of the population, versus 34% if a reactive quarantine policy is adopted; furthermore, the peak can be reduced to 14% if a proactive policy is enabled. For a highly restrictive quarantine ($r_{qrnt} = 0.8$), no-Q induces a similar peak of 43%, versus a similar 32% peak if react-Q is adopted, and a much smaller peak of 3.4% if proact-Q is adopted (see Figure 6).

The emergence of a residual infection and subsequent larger outbreaks is an ongoing and important debate in the recent literature on COVID-19 [51,52]. Within our simulation framework, we can reproduce this residual infection rate, as well as impactful secondary epidemic waves, based on the adopted recurrence scenario. Specifically, if we simulate a short recurrence, in which patients can reinfect, on average, in less than 6 months [26,27], we obtain approximately 1–12% remaining infected individuals after the first wave. This percentage decreases with an increase in the quarantine ratio, but even for highly restrictive quarantine, like $r_{qrnt} = 0.8$, we notice a larger secondary wave in Figure 7c. The epidemic recurrence phenomenon is induced by the emergent agent mobility modeled in our system. More precisely, recovered agents lose their immunity in time and travel to infected areas before the hotspots dissolve, and so, hotspots may be maintained for very long durations (e.g., several years). The suggested solution against these residual hotspots—based on our ABM results—is mobility reduction and proactive quarantine policies.

The general observations validated throughout the experiments are summarized as the following guidelines:

- The peak infected ratio (i.e., amplitude of the highest epidemic wave) is reduced in the following scenarios:
    1. By increasing the quarantine restrictiveness in our model through the r_{qrnt} parameter. Specifically, a phase transition is observed in the infected ratio when isolating more than 60% of the infected population. More precisely, the decreases in infection ratio, when increasing $r_{qrnt} = 0.6$ to $r_{qrnt} = 0.8$, range between 60-92% (when proactive quarantine is applied, see below).
    2. By combining the restrictiveness with a proactive quarantine policy, rather than a reactive one. This means isolating suspicious cases early, before they become contagious through population-wide contract tracing and testing. The proactive quarantine policy will reduce the peak infection ratio by 59–89% compared to the reactive policy.
    3. In less densely populated areas. This may be achieved again by partial isolation or relocation of the population.

- The total infected population (i.e., total number of new cases throughout the simulation) is reduced in the following scenarios:
    1. If the patient recurrence rate is longer: of one year, or more. In the short-R scenario (3 months), more than ×3 people become infected compared to long-R (one year). A long recurrence rate can be achieved through natural immunity or through vaccination, depending on the virus.
    2. By applying the previously described solutions: a proactive quarantine policy combined with a greater restrictiveness of more than 60% isolation.
    3. In less densely populated areas. This may be achieved again by partial isolation or relocation of the population.

- The residual infection ratio (i.e., the infected ratio that remains in the population for a very long time) is reduced in the following scenarios:
    1. Depends primarily on the recurrence rate; a short-R scenario can leave a 10–12% infected ratio in the population, while the long-R and never-R scenarios help to completely dissipate the epidemic in time.
    2. Depends on the population density combined with the quarantine policy. A react-Q policy and a high population density can lead to a residual infection ratio of 10–25%, while a proact-Q can keep the residual infection ratio within 1–5%.

In essence, the proactive quarantine in correlation to a higher quarantine ratio (i.e., stricter quarantine policy) triggers a phase transition reducing the total infected population by over 90%, compared to the reactive quarantine. Therefore, a proactive quarantine associated with a strict quarantine ratio can almost completely inhibit infectious spread, compared to a reactive quarantine which limits infectious spreading by about 20%, compared to no quarantine at all.

Some of the limitations of this study are discussed in more detail. First, our model is not able to fully capture the complexity of human behavior and interactions. However, while some ABM approaches focus on incorporating social aspects, such as agent age group [53], social similarity [36], or leisure activities [54], other models focus strictly on the complexity of human topology [32,33], mobility patterns [31] or computation [55]. In this sense, our model offers a contribution in regard to mobility patterns based on POIs, which have been proven to represent infection hotspots in dense areas [20]. Second, the closed space in which the agent population is allowed to move restricts the study of a very large-scale, heterogeneous, or hierarchical population model. However, a hierarchical ABM can be defined where each ABM corresponds to a neighborhood or a town in a larger geographic area, similar to the study in [56]. Third, the agents are considered identical; they travel with the same speed, each has one single point of interest (POI), and we did not add any demographic data like age groups or gender. However, our study focuses on the broader perspective of quarantine policies and recurrence rates, rather than on specific geo-social settings. Still, the ABM can be further developed to incorporate multiple POIs and some individual traits. Fourth, we tuned the SICARQD epidemic model with data specific to the SARS-CoV-2 virus. As such, the numerical estimations may change with other future viruses. A simple re-tuning of the model will suffice to adapt to new epidemic situations. Lastly, our SICARQD epidemic model does not consider asymptomatic cases. However, given the high number of unaccounted asymptomatic cases during the COVID-19 pandemic, the actual number of agents entering the aware state would be slightly reduced and the efficiency of the react-Q policy could drop (i.e., they cannot become aware without symptoms). This effect may stack up among recurrent infections.

In addition to the scientific potential of our proposed methodology, our results find immediate applicability in the real world in the context of COVID-19, as well as future pandemics with similar transmission mechanisms. Furthermore, the enumerated observations can be further adopted in social physics, by corroborating our epidemic control strategies with opinion injection strategies and competing influence dynamics [57–59], or other immunization strategies for viral outbreaks [60,61].

5. Conclusions

The design of effective strategies for the tracking and control of epidemics is a major concern for public health systems around the world and an ongoing scientific challenge [1–3]. Much of the current state-of-the-art work is either focused on *flattening the curve* solutions [4,5,14,62], or limited to specific geographic and demographic settings [16,17], or their epidemic models do not focus on various quarantine policies [18,19]. Therefore, our study integrates agent-based modeling with a novel epidemic model to provide a qualitative study on quarantine policies and recurrence rates in large populations. Overall, our research is innovative because it targets policy makers by providing a set of "epidemic control guidelines" based on our simulation results.

Specifically, in this article, we study the effect of control procedures, implemented through quarantine timing and strength of isolation, from the perspective of variable patient recurrence rates, to understand their impact on containing epidemics over closed populations. In our SICARQD epidemic model, we incorporate three infectious states (I, C, A) and a quarantine state (Q) which allow for implementing an early (proact-Q) quarantine policy and a late (react-Q) policy. Therefore, by specifying *when* and *how many* patients transit to Q, we simulate different levels of policy and restrictiveness in the population. Furthermore, we introduce three possible recurrence scenarios (short-R, long-R, never-R), which aim to demonstrate the effectiveness of the quarantine policies with respect to the natural response of patients to an infectious disease. Consequently, by corroborating all parameters into our ABM, we provide a discrete event simulation framework for current and future epidemic scenarios.

Regarding the remarkable amount of published literature on epidemic outbreaks during the COVID-19 period, we strongly believe that our study represents a promising direction of research aimed at better understanding the real-world impact of quarantine policies, their timing and strength, and the impact of the natural disease recurrence rate in patients, all by incorporating real epidemiological data. In general, we consider that our proposed ABM framework can trigger the creation of computational intelligence tools to further enhance strategies for the monitoring and control of large-scale public health systems in a safe and effective manner.

Future research directions may include (i) incorporating hierarchical agent-based models to replicate more realistic human populations, (ii) integrating a set of vaccination strategies on top of the existing policies, (iii) defining multiple categories of agent types and adding additional mobility patterns, (iv) and refining the SICARQD model with epidemic data characteristic to specific regions around the world.

Supplementary Materials: The following are available online at https://www.mdpi.com/article/10.3390/math11061336/s1, Figure S1: Time series data on daily COVID-19 cases in Romania, Figure S2: Fitting data on daily infectious cases, Figure S3: Fitting data on daily deaths, Table S1: Fitting accuracy of the epidemic model for COVID-19 data. Reference [63] is cited in Supplementary Materials.

Funding: This research received no external funding.

Data Availability Statement: All data used in this study is summarized in Table 1 and is represented by epidemiological data for the SARS-CoV-2 virus responsible for the COVID-19 pandemic. The parameters were used to tune the introduced SICARQD epidemic model for the experimental setup presented; these parameters were taken from cited resources as follows: infection rate [24,37], incubation period [38,39], delay between contagion and onset [37,38], delay between onset and recovery [40–42], death ratio [41,43], and recurrence rate [26,29].

Conflicts of Interest: The author declares no conflict of interest.

References

1. Keeling, M. The implications of network structure for epidemic dynamics. *Theor. Popul. Biol.* **2005**, *67*, 1–8. [CrossRef] [PubMed]
2. Keeling, M.J.; Rohani, P. *Modeling Infectious Diseases in Humans and Animals*; Princeton University Press: Princeton, NJ, USA, 2008.
3. Salathé, M.; Jones, J.H. Dynamics and control of diseases in networks with community structure. *PLoS Comput. Biol.* **2010**, *6*, e1000736. [CrossRef] [PubMed]
4. Hellewell, J.; Abbott, S.; Gimma, A.; Bosse, N.I.; Jarvis, C.I.; Russell, T.W.; Munday, J.D.; Kucharski, A.J.; Edmunds, W.J.; Funk, S.; et al. Feasibility of controlling COVID-19 outbreaks by isolation of cases and contacts. *Lancet Glob. Health* **2020**, *8*, e488–e496. [CrossRef] [PubMed]
5. Kucharski, A.J.; Russell, T.W.; Diamond, C.; Liu, Y.; Edmunds, J.; Funk, S.; Eggo, R.M.; Sun, F.; Jit, M.; Munday, J.D.; et al. Early dynamics of transmission and control of COVID-19: A mathematical modelling study. *Lancet Infect. Dis.* **2020**, *20*, 553–558. [CrossRef]
6. Koo, J.; Cook, A.; Park, M. Interventions to mitigate early spread of COVID-19 in Singapore: A modelling study. *Lancet Infect Dis.* **2020**, *20*, 678–688. [CrossRef]
7. Galea, S.; Riddle, M.; Kaplan, G.A. Causal thinking and complex system approaches in epidemiology. *Int. J. Epidemiol.* **2010**, *39*, 97–106. [CrossRef]
8. Siegenfeld, A.F.; Bar-Yam, Y. An introduction to complex systems science and its applications. *Complexity* **2020**, *2020*, 6105872. [CrossRef]
9. Adam, D. Special report: The simulations driving the world's response to COVID-19. *Nature* **2020**, *580*, 316–319. [CrossRef]
10. Hinch, R.; Probert, W.J.; Nurtay, A.; Kendall, M.; Wymant, C.; Hall, M.; Lythgoe, K.; Bulas Cruz, A.; Zhao, L.; Stewart, A.; et al. OpenABM-Covid19—An agent-based model for non-pharmaceutical interventions against COVID-19 including contact tracing. *PLoS Comput. Biol.* **2021**, *17*, e1009146. [CrossRef]
11. Topirceanu, A.; Udrescu, M.; Marculescu, R. Centralized and decentralized isolation strategies and their impact on the COVID-19 pandemic dynamics. *arXiv* **2020**, arXiv:2004.04222.
12. Diaz, P.; Constantine, P.; Kalmbach, K.; Jones, E.; Pankavich, S. A modified SEIR model for the spread of Ebola in Western Africa and metrics for resource allocation. *Appl. Math. Comput.* **2018**, *324*, 141–155. [CrossRef]
13. Arenas, A.; Cota, W.; Gomez-Gardenes, J.; Gómez, S.; Granell, C.; Matamalas, J.T.; Soriano-Panos, D.; Steinegger, B. A mathematical model for the spatiotemporal epidemic spreading of COVID19. *MedRxiv* **2020**. [CrossRef]
14. Ferguson, N.M.; Cummings, D.A.; Fraser, C.; Cajka, J.C.; Cooley, P.C.; Burke, D.S. Strategies for mitigating an influenza pandemic. *Nature* **2006**, *442*, 448–452. [CrossRef] [PubMed]

15. Mistry, D.; Litvinova, M.; y Piontti, A.P.; Chinazzi, M.; Fumanelli, L.; Gomes, M.F.; Haque, S.A.; Liu, Q.H.; Mu, K.; Xiong, X.; et al. Inferring high-resolution human mixing patterns for disease modeling. *Nat. Commun.* **2021**, *12*, 1–12. [CrossRef] [PubMed]
16. Hoertel, N.; Blachier, M.; Blanco, C.; Olfson, M.; Massetti, M.; Rico, M.S.; Limosin, F.; Leleu, H. A stochastic agent-based model of the SARS-CoV-2 epidemic in France. *Nat. Med.* **2020**, *26*, 1417–1421. [CrossRef] [PubMed]
17. Datta, A.; Winkelstein, P.; Sen, S. An agent-based model of spread of a pandemic with validation using COVID-19 data from New York State. *Phys. A Stat. Mech. Appl.* **2022**, *585*, 126401. [CrossRef] [PubMed]
18. Frias-Martinez, E.; Williamson, G.; Frias-Martinez, V. An agent-based model of epidemic spread using human mobility and social network information. In Proceedings of the 2011 IEEE Third International Conference on Privacy, Security, Risk and Trust and 2011 IEEE Third International Conference on Social Computing, Boston, MA, USA, 9–11 October 2011, pp. 57–64.
19. Alzu'bi, A.A.; Alasal, S.I.A.; Watzlaf, V.J. A simulation study of coronavirus as an epidemic disease using agent-based modeling. *Perspect. Health Inf. Manag.* **2021**, *18*, 1g. [PubMed]
20. Chang, S.; Pierson, E.; Koh, P.W.; Gerardin, J.; Redbird, B.; Grusky, D.; Leskovec, J. Mobility network models of COVID-19 explain inequities and inform reopening. *Nature* **2021**, *589*, 82–87. [CrossRef] [PubMed]
21. Nian, G.; Peng, B.; Sun, D.; Ma, W.; Peng, B.; Huang, T. Impact of COVID-19 on urban mobility during post-epidemic period in megacities: From the perspectives of taxi travel and social vitality. *Sustainability* **2020**, *12*, 7954. [CrossRef]
22. Li, Q.; Bessell, L.; Xiao, X.; Fan, C.; Gao, X.; Mostafavi, A. Disparate patterns of movements and visits to points of interest located in urban hotspots across US metropolitan cities during COVID-19. *R. Soc. Open Sci.* **2021**, *8*, 201209. [CrossRef] [PubMed]
23. Newman, M.E. Spread of epidemic disease on networks. *Phys. Rev. E* **2002**, *66*, 016128. [CrossRef] [PubMed]
24. Pastor-Satorras, R.; Castellano, C.; Van Mieghem, P.; Vespignani, A. Epidemic processes in complex networks. *Rev. Mod. Phys.* **2015**, *87*, 925. [CrossRef]
25. Tako, A.A.; Robinson, S. Comparing discrete-event simulation and system dynamics: Users' perceptions. In *System Dynamics*; Springer: Berlin, Germany, 2018; pp. 261–299.
26. Ward, H.; Cooke, G.; Atchison, C.J.; Whitaker, M.; Elliott, J.; Moshe, M.; Brown, J.C.; Flower, B.; Daunt, A.; Ainslie, K.E.; et al. Declining prevalence of antibody positivity to SARS-CoV-2: A community study of 365,000 adults. *MedRxiv* **2020**. [CrossRef]
27. Gudbjartsson, D.F.; Norddahl, G.L.; Melsted, P.; Gunnarsdottir, K.; Holm, H.; Eythorsson, E.; Arnthorsson, A.O.; Helgason, D.; Bjarnadottir, K.; Ingvarsson, R.F.; et al. Humoral immune response to SARS-CoV-2 in Iceland. *N. Engl. J. Med.* **2020**, *383*, 1724–1734. [CrossRef] [PubMed]
28. Zuo, J.; Dowell, A.; Pearce, H.; Verma, K.; Long, H.; Begum, J.; Aiano, F.; Amin-Chowdhury, Z.; Hallis, B.; Stapley, L.; et al. Robust SARS-CoV-2-specific T-cell immunity is maintained at 6 months following primary infection. *Nature Immunol.* **2021**, *22*, 620–626. [CrossRef]
29. Zayet, S.; Royer, P.Y.; Toko, L.; Pierron, A.; Gendrin, V.; Klopfenstein, T. Recurrence of COVID-19 after recovery? A case series in health care workers, France. *Microbes Infect.* **2021**, *23*, 104803. [CrossRef]
30. Kasereka, S.K.; Zohinga, G.N.; Kiketa, V.M.; Ngoie, R.B.M.; Mputu, E.K.; Kasoro, N.M.; Kyandoghere, K. Equation-Based Modeling vs. Agent-Based Modeling with Applications to the Spread of COVID-19 Outbreak. *Mathematics* **2023**, *11*, 253. [CrossRef]
31. De Oliveira, L.B.; Camponogara, E. Multi-agent model predictive control of signaling split in urban traffic networks. *Transp. Res. Part C Emerg. Technol.* **2010**, *18*, 120–139. [CrossRef]
32. Hackl, J.; Dubernet, T. Epidemic spreading in urban areas using agent-based transportation models. *Future Internet* **2019**, *11*, 92. [CrossRef]
33. Nadini, M.; Zino, L.; Rizzo, A.; Porfiri, M. A multi-agent model to study epidemic spreading and vaccination strategies in an urban-like environment. *Appl. Netw. Sci.* **2020**, *5*, 1–30. [CrossRef]
34. Girardi, P.; Gaetan, C. An SEIR Model with Time-Varying Coefficients for Analyzing the SARS-CoV-2 Epidemic. *Risk Anal.* **2021**, *43*, 144–155. [CrossRef]
35. Yin, K.; Mondal, A.; Ndeffo-Mbah, M.; Banerjee, P.; Huang, Q.; Gurarie, D. Bayesian Inference for COVID-19 Transmission Dynamics in India Using a Modified SEIR Model. *Mathematics* **2022**, *10*, 4037. [CrossRef]
36. Zhuge, C.; Shao, C.; Wei, B. An agent-based spatial urban social network generator: A case study of Beijing, China. *J. Comput. Sci.* **2018**, *29*, 46–58. [CrossRef]
37. Jones, N.R.; Qureshi, Z.U.; Temple, R.J.; Larwood, J.P.; Greenhalgh, T.; Bourouiba, L. Two metres or one: What is the evidence for physical distancing in covid-19? *BMJ* **2020**, *370*, m3223. [CrossRef] [PubMed]
38. Lauer, S.A.; Grantz, K.H.; Bi, Q.; Jones, F.K.; Zheng, Q.; Meredith, H.R.; Azman, A.S.; Reich, N.G.; Lessler, J. The incubation period of coronavirus disease 2019 (COVID-19) from publicly reported confirmed cases: Estimation and application. *Ann. Internal Med.* **2020**, *172*, 577–582. [CrossRef] [PubMed]
39. Li, Q.; Guan, X.; Wu, P.; Wang, X.; Zhou, L.; Tong, Y.; Ren, R.; Leung, K.S.; Lau, E.H.; Wong, J.Y.; et al. Early transmission dynamics in Wuhan, China, of novel coronavirus–infected pneumonia. *N. Engl. J. Med.* **2020**, *382*, 1199–1207. [CrossRef]
40. Eurosurveillance Editorial Team. Updated rapid risk assessment from ECDC on the novel coronavirus disease 2019 (COVID-19) pandemic: Increased transmission in the EU/EEA and the UK. *Euro Surveill. Bull. Eur. Sur Les Mal. Transm. Eur. Commun. Dis. Bull.* **2020**, *25*, 680.
41. Mission, W.C.J. *Report of the WHO-China Joint Mission on Coronavirus Disease 2019 (COVID-19)*; WHO: Geneva, Switzerland, 2020.

42. Linton, N.M.; Kobayashi, T.; Yang, Y.; Hayashi, K.; Akhmetzhanov, A.R.; Jung, S.M.; Yuan, B.; Kinoshita, R.; Nishiura, H. Incubation period and other epidemiological characteristics of 2019 novel coronavirus infections with right truncation: A statistical analysis of publicly available case data. *J. Clin. Med.* **2020**, *9*, 538. [CrossRef]
43. Wang, C.; Horby, P.W.; Hayden, F.G.; Gao, G.F. A novel coronavirus outbreak of global health concern. *Lancet* **2020**, *395*, 470–473. [CrossRef]
44. Barabási, A.L.; Pósfai, M. *Network Science*; Cambridge University Press: Cambridge, MA, USA, 2016.
45. Clauset, A.; Shalizi, C.R.; Newman, M.E. Power-law distributions in empirical data. *SIAM Rev.* **2009**, *51*, 661–703. [CrossRef]
46. Dunham, J.B. An agent-based spatially explicit epidemiological model in MASON. *J. Artif. Soc. Soc. Simul.* **2005**, *9*, 690.
47. Topîrceanu, A.; Udrescu, M. Statistical fidelity: A tool to quantify the similarity between multi-variable entities with application in complex networks. *Int. J. Comput. Math.* **2017**, *94*, 1787–1805. [CrossRef]
48. Kantner, M.; Koprucki, T. Beyond just "flattening the curve": Optimal control of epidemics with purely non-pharmaceutical interventions. *J. Math. Ind.* **2020**, *10*, 1–23. [CrossRef]
49. Topîrceanu, A.; Udrescu, M.; Udrescu, L.; Ardelean, C.; Dan, R.; Reisz, D.; Mihaicuta, S. SAS score: Targeting high-specificity for efficient population-wide monitoring of obstructive sleep apnea. *PLoS ONE* **2018**, *13*, e0202042. [CrossRef] [PubMed]
50. Udrescu, L.; Bogdan, P.; Chiş, A.; Sîrbu, I.O.; Topîrceanu, A.; Văruţ, R.M.; Udrescu, M. Uncovering New Drug Properties in Target-Based Drug–Drug Similarity Networks. *Pharmaceutics* **2020**, *12*, 879. [CrossRef]
51. Saadat, S.; Rawtani, D.; Hussain, C.M. Environmental perspective of COVID-19. *Sci. Total Environ.* **2020**, *728*, 138870. [CrossRef] [PubMed]
52. Rypdal, K.; Bianchi, F.M.; Rypdal, M. Intervention fatigue is the primary cause of strong secondary waves in the COVID-19 pandemic. *Int. J. Environ. Res. Public Health* **2020**, *17*, 9592. [CrossRef]
53. Mei, S.; Chen, B.; Zhu, Y.; Lees, M.H.; Boukhanovsky, A.; Sloot, P.M. Simulating city-level airborne infectious diseases. *Comput. Environ. Urban Syst.* **2015**, *51*, 97–105. [CrossRef]
54. Mao, L.; Bian, L. Spatial–temporal transmission of influenza and its health risks in an urbanized area. *Comput. Environ. Urban Syst.* **2010**, *34*, 204–215. [CrossRef]
55. Luo, W.; Gao, P.; Cassels, S. A large-scale location-based social network to understanding the impact of human geo-social interaction patterns on vaccination strategies in an urbanized area. *Comput. Environ. Urban Syst.* **2018**, *72*, 78–87. [CrossRef]
56. Topîrceanu, A.; Precup, R.E. A novel geo-hierarchical population mobility model for spatial spreading of resurgent epidemics. *Sci. Rep.* **2021**, *11*, 14341. [CrossRef]
57. Topîrceanu, A. Benchmarking Cost-Effective Opinion Injection Strategies in Complex Networks. *Mathematics* **2022**, *10*, 2067. [CrossRef]
58. Li, W.; Xue, X.; Pan, L.; Lin, T.; Wang, W. Competing spreading dynamics in simplicial complex. *Appl. Math. Comput.* **2022**, *412*, 126595. [CrossRef]
59. Topîrceanu, A. Competition-Based Benchmarking of Influence Ranking Methods in Social Networks. *Complexity* **2018**, *2018*, 4562609. [CrossRef]
60. Pastor-Satorras, R.; Vespignani, A. Immunization of complex networks. *Phys. Rev. E* **2002**, *65*, 036104. [CrossRef] [PubMed]
61. Topîrceanu, A. Immunization using a heterogeneous geo-spatial population model: A qualitative perspective on COVID-19 vaccination strategies. *Procedia Comput. Sci.* **2021**, *192*, 2095–2104. [CrossRef]
62. Matrajt, L.; Leung, T. Evaluating the effectiveness of social distancing interventions to delay or flatten the epidemic curve of coronavirus disease. *Emerg. Infect. Diseases* **2020**, *26*, 1740. [CrossRef]
63. European Centre for Disease Prevention and Control (An agency of the European Union). Data on the Daily Number of New Reported COVID-19 Cases and Deaths by EU/EEA country. Available online: https://www.ecdc.europa.eu/en/covid-19/data (accessed on 27 October 2022).

Disclaimer/Publisher's Note: The statements, opinions and data contained in all publications are solely those of the individual author(s) and contributor(s) and not of MDPI and/or the editor(s). MDPI and/or the editor(s) disclaim responsibility for any injury to people or property resulting from any ideas, methods, instructions or products referred to in the content.

Article

Advancing COVID-19 Understanding: Simulating Omicron Variant Spread Using Fractional-Order Models and Haar Wavelet Collocation

Zehba Raizah [1] and Rahat Zarin [2,*]

[1] Department of Mathematics, Faculty of Science, King Khalid University, Abha 62529, Saudi Arabia
[2] Department of Mathematics, Faculty of Science, King Mongkut's University of Technology Thonburi (KMUTT), 126 Pracha-Uthit Road, Bang Mod Thrung Khru, Bangkok 10140, Thailand
* Correspondence: rahat.zarin@uetpeshawar.edu.pk

Abstract: This study presents a novel approach for simulating the spread of the Omicron variant of the SARS-CoV-2 virus using fractional-order COVID-19 models and the Haar wavelet collocation method. The proposed model considers various factors that affect virus transmission, while the Haar wavelet collocation method provides an efficient and accurate solution for the fractional derivatives used in the model. This study analyzes the impact of the Omicron variant and provides valuable insights into its transmission dynamics, which can inform public health policies and strategies that are aimed at controlling its spread. Additionally, this study's findings represent a significant step forward in understanding the COVID-19 pandemic and its evolving variants. The results of the simulation showcase the effectiveness of the proposed method and demonstrate its potential to advance the field of COVID-19 research. The COVID epidemic model is reformulated by using fractional derivatives in the Caputo sense. The existence and uniqueness of the proposed model are illustrated in the model, taking into account some results of fixed point theory. The stability analysis for the system is established by incorporating the Hyers–Ulam method. For numerical treatment and simulations, we apply the Haar wavelet collocation method. The parameter estimation for the recorded COVID-19 cases in Pakistan from 23 June 2022 to 23 August 2022 is presented.

Keywords: parameter estimation; Haar wavelet; reproduction number; fractional modeling; COVID-19; numerical analysis

MSC: 34A08; 65P99; 49J15

1. Introduction

The global pandemic caused by SARS-CoV-2, a spike protein virus, has resulted in widespread infections of coronavirus. The severe acute respiratory syndrome (SARS) epidemic in China in 2003 and the Middle East respiratory syndrome (MERS) epidemic in Saudi Arabia in 2012 [1] were both coronavirus epidemics. Coronaviruses are a large family of viruses. The World Health Organization (WHO) first noted a SARS coronavirus 2 (COVID-19) outbreak in Wuhan, China, in December 2019. In March 2020, the WHO declared the outbreak to be a global pandemic. Over 6.6 million COVID-19 deaths [2] have been reported as of 8 January 2023, with over 659 million confirmed cases worldwide. Despite the fact that the disease is primarily transmitted through respiratory droplets produced by breathing, coughing, sneezing, and talking [3–5], more research indicates that it may also be transmitted through the air [6–9]. Additionally, COVID-19 can be acquired by coming into contact with contaminated objects. The virus can cause symptoms such as coughing, muscle pain, vertigo, high temperature, loss of smell, throat irritation, weakness, and nasal congestion and has an incubation period of 2–14 days.

The COVID-19 Omicron variant, which was first identified in South Africa in November 2021, has since raised serious concerns among public health authorities throughout the world, including in Pakistan. The COVID-19 virus, SARS-CoV-2, is thought to be more contagious and potentially more resistant to current vaccinations than earlier SARS-CoV-2 variants. The possibility of the virus spreading further in communities as a result has raised concerns, especially in regions with high population densities and low vaccination rates. The Omicron variant has been found in Pakistan's major cities such as Karachi, Lahore, and Islamabad as well as other regions of the nation. In response to the threat posed by the variant, the government has increased testing and contact tracing efforts, increased vaccination efforts, and implemented stricter public health measures such as social seclusion and mask use. Despite these initiatives, the situation is still worrying because COVID-19 cases are still on the rise across much of the nation. This emphasizes the requirement for ongoing watchfulness and a thorough, multifaceted strategy to stop the virus's spread. Along with public health initiatives, this also entails work intended to improve vaccination rates and access, as well as ongoing studies into the biology and epidemiology of the Omicron variant to better understand its potential effects on society. In the end, it will take a concerted effort from all facets of society, including the government, healthcare professionals, and the general public, to stop the spread of COVID-19 and its variants in Pakistan. By working together, it may be possible to reduce the impact of the Omicron variant and protect the health and well-being of communities in Pakistan.

Modeling the spread of the COVID-19 pandemic has been essential in guiding public health policies and interventions as it has been a significant global health crisis. Multiscale modeling techniques have become an effective method for understanding the intricate dynamics of virus pandemics and have been heavily utilized in the case of COVID-19 [10,11]. Epidemiological models have been extensively used to study the dynamics of COVID-19 transmission at the macroscopic level. These models span a spectrum ranging from straightforward compartmental models, such as the susceptible-infected-recovered (SIR) model, to more intricate models that take into account spatial and temporal heterogeneity, demographic factors, and interventions, such as vaccination and social isolation [12,13]. Additionally, the behavior of individual agents and their interactions during the COVID-19 outbreak have been simulated using agent-based models. Network models have been used to study the dissemination of COVID-19 within social networks and communities at the mesoscopic level. These models take into account the variation in contact patterns as well as the effects of social isolation policies on the dynamics of transmission. Large-scale datasets have also been analyzed to find patterns and trends in the spread of COVID-19 [14,15] using data-driven approaches, such as machine learning and artificial intelligence techniques. Models have been created to study the molecular interactions between the virus and host cells at the microscopic level. These models include the structural modeling of virus proteins and their interactions with host cell receptors (see, for example, [16,17]), as well as molecular dynamics simulations. These methods have been applied to comprehend the molecular mechanisms of virus replication and infection and to find potential therapeutic targets. Along with these conventional modeling techniques, there is growing interest in the use of multiscale models, which combine various modeling scales to capture the intricate relationships between virus spread, individual behavior, and public health policies. These models seek to offer a more thorough understanding of the COVID-19 dynamics and assist in the creation of efficient interventions and policies. Multiscale modeling has generally been shown to be an effective method for analyzing the COVID-19 epidemic and informing public health policies and interventions. Multiscale modeling techniques offer a promising way to deepen our understanding of this intricate global health crisis, even though there is still much to learn about the dynamics of COVID-19.

Fractional order differential equations (FODs) have been used to gain a deeper understanding of diseases, and mathematical models have been formulated and studied for various diseases [18–23]. The process of differentiation and integration in fractional calculus is generalized to non-integer orders, making it a valuable tool for research in various fields [24]. Caputo–Fabrizio (CF) operators [25], which use non-singular kernels, are one type of fractional derivative that has been developed to overcome the limitations of the ordinary operator. However, the CF operator has a locality problem, leading to the proposal of Mittag-Leffler kernels as a novel type of fractional derivative by Atangana and Baleanu [26]. Overall, the use of fractional calculus and mathematical models can provide insights into the dynamics of infectious diseases and inform public health policy. Omame et al. [27] explore the potential impact of COVID-19 on the dynamics of dengue and HIV transmission in their paper titled "Assessing the impact of SARS-CoV-2 infection on the dynamics of dengue and HIV via fractional derivatives". Using fractional calculus, the study creates mathematical models and suggests that the pandemic could significantly affect the transmission and control of these diseases. The authors emphasize the importance of considering multiple infectious diseases during pandemics. In their paper "Backward bifurcation and optimal control in a co-infection model for SARS-CoV-2 and ZIKV" [28], the authors present a mathematical model for the co-infection of SARS-CoV-2 and ZIKV. The study examines the role of backward bifurcation in the dynamics of the co-infection and evaluates the potential benefits of optimal control strategies. The study highlights the importance of considering multiple infectious diseases and optimal control strategies in public health policy.

Furthermore, fractional operators with non-singular and singular kernels have been proposed in several works [29–31], and research related to these topics and their applications can be found in a number of recent publications [32–36]. As of late, a number of studies have appeared in mathematical modeling that investigate addressing social issues, such as criminal issues, using FCs. A time lag exists between the individual's offense and the judgment, which is why Bansal et al. [37] introduced the time-delay coefficient to extend the proposed fractional-order crime transmission model to the delayed model. With regard to analyzing crime congestion, in their study, Pritam et al. [38] examined a mathematical model of crime transmission using a fractional-order derivative that includes memory effects, allowing for the previous input's impact to be taken into account when forecasting the growth rate of crime. Using the iterative fractional-order Adams–Bashforth approach, ref. [39] found the approximate solution and numerically simulated it for various control strategies in different fractional orders.

In recent years, there has been a growing interest in the use of Haar wavelet numerical methods for solving problems related to the COVID-19 pandemic. These methods have been used to model the spread of the virus over time, taking into account the various factors that influence its transmission. The results of these studies have provided valuable insights into the dynamics of the pandemic and its evolution and have informed public health policies and strategies aimed at controlling its impact. Overall, Haar wavelet numerical methods have proven to be a valuable tool for solving problems with fractional derivatives and have been widely adopted in various fields due to their efficiency and accuracy. With ongoing research and development, they are expected to play an increasingly important role in solving complex problems in the future.

Similarly, wavelet analyses have been extensively applied in numerical analyses, statistical applications, image digital processing, quantum field theory, and many other fields. A wide range of applications has been made for Haar wavelets, including in communication and physics research as well as more mathematically-based research on differential equations and nonlinear problems [40]. There is an emphasis on Haar wavelets among all wavelet families. In mathematics, they are the simplest wavelet family because they consist of pairs of piecewise constant functions. In addition, Haar wavelets can also be integrated analytically in random times. Recently, researchers applied this technique for solving different fractional-order mathematical models [41,42]. In addition to being a fast method, the method is also more stable.

2. Preliminaries

The definition of the fractional derivative has developed greatly in recent years [43,44], ranging from the non-singular kernel derivatives and Riemann–Liouville (RL) fractional derivative without a singular kernel to the two-parameter derivative with non-singular and non-local kernels [45]. The two most commonly used definitions are described below:

Definition 1. *The fractional derivative of y of order δ is defined by Riemann–Liouville as follows [42]:*

$$\mathbb{D}_*^\delta \mathcal{F}(t) = \begin{cases} \frac{1}{\Gamma(s-\delta)} \left(\frac{d}{dt}\right)^s \int_0^t \frac{\mathcal{F}(v)}{(t-v)^{\delta-s+1}} dv, & 0 \leq s-1 < \delta < s, \quad s \in \mathbb{N}, \\ (d/dt)^s \mathcal{F}(t), & \delta = s, \quad s \in \mathbb{N}. \end{cases} \quad (1)$$

Definition 2. *The Caputo fractional derivative of the function $\mathcal{F}$ of order δ is defined as follows [42]:*

$$\mathbb{D}_*^\delta \mathcal{F}(t) = \begin{cases} \frac{1}{\Gamma(s-\delta)} \int_0^t \frac{(d/dv)^s \mathcal{F}(v)}{(t-v)^{\delta-s+1}} dv, & 0 \leq s-1 < \delta < s, \quad s \in \mathbb{N}, \\ (d/dt)^s \mathcal{F}(t), & \delta = s, \quad s \in \mathbb{N}. \end{cases} \quad (2)$$

Also used in this study is the RL form of the fractional integral operator $\mathbb{D}_*^{-\delta}$ of order δ, which is defined as follows:

$$\mathbb{D}_*^{-\delta} \mathcal{F}(t) = \frac{1}{\Gamma(\delta)} \int_0^t \mathcal{F}(v)(t-v)^{\delta-1} dv \quad (3)$$

Haar Wavelets

If $\psi(t)$ and $\tilde{\psi}_0(t)$ represent the mother Haar wavelet function (on the real line) and Haar scaling function, respectively, then they are given by [46,47]:

$$\psi(t) = \begin{cases} 1, & \text{if } t \in \left[0, \frac{1}{2}\right), \\ -1, & \text{if } t \in \left[\frac{1}{2}, 1\right), \\ 0, & \text{elsewhere,} \end{cases} \quad (4)$$

$$\tilde{\psi}_0(t) = 1, \text{ if } t \in [0,1). \quad (5)$$

As a result, if the various Haar wavelets that are produced on the interval $[0,1)$ using multiresolution analysis are $\tilde{\psi}_m(t)$, then:

$$\tilde{\psi}_m(t) = 2^{j/2} \psi\left(2^j t - p\right), m = 1, 2, \ldots; \quad (6)$$

where $m = 2^j + p : p = 0, 1, \ldots, 2^j - 1; j = 0, 1, \ldots$. Furthermore, we can translate the Haar functions on $u - 1 \leq t < u$ as

$$\tilde{\psi}_{u,m}(t) = \tilde{\psi}_m(t+1-u), m = 0,1,2,\ldots, \quad u = 1,2,\ldots,\varrho, \quad \varrho \in \mathbb{N}. \quad (7)$$

The resulting sequence $\{\tilde{\psi}_m(t)\}_{m=0}^\infty$ forms a complete orthonormal system [47] in $\mathcal{L}^2[0,1)$. Similarly, the sequence $\{\tilde{\psi}_{u,m}(t)\}_{m=0}^\infty$, $u = 1, 2, \ldots, \varrho$, forms a complete orthonormal system in $\mathcal{L}^2[0,\varrho)$. Therefore, any function $\mathcal{F}(t) \in \mathcal{L}^2[0,\varrho)$ can be expanded in terms of Haar orthonormal basis functions as

$$\mathcal{F}(t) = \sum_{u=1}^\varrho \sum_{m=0}^\infty \mathcal{G}_{u,m} \tilde{\psi}_{u,m}(t). \quad (8)$$

Additionally, after truncating the series $\mathcal{F}(t)$, we obtain the equivalent approximation $y_p(t)$ of $\mathcal{F}(t)$ as

$$\mathcal{F}(t) \approx y_p(t) = \sum_{u=1}^{\varrho} \sum_{m=0}^{p-1} \mathcal{G}_{u,m} \tilde{\psi}_{u,m}(t) = B_{\varrho p \times 1}^T \tilde{\psi}_{\varrho p \times 1}(t), \tag{9}$$

where the coefficients $\mathcal{G}_{u,m}$ can be expressed by the inner product

$$\begin{aligned} \langle \mathcal{F}(t), \tilde{\psi}_{u,m}(t) \rangle &= \int_{u-1}^{u} \mathcal{F}(t) \tilde{\psi}_{u,m}(t) dt, \quad m = 1, 2, \ldots, (p-1), u = 1, 2, \ldots, \varrho, \\ B_{\varrho p \times 1} &= [\mathcal{G}_{1,0}, \ldots, \mathcal{G}_{1,p-1}, \mathcal{G}_{2,0}, \ldots, \mathcal{G}_{2,p-1}, \ldots, \mathcal{G}_{\varrho,0}, \ldots, \mathcal{G}_{\varrho,p-1}]^T, \\ \tilde{\psi}_{\varrho p \times 1} &= [\tilde{\psi}_{1,0}, \ldots, \tilde{\psi}_{1,p-1}, \tilde{\psi}_{2,0}, \ldots, \tilde{\psi}_{2,p-1}, \ldots, \tilde{\psi}_{\varrho,0}, \ldots, \tilde{\psi}_{\varrho,p-1}]^T, \end{aligned} \tag{10}$$

and superscript T indicates the transpose of a matrix.

3. Mathematical Model

Understanding and forecasting the spread of infectious illnesses are greatly aided by mathematical modeling. Researchers can model how illnesses behave and spread within communities using mathematical equations and algorithms. As a result, they can forecast how the disease will develop in the future and can try different intervention techniques. In addition to helping to identify risk factors and guide public health policy, mathematical models can offer important insights into the fundamental mechanisms of disease transmission. It is crucial to keep in mind that the correctness of these models depends on the caliber of the data input and model assumptions, and they should always be utilized in conjunction with other information sources. Regarding the work of [6,7], the model used consists of the following ODEs:

$$\begin{cases} \dfrac{d\mathbf{S}(t)}{dt} = B - \theta \mathbf{S}(t) \mathbf{I}(t)(1 + \tau \mathbf{I}(t)) - (\varepsilon_1 + \rho + \eta) \mathbf{S}(t) + \mathcal{K} \mathbf{V}(t), \\[4pt] \dfrac{d\mathbf{V}(t)}{dt} = \pi \mathbf{S}(t) - \psi \mathbf{V}(t) - (\mathcal{K} + \rho) \mathbf{V}(t), \\[4pt] \dfrac{d\mathbf{E}(t)}{dt} = \theta \mathbf{S}(t) \mathbf{I}(t)(1 + \tau \mathbf{I}(t)) + \psi \mathbf{V}(t) - (\varepsilon_2 + \rho + \varphi) \mathbf{E}(t), \\[4pt] \dfrac{d\mathbf{I}(t)}{dt} = \varphi \mathbf{E}(t) - (\lambda + \epsilon + \rho + \varepsilon_3) \mathbf{I}(t), \\[4pt] \dfrac{d\mathbf{Q}(t)}{dt} = \varepsilon_1 \mathbf{S}(t) + \varepsilon_2 \mathbf{E}(t) + \varepsilon_3 \mathbf{I}(t) - (\rho + \sigma) \mathbf{Q}(t), \\[4pt] \dfrac{d\mathbf{R}(t)}{dt} = \eta \mathbf{S}(t) + \sigma \mathbf{Q}(t) + \lambda \mathbf{I}(t) - \rho \mathbf{R}(t), \\[4pt] \mathbf{S}(t) \geq 0, \quad \mathbf{V}(t) \geq 0, \quad \mathbf{E}(t) \geq 0, \quad \mathbf{I}(t) \geq 0, \quad \mathbf{Q}(t) \geq 0, \quad \mathbf{R}(t) \geq 0. \end{cases} \tag{11}$$

A representation of COVID-19 transmission is provided by the system of ordinary differential equations in mathematical model refe1; the model is divided into six compartments, each representing a different group of people in the population that is affected by the disease. These compartments are as follows: susceptible people $\mathbf{S}(t)$, vaccinated people $\mathbf{V}(t)$, exposed people $\mathbf{E}(t)$, infectious people $\mathbf{I}(t)$, quarantined people $\mathbf{Q}(t)$, and recovered people $\mathbf{R}(t)$. Susceptible people have not been infected but can contract the virus, whereas vaccinated people have received a vaccine and are immune to the virus. Exposed people have been infected but have not yet developed symptoms, whereas infectious people are infected and can spread the virus. Individuals quarantined have been identified as infected and are being isolated to prevent further transmission. Finally, recovered individuals have recovered from the disease and are now immune.

Numerous parameters that affect the dynamics of the disease are included in the model for predicting the spread of COVID-19. The following parameters make up this list: birth rate (B), transmission rate (θ), infectivity coefficient (τ), quarantine rates for exposed and susceptible individuals (ε_2) and (ε_1) and for natural deaths (ρ), vaccination efficacy rate (η), vaccination rate ($\mathcal{K}$), vaccination loss rate (ψ), rate of susceptibility to vaccination (π), progression rate (φ), recovery rate (λ), disease-induced death rate (ϵ), quarantine rate for infectious individuals (ε_3), and quarantine loss rate (σ). The model's differential equations predict how these parameters and variables will change over time to affect how many people are in each compartment. It is possible to forecast the spread of COVID-19 and assess the effects of different interventions, such as vaccination and quarantine, by simulating the model.

3.1. Formulation of Fractional Model

The ability to model with fractional-order derivatives, which can more precisely capture the dynamics of some diseases that exhibit non-integer order behavior, is one benefit of using the Caputo derivative over the classical derivative in the context of disease modeling. Because the fractional-order derivatives can capture both memory effects and the power-law decay characteristic of many disease models, this can result in more accurate predictions and better control strategies. The Caputo derivative is also a more economical option for disease modeling because it frequently requires fewer computational resources and data. Additionally, the majority of natural phenomena, including epidemiological dynamics, exhibit the time memory effect. Model (11) is expressed in integral form as:

$$\begin{cases} \dfrac{d\mathbf{S}(t)}{dt} = \int_{t_0}^{t} \varsigma(t-\omega)[B - \theta \mathbf{S}(t)\mathbf{I}(t)(1+\tau \mathbf{I}(t)) - (\varepsilon_1+\rho+\eta)\mathbf{S}(t) + \mathcal{K}\mathbf{V}(t)]d\omega, \\ \dfrac{d\mathbf{V}(t)}{dt} = \int_{t_0}^{t} \varsigma(t-\omega)[\pi \mathbf{S}(t) - \psi \mathbf{V}(t) - (\mathcal{K}+\rho)\mathbf{V}(t)]d\omega, \\ \dfrac{d\mathbf{E}(t)}{dt} = \int_{t_0}^{t} \varsigma(t-\omega)[\theta \mathbf{S}(t)\mathbf{I}(t)(1+\tau \mathbf{I}(t)) + \psi \mathbf{V}(t) - (\varepsilon_2+\rho+\varphi)\mathbf{E}(t)]d\omega, \\ \dfrac{d\mathbf{I}(t)}{dt} = \int_{t_0}^{t} \varsigma(t-\omega)[\varphi \mathbf{E}(t) - (\lambda+\epsilon+\rho+\varepsilon_3)\mathbf{I}(t)]d\omega, \\ \dfrac{d\mathbf{Q}(t)}{dt} = \int_{t_0}^{t} \varsigma(t-\omega)[\varepsilon_1 \mathbf{S}(t) + \varepsilon_2 \mathbf{E}(t) + \varepsilon_3 \mathbf{I}(t) - (\rho+\sigma)\mathbf{Q}(t)]d\omega, \\ \dfrac{d\mathbf{R}(t)}{dt} = \int_{t_0}^{t} \varsigma(t-\varsigma)[\eta \mathbf{S}(t) + \sigma \mathbf{Q}(t) + \lambda \mathbf{I}(t) - \rho \mathbf{R}(t)]d\omega. \end{cases} \quad (12)$$

Incorporating the Caputo derivative, we get

$$\begin{cases} {}^C D_t^{\delta-1}\left[\dfrac{d\mathbf{S}(t)}{dt}\right] = {}^C D_t^{\delta-1} I^{-(\delta-1)}[B - \theta \mathbf{S}(t)\mathbf{I}(t)(1+\tau \mathbf{I}(t)) - (\varepsilon_1+\rho+\eta)\mathbf{S}(t) + \mathcal{K}\mathbf{V}(t)], \\ {}^C D_t^{\delta-1}\left[\dfrac{d\mathbf{V}(t)}{dt}\right] = {}^C D_t^{\delta-1} I^{-(\delta-1)}[\pi \mathbf{S}(t) - \psi \mathbf{V}(t) - (\mathcal{K}+\rho)\mathbf{V}(t)], \\ {}^C D_t^{\delta-1}\left[\dfrac{d\mathbf{E}(t)}{dt}\right] = {}^C D_t^{\delta-1} I^{-(\delta-1)}[\theta \mathbf{S}(t)\mathbf{I}(t)(1+\tau \mathbf{I}(t)) + \psi \mathbf{V}(t) - (\varepsilon_2+\rho+\varphi)\mathbf{E}(t)], \\ {}^C D_t^{\delta-1}\left[\dfrac{d\mathbf{I}(t)}{dt}\right] = {}^C D_t^{\delta-1} I^{-(\delta-1)}[\varphi \mathbf{E}(t) - (\lambda+\epsilon+\rho+\varepsilon_3)\mathbf{I}(t)], \\ {}^C D_t^{\delta-1}\left[\dfrac{d\mathbf{Q}(t)}{dt}\right] = {}^C D_t^{\delta-1} I^{-(\delta-1)}[\varepsilon_1 \mathbf{S}(t) + \varepsilon_2 \mathbf{E}(t) + \varepsilon_3 \mathbf{I}(t) - (\rho+\sigma)\mathbf{Q}(t)], \\ {}^C D_t^{\delta-1}\left[\dfrac{d\mathbf{R}(t)}{dt}\right] = {}^C D_t^{\delta-1} I^{-(\delta-1)}[\eta \mathbf{S}(t) + \sigma \mathbf{Q}(t) + \lambda \mathbf{I}(t) - \rho \mathbf{R}(t)]. \end{cases} \quad (13)$$

After calculations, we reach

$$\begin{cases} {}^C D_t^\delta \mathbf{S}(t) = B - \theta \mathbf{S}(t)\mathbf{I}(t)(1 + \tau \mathbf{I}(t)) - (\varepsilon_1 + \rho + \eta)\mathbf{S}(t) + \mathcal{K}\mathbf{V}(t), \\ {}^C D_t^\delta \mathbf{V}(t) = \pi \mathbf{S}(t) - \psi \mathbf{V}(t) - (\mathcal{K} + \rho)\mathbf{V}(t), \\ {}^C D_t^\delta \mathbf{E}(t) = \theta \mathbf{S}(t)\mathbf{I}(t)(1 + \tau \mathbf{I}(t)) + \psi \mathbf{V}(t) - (\varepsilon_2 + \rho + \varphi)\mathbf{E}(t), \\ {}^C D_t^\delta \mathbf{I}(t) = \varphi \mathbf{E}(t) - (\lambda + \epsilon + \rho + \varepsilon_3)\mathbf{I}(t), \\ {}^C D_t^\delta \mathbf{Q}(t) = \varepsilon_1 \mathbf{S}(t) + \varepsilon_2 \mathbf{E}(t) + \varepsilon_3 \mathbf{I}(t) - (\rho + \sigma)\mathbf{Q}(t), \\ {}^C D_t^\delta \mathbf{R}(t) = \eta \mathbf{S}(t) + \sigma \mathbf{Q}(t) + \lambda \mathbf{I}(t) - \rho \mathbf{R}(t). \end{cases} \quad (14)$$

3.2. Basic Reproductive Number R_0

The basic reproductive number R_0, which expresses the typical number of secondary infections caused by a single infected person in a population that is fully susceptible, is an essential component of epidemiological modeling. The advantages of R_0 in epidemiological modeling are extensive: it offers a precise measurement of a disease's transmissibility, where a disease is more contagious if its R_0 is high, whereas a low R_0 suggests that it is less contagious; predicting the potential spread of an outbreak, which helps epidemiologists forecast the size and length of an outbreak and create the most efficient control strategies by calculating R_0; it can be employed to assess the efficacy of interventions, and epidemiologists can assess the success of an intervention in reducing transmission by comparing the R_0 before and after the implementation of a vaccine or quarantine measures; it is useful for locating the crucial control points, and the amount of infected people in a population, for example, is one of the crucial factors in the transmission of an infection that R_0 can be used to pinpoint; and the herd immunity threshold can be predicted using the number, and the percentage of the population that must be immune to a disease for there to be herd immunity is known as the herd immunity threshold. The herd immunity threshold and subsequently the overall efficacy of vaccination programs are predicted with the aid of R_0. Overall, R_0 is a useful tool for understanding the spread of infectious diseases and for designing effective public health interventions.

The DFE of Model (14) is denoted by $E^0 = (\mathbf{S}_0, \mathbf{V}_0, 0, 0, \mathbf{Q}_0, \mathbf{R}_0)$, where

$$\mathbf{S}_0 = \frac{B}{\eta + \rho + \varepsilon_1}, \quad \mathbf{V}_0 = \frac{\pi \mathbf{S}_0}{\mathcal{K} + \rho}, \quad \mathbf{Q}_0 = \frac{\varepsilon_1 \mathbf{S}_0}{\rho + \tau}, \quad \mathbf{R}_0 = \frac{\eta \mathbf{S}_0 + \tau \mathbf{Q}_0}{\rho}.$$

Our proposed model is split into two matrices [48].

$$\tilde{U} = \begin{bmatrix} 0 & \theta \mathbf{S}_0 \\ 0 & 0 \end{bmatrix}, \quad \tilde{V} = \begin{bmatrix} \varphi + \rho + \varepsilon_2 & 0 \\ -\varphi & \rho + \epsilon + \lambda + \varepsilon_3 \end{bmatrix},$$

$$\tilde{V}^{-1} = \frac{1}{(\varphi + \rho + \varepsilon_2)(\rho + \epsilon + \lambda + \varepsilon_3)} \begin{bmatrix} \rho + \epsilon + \lambda + \varepsilon_3 & 0 \\ \varphi & \varphi + \rho + \varepsilon_2 \end{bmatrix},$$

$$\tilde{U}\tilde{V}^{-1} = \begin{bmatrix} \frac{\theta \mathbf{S}_0 \varphi}{(\varphi + \rho + \varepsilon_2)(\rho + \epsilon + \lambda + \varepsilon_3)} & \frac{\theta \mathbf{S}_0}{\rho + \epsilon + \lambda + \varepsilon_3} \\ 0 & 0 \end{bmatrix}.$$

Hence,

$$R_0 = \frac{\varphi \theta B}{(\eta + \rho + \varepsilon_1)(\varphi + \rho + \varepsilon_2)(\rho + \epsilon + \lambda + \varepsilon_3)}.$$

4. Existence and Uniqueness

The solution for the system (14) using the Caputo operator will be described below, along with its existence and uniqueness. Assume that the continuous real-valued function $\mathcal{A}(Y)$, which has the sup-norm property, is a Banach space on $J = [0, b]$ and that $Y = [0, \kappa]$ and $P = \mathcal{A}(Y) \times \mathcal{A}(Y) \times \mathcal{A}(Y) \times \mathcal{A}(Y) \times \mathcal{A}(Y)$ with norm $\|(\mathbf{S}, \mathbf{V}, \mathbf{E}, \mathbf{Q}, \mathbf{I}, \mathbf{R})\| = \|\mathbf{S}\| + \|\mathbf{V}\| + \|\mathbf{E}\| + \|\mathbf{Q}\| + \|\mathbf{I}\| + \|\mathbf{R}\|$, where $\|\mathbf{S}\| = \sup_{t \in Y} |\mathbf{S}(t)|, \|\mathbf{V}(t)\| = \sup_{t \in Y} |\mathbf{V}(t)|, \|\mathbf{E}(t)\| =$

$\sup_{t\in Y}|\mathbf{E}(t)|, \|\mathbf{Q}\| = \sup_{t\in Y}|\mathbf{Q}(t)|, \|\mathbf{I}\| = \sup_{t\in Y}|\mathbf{I}(t)|, \|\mathbf{R}\| = \sup_{t\in Y}|\mathbf{R}(t)|$. The following equation is obtain by using the Caputo fractional integral operator on both sides of (14):

$$\begin{cases} \mathbf{S}(t) - \mathbf{S}(0) =^C \mathbb{D}_{0,t}^{\delta}\mathbf{S}(t)\{B - \theta\mathbf{S}\mathbf{I}(t)(1+\tau\mathbf{I}) - (\varepsilon_1 + \rho + \eta)\mathbf{S}(t) + \mathcal{K}\mathbf{V}(t)\}, \\ \mathbf{V}(t) - \mathbf{V}(0) =^C \mathbb{D}_{0,t}^{\delta}\mathbf{V}(t)\{\pi\mathbf{S}(t) - \psi\mathbf{V}(t) - (\mathcal{K}+\rho)\mathbf{V}(t)\}, \\ \mathbf{E}(t) - \mathbf{E}(0) =^C \mathbb{D}_{0,t}^{\delta}\mathbf{E}(t)\{\theta\mathbf{S}(t)\mathbf{I}(t)(1+\tau\mathbf{I}(t)) + \psi\mathbf{V}(t) - (\varepsilon_2 + \rho + \varphi)\mathbf{E}(t)\}, \\ \mathbf{I}(t) - \mathbf{I}(0) =^C \mathbb{D}_{0,t}^{\delta}\mathbf{I}(t)\{\varphi\mathbf{E}(t) - (\lambda + \epsilon + \rho + \varepsilon_3)\mathbf{I}(t)\}, \\ \mathbf{Q}(t) - \mathbf{Q}(0) =^C \mathbb{D}_{0,t}^{\delta}\mathbf{Q}(t)\{\varepsilon_1\mathbf{S}(t) + \varepsilon_2\mathbf{E}(t) + \varepsilon_3\mathbf{I}(t) - (\rho+\sigma)\mathbf{Q}(t)\}, \\ \mathbf{R}(t) - \mathbf{R}(0) =^C \mathbb{D}_{0,t}^{\delta}\mathbf{R}(t)\{\eta\mathbf{S}(t) + \sigma\mathbf{Q}(t) + \lambda\mathbf{I}(t) - \rho\mathbf{R}(t)\}. \end{cases} \quad (15)$$

After calculation,

$$\mathbf{S}(t) - \mathbf{S}(0) = \mathcal{H}(\delta)\int_0^t (t-\varpi)^{-\delta}\mathcal{B}_1(\delta,\varpi,\mathbf{S}(\varpi))d\varpi,$$

$$\mathbf{V}(t) - \mathbf{V}(0) = \mathcal{H}(\delta)\int_0^t (t-\varpi)^{-\delta}\mathcal{B}_2(\delta,\varpi,\mathbf{V}(t)(\varpi))d\varpi,$$

$$\mathbf{E}(t) - \mathbf{E}(0) = \mathcal{H}(\delta)\int_0^t (t-\varpi)^{-\delta}\mathcal{B}_2(\delta,\varpi,\mathbf{E}(t)(\varpi))d\varpi, \quad (16)$$

$$\mathbf{I}(t) - \mathbf{I}(0) = \mathcal{H}(\delta)\int_0^t (t-\varpi)^{-\delta}\mathcal{B}_4(\delta,\varpi,\mathbf{I}(t)(\varpi))d\varpi,$$

$$\mathbf{Q}(t) - \mathbf{Q}(0) = \mathcal{H}(\delta)\int_0^t (t-\varpi)^{-\delta}\mathcal{B}_3(\delta,\varpi,\mathbf{Q}(t)(\varpi))d\varpi,$$

$$\mathbf{R}(t) - \mathbf{R}(0) = \mathcal{H}(\delta)\int_0^t (t-\varpi)^{-\delta}\mathcal{B}_5(\delta,\varpi,\mathbf{R}(t)(\varpi))d\varpi,$$

where

$$\begin{cases} \mathbf{S}(t) - \mathbf{S}(0) =^C \mathbb{D}_{0,t}^{\delta}\mathbf{S}(t)\{B - \theta\mathbf{S}(t)\mathbf{I}(t)(1+\tau\mathbf{I}(t)) - (\varepsilon_1 + \rho + \eta)\mathbf{S}(t) + \mathcal{K}\mathbf{V}(t)\}, \\ \mathbf{V}(t) - \mathbf{V}(0) =^C \mathbb{D}_{0,t}^{\delta}\mathbf{V}(t)\{\pi\mathbf{S}(t) - \psi\mathbf{V}(t) - (\mathcal{K}+\rho)\mathbf{V}(t)\}, \\ \mathbf{E}(t) - \mathbf{E}(0) =^C \mathbb{D}_{0,t}^{\delta}\mathbf{E}(t)\{\theta\mathbf{S}(t)\mathbf{I}(t)(1+\tau\mathbf{I}(t)) + \psi\mathbf{V}(t) - (\varepsilon_2 + \rho + \varphi)\mathbf{E}(t)\}, \\ \mathbf{I}(t) - \mathbf{I}(0) =^C \mathbb{D}_{0,t}^{\delta}\mathbf{I}(t)\{\varphi\mathbf{E}(t) - (\lambda + \epsilon + \rho + \varepsilon_3)\mathbf{I}(t)\}, \\ \mathbf{Q}(t) - \mathbf{Q}(0) =^C \mathbb{D}_{0,t}^{\delta}\mathbf{Q}(t)\{\varepsilon_1\mathbf{S}(t) + \varepsilon_2\mathbf{E}(t) + \varepsilon_3\mathbf{I}(t) - (\rho+\sigma)\mathbf{Q}(t)\}, \\ \mathbf{R}(t) - \mathbf{R}(0) =^C \mathbb{D}_{0,t}^{\delta}\mathbf{R}(t)\{\eta\mathbf{S}(t) + \sigma\mathbf{Q}(t) + \lambda\mathbf{I}(t) - \rho\mathbf{R}(t)\}. \end{cases} \quad (17)$$

$$\mathcal{B}_1(\delta,t,\mathbf{S}(t)) = B - \theta\mathbf{S}\mathbf{I}(t)(1+\tau\mathbf{I}(t)) - (\varepsilon_1 + \rho + \eta)\mathbf{S}(t) + \mathcal{K}\mathbf{V}(t),$$

$$\mathcal{B}_2(\delta,t,\mathbf{V}(t)) = \pi\mathbf{S}(t) - \psi\mathbf{V}(t) - (\mathcal{K}+\rho)\mathbf{V}(t),$$

$$\mathcal{B}_2(\delta,t,\mathbf{E}(t)) = \theta\mathbf{S}\mathbf{I}(t)(1+\tau\mathbf{I}(t)) + \psi\mathbf{V}(t) - (\varepsilon_2 + \rho + \varphi)\mathbf{E}(t), \quad (18)$$

$$\mathcal{B}_3(\delta,t,\mathbf{I}(t)) = \varphi\mathbf{E}(t) - (\lambda + \epsilon + \rho + \varepsilon_3)\mathbf{I}(t),$$

$$\mathcal{B}_4(\delta,t,\mathbf{Q}(t)) = \varepsilon_1\mathbf{S}(t) + \varepsilon_2\mathbf{E}(t) + \varepsilon_3\mathbf{I}(t) - (\rho+\sigma)\mathbf{Q}(t),$$

$$\mathcal{B}_5(\delta,t,\mathbf{R}(t)) = \eta\mathbf{S}(t) + \sigma\mathbf{Q}(t) + \lambda\mathbf{I}(t) - \rho\mathbf{R}(t).$$

If $\mathbf{S}(t), \mathbf{V}(t), \mathbf{E}(t), \mathbf{Q}(t), \mathbf{I}(t),$ and $\mathbf{R}(t)$ have an upper bound, then the symbols $\mathcal{B}_1, \mathcal{B}_2, \mathcal{B}_3, \mathcal{B}_4, \mathcal{B}_5,$ and $\mathcal{B}_6$ are necessary for the Lipschitz condition. It should be noted that $\mathbf{S}(t)$ and $\mathbf{S}^*(t)$ are paired functions, and we reach

$$\|\mathcal{B}_1(\delta,t,\mathbf{S}(t)) - \mathcal{B}_1(\delta,t,\mathbf{S}^*(t))\| = \|-(\theta\mathbf{I}(t)(1+\varpropto\mathbf{I}(t)) + \eta + \rho + \varepsilon_1)(\mathbf{S}(t) - \mathbf{S}^*(t))\|. \quad (19)$$

Taking into account $\Lambda_1 = \|-(\theta\mathbf{I}(t)(1+\tau\mathbf{I}(t))+\eta+\rho+\varepsilon_1)\|$, one reaches

$$\|\mathcal{B}_1(\delta,t,\mathbf{S}(t)) - \mathcal{B}_1(\delta,t,\mathbf{S}^*(t))\| \leq \Lambda_1\|\mathbf{S}(t) - \mathbf{S}^*(t)\|. \tag{20}$$

Similarly,

$$\|\mathcal{B}_2(\delta,t,\mathbf{V}(t)) - \mathcal{B}_2(\delta,t,\mathbf{V}^*(t))\| \leq \Lambda_2\|\mathbf{V}(t) - \mathbf{V}^*(t)\|,$$

$$\|\mathcal{B}_3(\delta,t,\mathbf{E}(t)) - \mathcal{B}_3(\delta,t,\mathbf{E}^*(t))\| \leq \Lambda_3\|\mathbf{E}(t) - \mathbf{E}^*(t)\|,$$

$$\|\mathcal{B}_4(\delta,t,\mathbf{Q}(t)) - \mathcal{B}_4(\delta,t,\mathbf{Q}^*(t))\| \leq \Lambda_4\|\mathbf{Q}(t) - \mathbf{Q}^*(t)\|, \tag{21}$$

$$\|\mathcal{B}_5(\delta,t,\mathbf{I}(t)) - \mathcal{B}_5(\delta,t,\mathbf{I}^*(t))\| \leq \Lambda_5\|\mathbf{I}(t) - \mathbf{I}^*(t)\|,$$

$$\|\mathcal{B}_6(\delta,t,\mathbf{R}(t)) - \mathcal{B}_6(\delta,t,\mathbf{R}^*(t))\| \leq \Lambda_6\|\mathbf{R}(t) - \mathbf{R}^*(t)\|.$$

where

$$\Lambda_3 = \|-(\varphi+\rho+\varepsilon_2)\|$$

$$\Lambda_4 = \|-(\rho+\epsilon+\lambda+\varepsilon_3)\|$$

$$\Lambda_5 = \|-(\rho+\tau)\|$$

$$\Lambda_6 = \|-(\rho)\|.$$

This indicates that for each of the five functions, the Lipschitz condition is true. Recursively applying the expressions in (16), we obtain

$$\mathbf{S}_n(t) = \mathcal{H}(\delta) \int_0^t (t-\omega)^{-\delta} \mathcal{B}_1(\delta,\omega,\mathbf{S}_{n-1}(\omega)) d\omega,$$

$$\mathbf{V}(t)_n(t) = \mathcal{H}(\delta) \int_0^t (t-\omega)^{-\delta} \mathcal{B}_2(\delta,\omega,\mathbf{V}(t)_{n-1}(\omega)) d\omega,$$

$$\mathbf{E}(t)_n(t) = \mathcal{H}(\delta) \int_0^t (t-\omega)^{-\delta} \mathcal{B}_3(\delta,\omega,\mathbf{E}(t)_{n-1}(\omega)) d\omega, \tag{22}$$

$$\mathbf{Q}(t)_n(t) = \mathcal{H}(\delta) \int_0^t (t-\omega)^{-\delta} \mathcal{B}_4(\delta,\omega,\mathbf{Q}(t)_{n-1}(\omega)) d\omega,$$

$$\mathbf{I}(t)_n(t) = \mathcal{H}(\delta) \int_0^t (t-\omega)^{-\delta} \mathcal{B}_5(\delta,\omega,\mathbf{I}(t)_{n-1}(\omega)) d\omega,$$

$$\mathbf{R}(t)_n(t) = \mathcal{H}(\delta) \int_0^t (t-\omega)^{-\delta} \mathcal{B}_6(\delta,\omega,\mathbf{R}(t)_{n-1}(\omega)) d\omega,$$

We can obtain the successive terms difference by considering $\mathbf{S}_0(t) = \mathbf{S}_0, \mathbf{V}_0(t) = \mathbf{V}_0, \mathbf{E}_0(t) = \mathbf{E}_0, \mathbf{Q}_0(t) = \mathbf{Q}_0, \mathbf{I}_0(t) = \mathbf{I}_0$ and $\mathbf{R}_0(t) = \mathbf{R}_0$ in conjunction with other relevant information.

$$\begin{aligned}
\Xi_{\mathbf{S},n}(t) &= \mathbf{S}_n(t) - \mathbf{S}_{n-1}(t) \\
&= \mathcal{H}(\delta) \int_0^t (t-\omega)^{-\delta} (\mathcal{B}_1(\delta,\omega,\mathbf{S}_{n-1}(\omega)) - \mathcal{B}_1(\delta,\omega,\mathbf{S}_{n-2}(\omega))) d\omega, \\
\Xi_{\mathbf{V},n}(t) &= \mathbf{V}_n(t) - \mathbf{V}_{n-1}(t) \\
&= \mathcal{H}(\delta) \int_0^t (t-\omega)^{-\delta} (\mathcal{B}_2(\delta,\omega,\mathbf{V}_{n-1}(\omega)) - \mathcal{B}_2(\delta,\omega,\mathbf{V}_{n-2}(\omega))) d\omega, \\
\Xi_{\mathbf{E}(t),n}(t) &= \mathbf{E}(t)_n - \mathbf{E}(t)_{n-1} \\
&= \mathcal{H}(\delta) \int_0^t (t-\omega)^{-\delta} (\mathcal{B}_3(\delta,\omega,\mathbf{E}(t)_{n-1}(\omega)) - \mathcal{B}_3(\delta,\omega,\mathbf{E}(t)_{n-2}(\omega))) d\omega, \\
\Xi_{\mathbf{I}(t),n}(t) &= \mathbf{I}(t)_{2n} - \mathbf{I}(t)_{n-1} \\
&= \mathcal{H}(\delta) \int_0^t (t-\omega)^{-\delta} (\mathcal{B}_4(\delta,\omega,\mathbf{I}(t)_{n-1}(\omega)) - \mathcal{B}_4(\delta,\omega,\mathbf{I}(t)_{n-2}(\omega))) d\omega, \\
\Xi_{\mathbf{Q}(t),n}(t) &= \mathbf{Q}(t)_{1n} - \mathbf{Q}(t)_{n-1} \\
&= \mathcal{H}(\delta) \int_0^t (t-\omega)^{-\delta} (\mathcal{B}_5(\delta,\omega,\mathbf{Q}(t)_{n-1}(\omega)) - \mathcal{B}_5(\delta,\omega,\mathbf{Q}(t)_{n-2}(\omega))) d\omega, \\
\Xi_{\mathbf{R}(t),n}(t) &= \mathbf{R}(t)_n - \mathbf{R}(t)_{n-1} \\
&= \mathcal{H}(\delta) \int_0^t (t-\omega)^{-\delta} (\mathcal{B}_6(\delta,\omega,\mathbf{R}(t)_{n-1}(\omega)) - \mathcal{B}_6(\delta,\omega,\mathbf{R}(t)_{n-2}(\omega))) d\omega.
\end{aligned} \quad (23)$$

It is vital to observe that

$$\mathbf{S}_n(t) = \sum_{m=0}^n \Xi_{\mathbf{S},m}(t), \quad \mathbf{V}_n(t) = \sum_{m=0}^n \Xi_{\mathbf{V},m}(t), \quad \mathbf{V}_n(t) = \sum_{m=0}^n \Xi_{\mathbf{E},m}(t),$$

$$\mathbf{Q}_n(t) = \sum_{m=0}^n \Xi_{\mathbf{Q},m}(t), \quad \mathbf{I}_n(t) = \sum_{m=0}^n \Xi_{\mathbf{I},m}(t), \quad \mathbf{R}_n(t) = \sum_{m=0}^n \Xi_{\mathbf{R},m}(t).$$

Additionally, by using Equations (20) and (21) and considering that

$$\Xi_{\mathbf{S},n-1}(t) = \mathbf{S}_{n-1}(t) - \mathbf{S}_{n-2}(t), \quad \Xi_{\mathbf{V},n-1}(t) = \mathbf{V}_{n-1}(t) - \mathbf{V}_{n-2}(t), \quad \Xi_{\mathbf{E},n-1}(t) = \mathbf{E}_{n-1}(t) - \mathbf{E}_{n-2}(t),$$
$$\Xi_{\mathbf{Q},n-1}(t) = \mathbf{Q}_{n-1}(t) - \mathbf{Q}_{n-2}(t), \quad \Xi_{\mathbf{I},n-1}(t) = \mathbf{I}_{n-1}(t) - \mathbf{I}_{n-2}(t), \quad \Xi_{\mathbf{R},n-1}(t) = \mathbf{R}(t)_{n-1} - \mathbf{R}_{n-2}(t),$$

we reach

$$\begin{aligned}
\|\Xi_{\mathbf{S},n}(t)\| &\leq \mathcal{H}(\delta) \Lambda_1 \int_0^t (t-\omega)^{-\delta} \|\Xi_{\mathbf{S},n-1}(\omega)\| d\omega, \\
\|\Xi_{\mathbf{V},n}(t)\| &\leq \mathcal{H}(\delta) \Lambda_2 \int_0^t (t-\omega)^{-\delta} \|\Xi_{\mathbf{V},n-1}(\omega)\| d\omega, \\
\|\Xi_{\mathbf{E},n}(t)\| &\leq \mathcal{H}(\delta) \Lambda_3 \int_0^t (t-\omega)^{-\delta} \|\Xi_{\mathbf{E},n-1}(\omega)\| d\omega, \\
\|\Xi_{\mathbf{I},n}(t)\| &\leq \mathcal{H}(\delta) \Lambda_4 \int_0^t (t-\omega)^{-\delta} \|\Xi_{\mathbf{I},n-1}(\omega)\| d\omega, \\
\|\Xi_{\mathbf{Q},n}(t)\| &\leq \mathcal{H}(\delta) \Lambda_5 \int_0^t (t-\omega)^{-\delta} \|\Xi_{\mathbf{Q},n-1}(\omega)\| d\omega, \\
\|\Xi_{\mathbf{R},n}(t)\| &\leq \mathcal{H}(\delta) \Lambda_6 \int_0^t (t-\omega)^{-\delta} \|\Xi_{\mathbf{R},n-1}(\omega)\| d\omega.
\end{aligned} \quad (24)$$

Theorem 1. *If the following condition holds,*

$$\frac{\mathcal{H}(\delta)}{\delta} \kappa^\delta \Lambda_m < 1, \, m = 1, 2, \ldots, 6. \quad (25)$$

Then, (14) has a unique solution for $t \in [0, \kappa]$.

Proof. It has been demonstrated that the functions $\mathbf{S}(t), \mathbf{V}(t), \mathbf{E}(t), \mathbf{Q}(t), \mathbf{I}(t)$, and $\mathbf{R}(t)$ are bounded. Furthermore, it can be observed from Equations (20) and (21) that the symbols $\mathcal{B}_1, \mathcal{B}_2, \mathcal{B}_3, \mathcal{B}_4, \mathcal{B}_5$, and $\mathcal{B}_6$ are applicable to the Lipschitz condition. Thus, by using Equation (24) along with a recursive hypothesis, we can derive:

$$\|\Xi_{\mathbf{S},n}(t)\| \leq \|\mathbf{S}_0(t)\| \left(\frac{\mathcal{H}(\delta)}{\delta}\kappa^\delta \Lambda_1\right)^n,$$

$$\|\Xi_{\mathbf{V},n}(t)\| \leq \|\mathbf{V}_0(t)\| \left(\frac{\mathcal{H}(\delta)}{\delta}\kappa^\delta \Lambda_2\right)^n,$$

$$\|\Xi_{\mathbf{E},n}(t)\| \leq \|\mathbf{E}_0(t)\| \left(\frac{\mathcal{H}(\delta)}{\delta}\kappa^\delta \Lambda_3\right)^n,$$

$$\|\Xi_{\mathbf{I},n}(t)\| \leq \|\mathbf{I}_0(t)\| \left(\frac{\mathcal{H}(\delta)}{\delta}\kappa^\delta \Lambda_4\right)^n, \quad (26)$$

$$\|\Xi_{\mathbf{Q},n}(t)\| \leq \|\mathbf{Q}_0(t)\| \left(\frac{\mathcal{H}(\delta)}{\delta}\kappa^\delta \Lambda_5\right)^n,$$

$$\|\Xi_{\mathbf{R},n}(t)\| \leq \|\mathbf{R}_0(t)\| \left(\frac{\mathcal{H}(\delta)}{\delta}\kappa^\delta \Lambda_6\right)^n.$$

As a result, it is evident that the sequences fulfill and exist

$$\|\Xi_{\mathbf{S},n}(t)\| \to 0, \|\Xi_{\mathbf{V},n}(t)\| \to 0, \|\Xi_{\mathbf{E},n}(t)\| \to 0, \|\Xi_{\mathbf{I},n}(t)\| \to 0, \|\Xi_{\mathbf{Q},n}(t)\| \to 0, \quad \|\Xi_{\mathbf{R},n}(t)\| \to 0 \text{ as } n \to \infty.$$

Additionally, using Equation (26) and the triangle inequality, for any s, we have

$$\|\mathbf{S}_{n+s}(t) - \mathbf{S}_n(t)\| \leq \sum_{j=n+1}^{n+s} X_1^j = \frac{X_1^{n+1} - X_1^{n+s+1}}{1 - X_1},$$

$$\|\mathbf{V}_{n+s}(t) - \mathbf{V}_n(t)\| \leq \sum_{j=n+1}^{n+s} X_2^j = \frac{X_2^{n+1} - X_2^{n+s+1}}{1 - X_2},$$

$$\|\mathbf{E}_{n+s}(t) - \mathbf{E}_n(t)\| \leq \sum_{j=n+1}^{n+s} X_2^j = \frac{X_3^{n+1} - X_3^{n+s+1}}{1 - X_3},$$

$$\|\mathbf{I}_{n+s}(t) - \mathbf{I}_n(t)\| \leq \sum_{j=n+1}^{n+s} X_4^j = \frac{X_4^{n+1} - X_4^{n+s+1}}{1 - X_4}, \quad (27)$$

$$\|\mathbf{Q}_{n+s}(t) - \mathbf{Q}_n(t)\| \leq \sum_{j=n+1}^{n+s} X_5^j = \frac{X_5^{n+1} - X_5^{n+s+1}}{1 - X_5},$$

$$\|\mathbf{R}_{n+s}(t) - \mathbf{R}_n(t)\| \leq \sum_{m=n+1}^{n+s} X_6^j = \frac{X_6^{n+1} - X_6^{n+s+1}}{1 - X_6},$$

with $X_m = \frac{\mathcal{H}(\delta)}{\delta}\kappa^\delta \Lambda_m < 1$ by hypothesis. Therefore, with $\mathbf{S}_n, \mathbf{V}_n, \mathbf{E}_n, \mathbf{I}_n, \mathbf{Q}_n$, it is possible to think of $\mathbf{R}(t)_n$ as a Cauchy sequence in the $\mathcal{A}(Y)$ Banach space. This has demonstrated that they are uniformly convergent [49]. □

5. Parameter Estimation

Parameter estimation in epidemic models involves identifying the model parameters that best fit the observed data. One widely used approach for parameter estimation is least-squared curve-fitting, which entails minimizing the difference between the observed data and the model predictions by adjusting the model parameters. This method assumes that the residuals of the model are normally distributed and have a constant variance. Least-squared curve-fitting tools estimate the parameters of the epidemic model by minimizing the sum of squared residuals. This method is computationally efficient and provides reliable estimates when the assumptions of normality and constant variance are met. However, it is critical to verify that the model assumptions are satisfied and that the goodness-of-fit is assessed using appropriate statistical tests. In this study, we employed least-squared curve-fitting methods to analyze the COVID-19 cases reported in Pakistan between 23 June 2022 and 23 August 2022 [50]. The estimated parameters of the system were based on Pakistan's confirmed cases and fatalities. The ordinary least square solution (OLS) was used to minimize the error terms in the daily reports, and the goodness-of-fit was evaluated using the relative error.

$$\min\left(\frac{\sum_{i=1}^{n}(\mathbf{I}_i - \hat{\mathbf{I}}_i)^2}{\sum_{i=1}^{n}\mathbf{I}_i^2}\right). \tag{28}$$

The simulated cumulative number of infected individuals, denoted by $\mathbf{I}_i$, is computed by adding the total number of individuals transitioning from the infected compartment to the recovered compartment each day. Meanwhile, the reported total number of infected individuals is represented by $\hat{\mathbf{I}}_i$. Table 1 presents the estimated parameter values, while a comparison between the model predictions and the reported cases is illustrated in Figure 1.

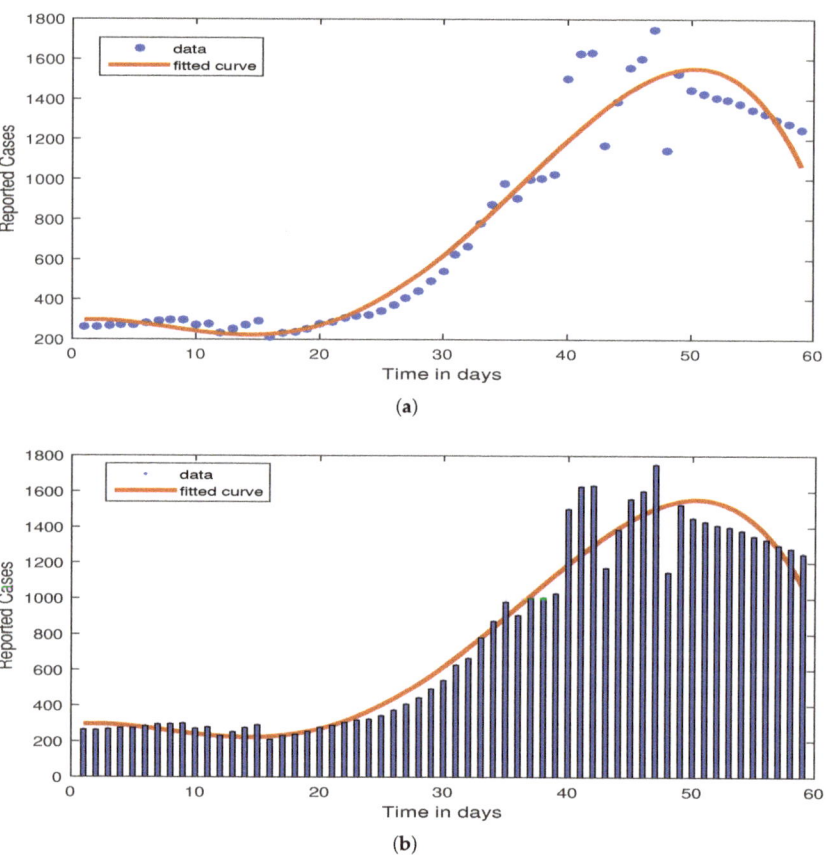

Figure 1. Cont.

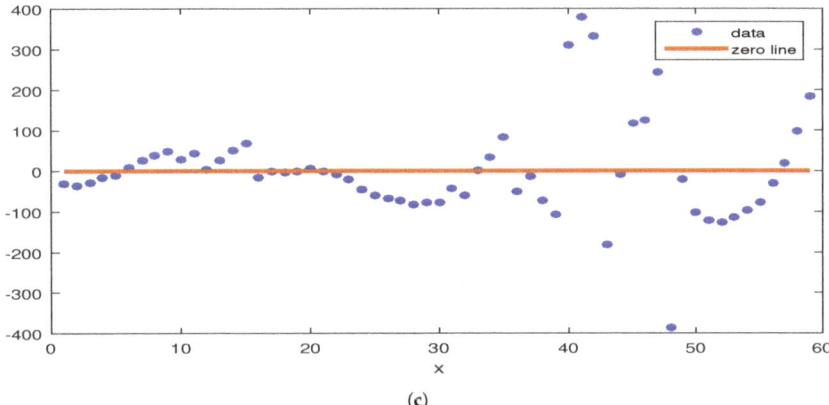

(c)

Figure 1. The figure illustrates the profiles of the best-fitted curve and its corresponding residuals for the daily cumulative cases of COVID-19 in Pakistan from 23 Jun 2022 to 23 August 2022. Furthermore, the figures (**a**,**b**) represent the model fit with reported infected cases and (**c**) represents the residuals.

Table 1. The table contains descriptions and estimated values for the parameters.

Symbol of Parameters	Values of Parameters	References
B	50.4057	[51]
θ	0.0019	Estimated
τ	0.0205	Estimated
ϵ	0.1571	Estimated
ϵ_1	0.01772	Estimated
ϵ_2	0.0805	Estimated
ϵ_3	0.0876	Estimated
ρ	0.0205	[51]
η	0.1460	Estimated
α	0.3506	Estimated
$\mathcal{K}$	0.1530	Estimated
σ	0.0059	Estimated
π	0.0029	Estimated
ψ	0.0532	Estimated
λ	0.0105	Estimated

6. Sensitivity Analysis

A sensitivity analysis identifies the parameters that are most effective in curbing COVID-19 spread. Even though forward sensitivity analysis becomes tedious for complex biological models, it is an essential component of phenomena modeling. Ecologists and epidemiologists have taken an active interest in R_0 sensitivity analysis.

Definition 3. *The normalized forward sensitivity index of the R_0 that depends differentiability on a parameter $\varkappa$ is defined as*

$$\Upsilon_\varkappa = \frac{\varkappa}{R_0} \frac{\partial R_0}{\partial \varkappa}.$$

A vital tool for assessing how uncertain input parameters affect a system's output is sensitivity analysis. Sensitivity indices can be computed using a variety of techniques, including direct differentiation, Latin hypercube sampling, and system linearization. In this study, the analytical expressions for the indices were provided using the direct differentiation method, allowing for a more thorough understanding of the behavior of the system. This study investigated the effects of various factors and gained a critical understanding of the comparative variability of the fundamental reproduction number, R_0, and other parameters by applying this method to a COVID-19 model. These sensitivity indices provide policymakers and law enforcement agencies with useful data to create strategies that effectively combat COVID-19, and they are therefore essential in battling the pandemic. The graphical results are displayed in Figures 2 and 3.

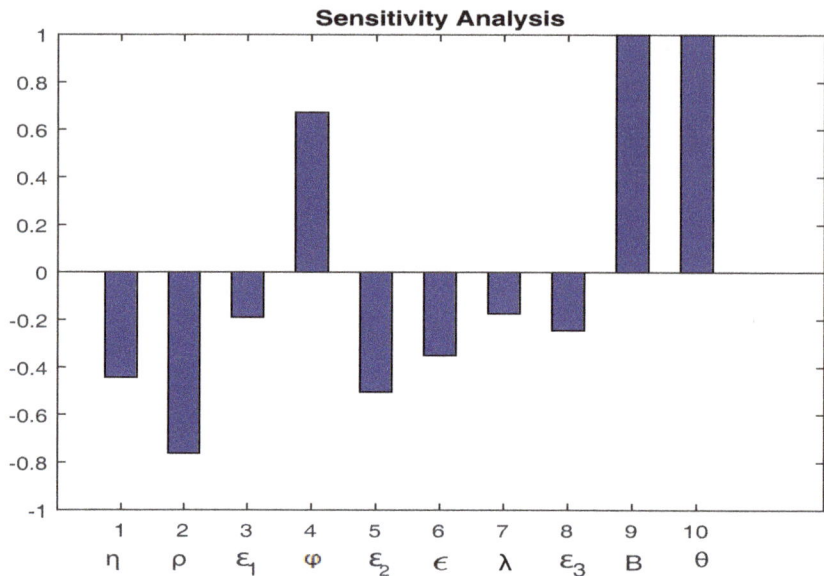

Figure 2. Global sensitivity.

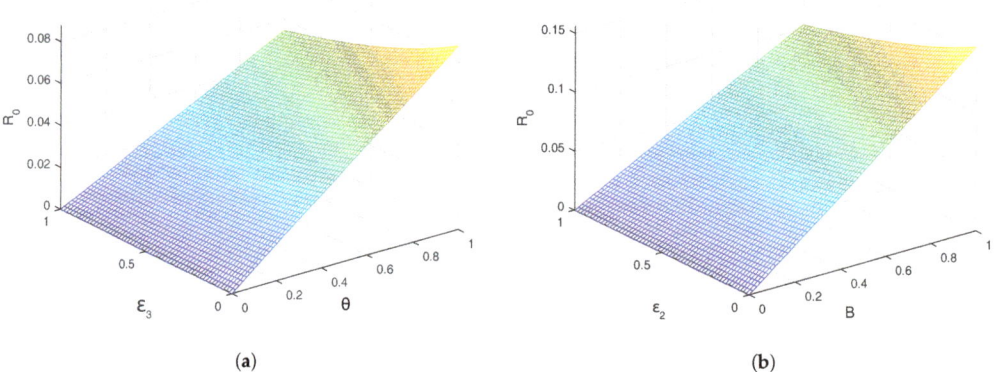

(**a**)   (**b**)

Figure 3. *Cont.*

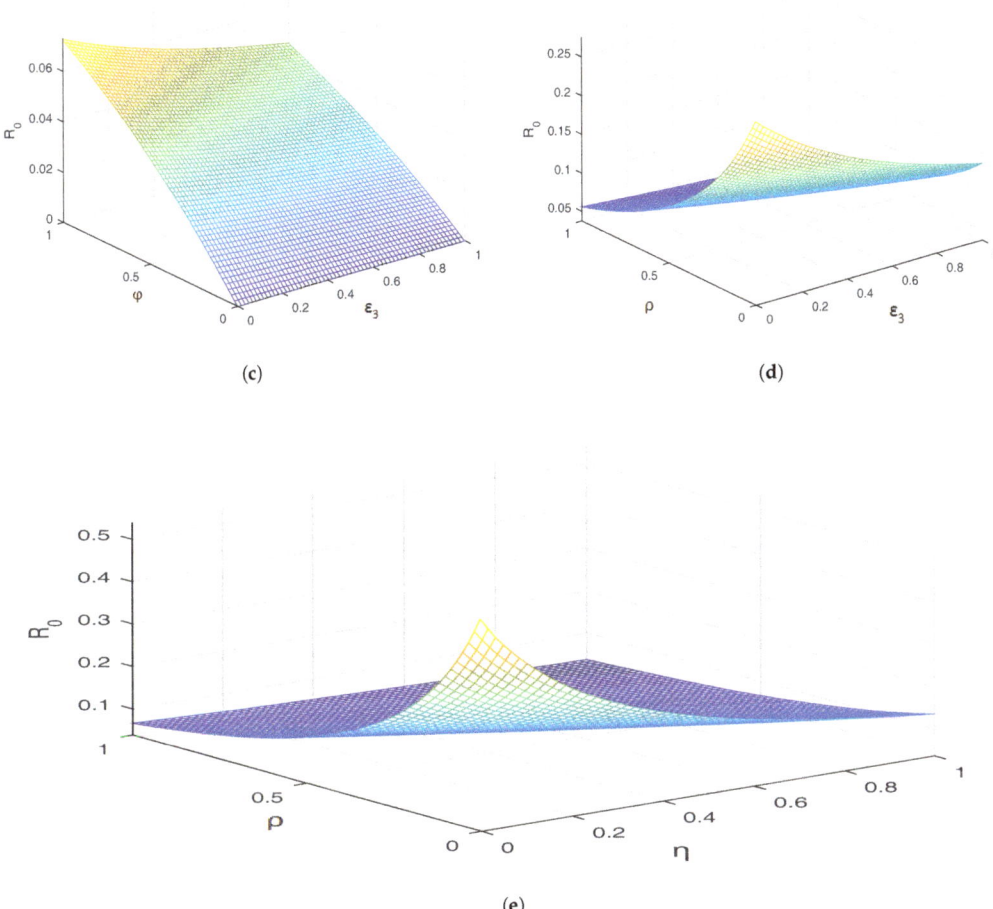

Figure 3. Sensitivity of various parameters vs. R_0 are displayed in (**a**–**e**).

7. Numerical Scheme and Graphical Results

The Haar wavelet collocation method is a powerful numerical technique used for solving various types of differential equations. One of the main advantages of this method is its high accuracy, as it is capable of producing highly precise solutions for a wide range of problems. Additionally, the Haar wavelet collocation method is highly efficient, having a relatively low computational cost compared with other numerical methods. This is due to the compact support of the Haar wavelet, which allows for the use of fewer basis functions to represent the solution, resulting in faster calculations. Overall, the Haar wavelet collocation method is a reliable and efficient tool for numerical analysis, making it a popular choice among researchers and practitioners in various fields of science and engineering.

Consider the square integrable function space $L_2[0, 1)$, where $\dot{\mathbf{S}}(t), \dot{\mathbf{V}}(t), \dot{\mathbf{E}}(t), \dot{\mathbf{I}}(t), \dot{\mathbf{Q}}(t)$, and $\dot{\mathbf{R}}(t)$ can be expressed as a Haar series, given by:

$$\dot{\mathbf{S}}(t) = \sum_{m=1}^{N} \alpha_m \tilde{\psi}_m(t)$$

$$\dot{\mathbf{V}}(t) = \sum_{m=1}^{N} \gamma_m \tilde{\psi}_m(t)$$

$$\dot{\mathbf{E}}(t) = \sum_{m=1}^{N} \theta_m \tilde{\psi}_m(t)$$

$$\dot{\mathbf{I}}(t) = \sum_{m=1}^{N} \lambda_m \tilde{\psi}_m(t),$$

$$\dot{\mathbf{Q}}(t) = \sum_{m=1}^{N} \omega_m \tilde{\psi}_m(t)$$

$$\dot{\mathbf{R}}(t) = \sum_{m=1}^{N} \sigma_m \tilde{\psi}_m(t).$$

Here, $\mathbf{S}(0) = \mathbf{S}_0$ represents the initial susceptible compartment, $\mathbf{V}(0) = \mathbf{V}_0$ represents the initial vaccinated compartment, $\mathbf{E}(0) = \mathbf{E}_0$ represents the initial exposed compartment, $\mathbf{I}(0) = \mathbf{I}_0$ represents the initial infected compartment, $\mathbf{Q}(0) = \mathbf{Q}_0$ represents the initial infected compartment, and $\mathbf{R}(0) = \mathbf{R}_0$ represents the initial recovered compartment. Integration of these equations leads to the following relation:

$$\begin{aligned} \mathbf{S}(t) = \mathbf{S}_0 + \sum_{m=1}^{N} \alpha_m \vartheta_{m,1}(t), \mathbf{V}(t) = \mathbf{V}_0 + \sum_{m=1}^{N} \gamma_m \vartheta_{m,1}(t), \mathbf{E}(t) = \mathbf{E}_0 + \sum_{m=1}^{N} \theta_m \vartheta_{m,1}(t), \\ \mathbf{I}(t) = \mathbf{I}_0 + \sum_{m=1}^{N} \lambda_m \vartheta_{m,1}(t), \mathbf{Q}(t) = \mathbf{Q}_0 + \sum_{m=1}^{N} \omega_m \vartheta_{m,1}(t), \mathbf{R}(t) = \mathbf{R}_0 + \sum_{m=1}^{N} \sigma_m \vartheta_{m,1}(t). \end{aligned} \quad (29)$$

By using the Caputo derivative, we have

$$\begin{cases} \frac{1}{\Gamma(n-\delta)} \int_0^t \mathbf{S}^{(n)}(\omega)(t-\omega)^{n-\delta-1} d\omega = B - \theta \mathbf{S}(t)\mathbf{I}(t)(1+\tau\mathbf{I}(t)) - (\varepsilon_1 + \rho + \eta)\mathbf{S}(t) + \mathcal{K}\mathbf{V}(t), \\ \frac{1}{\Gamma(n-\delta)} \int_0^t \mathbf{V}^{(n)}(\omega)(t-\omega)^{n-\delta-1} d\omega = \pi \mathbf{S}(t) - \psi\mathbf{V}(t) - (\mathcal{K}+\rho)\mathbf{V}(t), \\ \frac{1}{\Gamma(n-\delta)} \int_0^t \mathbf{E}^{(n)}(\omega)(t-\omega)^{n-\delta-1} d\omega = \theta \mathbf{S}(t)\mathbf{I}(t)(1+\tau\mathbf{I}(t)) + \psi\mathbf{V}(t) - (\varepsilon_2 + \rho + \not{2})\mathbf{E}(t), \\ \frac{1}{\Gamma(n-\delta)} \int_0^t \mathbf{I}^{(n)}(\omega)(t-\omega)^{n-\delta-1} d\omega = \varphi \mathbf{E}(t) - (\lambda + \epsilon + \rho + \varepsilon_3)\mathbf{I}(t), \\ \frac{1}{\Gamma(n-\delta)} \int_0^t \mathbf{Q}^{(n)}(\omega)(t-\omega)^{n-\delta-1} d\omega = \varepsilon_1 \mathbf{S}(t) + \varepsilon_2 \mathbf{E}(t) + \varepsilon_3 \mathbf{I}(t) + -(\rho + \sigma)\mathbf{Q}(t), \\ \frac{1}{\Gamma(n-\delta)} \int_0^t \mathbf{R}^{(n)}(\omega)(t-\omega)^{n-\delta-1} d\omega = \eta \mathbf{S}(t) + \sigma \mathbf{Q}(t) + \lambda \mathbf{I}(t) - \rho \mathbf{R}(t). \end{cases} \quad (30)$$

As we have assumed that $0 < \delta < 1$, therefore $n = 1$, and we have

$$\begin{cases} \dfrac{1}{\Gamma(1-\delta)}\int_0^t \dot{\mathbf{S}}(t)(\varpi)(t-\varpi)^{-\delta}d\varpi = B - \theta \mathbf{S}(t)\mathbf{I}(t)(1+\tau \mathbf{I}(t)) - (\varepsilon_1 + \rho + \eta)\mathbf{S}(t) + \mathcal{K}\mathbf{V}(t), \\ \dfrac{1}{\Gamma(1-\delta)}\int_0^t \dot{\mathbf{V}}(t)(\varpi)(t-\varpi)^{-\delta}d\varpi = \pi \mathbf{S}(t) - \psi \mathbf{V}(t) - (\rho + \mathcal{K})\mathbf{V}(t), \\ \dfrac{1}{\Gamma(1-\delta)}\int_0^t \dot{\mathbf{E}}(t)(\varpi)(t-\varpi)^{-\delta}d\varpi = \theta \mathbf{S}(t)\mathbf{I}(t)(1+\tau \mathbf{I}(t)) + \psi \mathbf{V}(t) - (\varepsilon_2 + \rho + \rightleftarrows)\mathbf{E}(t), \\ \dfrac{1}{\Gamma(1-\delta)}\int_0^t \dot{\mathbf{I}}(t)(\varpi)(t-\varpi)^{-\delta}d\varpi = \varphi \mathbf{E}(t) - (\lambda + \epsilon + \rho + \varepsilon_3)\mathbf{I}(t), \\ \dfrac{1}{\Gamma(1-\delta)}\int_0^t \dot{\mathbf{Q}}(t)(\varpi)(t-\varpi)^{-\delta}d\varpi = \varepsilon_1 \mathbf{S}(t) + \varepsilon_2 \mathbf{E}(t) + \varepsilon_3 \mathbf{I}(t) - (\rho + \sigma)\mathbf{Q}(t), \\ \dfrac{1}{\Gamma(1-\delta)}\int_0^t \dot{\mathbf{R}}(t)(\varpi)(t-\varpi)^{-\delta}d\varpi = \eta \mathbf{S}(t) + \sigma \mathbf{Q}(t) + \lambda \mathbf{I}(t) - \rho \mathbf{R}(t). \end{cases} \quad (31)$$

Haar approximations are used, and we have

$$\begin{aligned}
\dfrac{1}{\Gamma(1-\delta)} & \int_0^t \sum_{m=1}^N \alpha_m \tilde{\psi}_m(\varpi)(t-\varpi)^{-\delta}d\varpi = B - \theta\left(\mathbf{I}_0 + \sum_{m=1}^N \lambda_m \vartheta_{m,1}(t)\right)\left(\mathbf{S}_0 + \sum_{m=1}^N \alpha_m \vartheta_{m,1}(t)\right) \\
& \left(1 + \tau\left(\mathbf{I}_0 + \sum_{m=1}^N \lambda_m \vartheta_{m,1}(t)\right)\right) - (\varepsilon_1 + \rho + \eta)\left(\mathbf{S}_0 + \sum_{m=1}^N \alpha_m \vartheta_{m,1}(t)\right) + \mathcal{K}\left(\mathbf{S}_0 + \sum_{m=1}^N \gamma_m \vartheta_{m,1}(t)\right) \\
\dfrac{1}{\Gamma(1-\delta)} & \int_0^t \sum_{m=1}^N \gamma_m \tilde{\psi}_m(\varpi)(t-\varpi)^{-\delta}d\varpi = \pi\left(\mathbf{S}_0 + \sum_{m=1}^N \alpha_m \vartheta_{m,1}(t)\right) \\
& - \psi\left(\mathbf{V}_0 + \sum_{m=1}^N \gamma_m \vartheta_{m,1}(t)\right) - (\rho + \mathcal{K})\left(\mathbf{V}_0 + \sum_{m=1}^N \gamma_m \vartheta_{m,1}(t)\right) \\
\dfrac{1}{\Gamma(1-\delta)} & \int_0^t \sum_{m=1}^N \theta_m \tilde{\psi}_m(\varpi)(t-\varpi)^{-\delta}d\varpi = \theta\left(\mathbf{I}_0 + \sum_{m=1}^N \lambda_m \vartheta_{m,1}(t)\right)\left(\mathbf{S}_0 + \sum_{m=1}^N \alpha_m \vartheta_{m,1}(t)\right) + \psi\left(\mathbf{V}_0 + \sum_{m=1}^N \gamma_m \vartheta_{m,1}(t)\right) \\
& \left(1 + \tau\left(\mathbf{I}_0 + \sum_{m=1}^N \lambda_m \vartheta_{m,1}(t)\right)\right) - (\varepsilon_2 + \rho + \varphi)\left(\mathbf{E}_0 + \sum_{m=1}^N \theta_m \vartheta_{m,1}(t)\right) \\
\dfrac{1}{\Gamma(1-\delta)} & \int_0^t \sum_{m=1}^N \lambda_m \tilde{\psi}_m(\varpi)(t-\varpi)^{-\delta}d\varpi = \\
& \varphi\left(\mathbf{E}_0 + \sum_{m=1}^N \theta_m \vartheta_{m,1}(t)\right) - (\lambda + \epsilon + \rho + d_1)\left(\mathbf{I}_0 + \sum_{m=1}^N \lambda_m \vartheta_{m,1}(t)\right) \\
\dfrac{1}{\Gamma(1-\delta)} & \int_0^t \sum_{m=1}^N \varpi_m \tilde{\psi}_m(\varpi)(t-\subsetneq)^{-\delta}d\varpi = \varepsilon_3\left(\mathbf{I}(t)_0 + \sum_{m=1}^N \lambda_m \vartheta_{m,1}(t)\right) + \varepsilon_2\left(\mathbf{E}_0 + \sum_{m=1}^N \theta_m \vartheta_{m,1}(t)\right) \\
& + \varepsilon_1\left(\mathbf{S}_0 + \sum_{m=1}^N \alpha_m \vartheta_{m,1}(t)\right) - (\rho + \sigma)\left(\mathbf{Q}_0 + \sum_{m=1}^N \varpi_m \vartheta_{m,1}(t)\right) \\
\dfrac{1}{\Gamma(1-\delta)} & \int_0^t \sum_{m=1}^N \sigma_m \tilde{\psi}_m(\varpi)(t-\varpi)^{-\delta}d\varpi = \eta\left(\mathbf{S}_0 + \sum_{m=1}^N \alpha_m \vartheta_{m,1}(t)\right) \\
& + \sigma\left(\mathbf{Q}_0 + \sum_{m=1}^N \sigma_m \vartheta_{m,1}(t)\right) + \lambda\left(\mathbf{I}_0 + \sum_{m=1}^N \lambda_m \vartheta_{m,1}(t)\right) + \rho\left(\mathbf{R}_0 + \sum_{m=1}^N \zeta_m \vartheta_{m,1}(t)\right).
\end{aligned} \quad (32)$$

Upon simplification, we have

$$\left\{\begin{array}{l}\frac{1}{\Gamma(1-\delta)}\sum_{m=1}^{N}\alpha_m\tilde{\psi}_m(\varpi)(t-\varpi)^{-\delta}d\varpi - B + \theta(1+\tau\mathbf{I}_0)\times \\[4pt] \left(\mathbf{I}_0\mathbf{S}_0 + \mathbf{I}_0\sum_{m=1}^{N}\alpha_m\vartheta_{m,1}(t) + \mathbf{S}_0\sum_{m=1}^{N}\theta_m\vartheta_{m,1}(t) + \sum_{m=1}^{N}\alpha_m\vartheta_{m,1}(t)\sum_{m=1}^{N}\theta_m\vartheta_{m,1}(t)\right) \\[4pt] +\theta\left[\mathbf{I}_0\mathbf{S}_0\tau\sum_{m=1}^{N}\theta_m\vartheta_{m,1}(t) + \mathbf{I}_0\tau\sum_{m=1}^{N}\alpha_m\vartheta_{m,1}(t)\sum_{m=1}^{N}\theta_m\vartheta_{m,1}(t) + \mathbf{S}_0\tau\left(\sum_{m=1}^{N}\theta_m\vartheta_{m,1}(t)\right)^2\right. \\[4pt] \left.+\tau\sum_{m=1}^{N}\alpha_m\vartheta_{m,1}(t)\left(\sum_{m=1}^{N}\theta_m\vartheta_{m,1}(t)\right)^2\right] + (\epsilon_1+\rho+\eta)\mathbf{S}_0 + (\epsilon_1+\rho+\eta)\sum_{m=1}^{N}\alpha_m\vartheta_{m,1}(t)\end{array}\right\} = 0, \quad (33)$$

$$\begin{aligned}\frac{1}{\Gamma(1-\delta)}\int_0^t\sum_{m=1}^{N}\gamma_m\tilde{\psi}_m(\varpi)(t-\varpi)^{-\delta}d\varpi - \pi\left(\mathbf{S}_0+\sum_{m=1}^{N}\alpha_m\vartheta_{m,1}(t)\right) \\ + \psi\left(\mathbf{V}_0+\sum_{m=1}^{N}\gamma_m\vartheta_{m,1}(t)\right) + (\rho+\mathcal{K})\left(\mathbf{V}_0+\sum_{m=1}^{N}\gamma_m\vartheta_{m,1}(t)\right) = 0,\end{aligned} \quad (34)$$

$$\left\{\begin{array}{l}\frac{1}{\Gamma(1-\delta)}\sum_{m=1}^{N}\theta_m\tilde{\psi}_m(\varpi)(t-\varpi)^{-\delta}d\varpi + \theta(1+\tau\mathbf{I}_0)\times \\[4pt] \left(\mathbf{I}_0\mathbf{S}_0 + \mathbf{I}_0\sum_{m=1}^{N}\alpha_m\vartheta_{m,1}(t) + \mathbf{S}_0\sum_{m=1}^{N}\theta_m\vartheta_{m,1}(t) + \sum_{m=1}^{N}\alpha_m\vartheta_{m,1}(t)\sum_{m=1}^{N}\theta_m\vartheta_{m,1}(t)\right) \\[4pt] +\theta\left[\mathbf{I}_0\mathbf{S}_0\tau\sum_{m=1}^{N}\theta_m\vartheta_{m,1}(t) + \mathbf{I}_0\tau\sum_{m=1}^{N}\alpha_m\vartheta_{m,1}(t)\sum_{m=1}^{N}\theta_m\vartheta_{m,1}(t) + \mathbf{S}_0\tau\left(\sum_{m=1}^{N}\theta_m\vartheta_{m,1}(t)\right)^2\right. \\[4pt] \left.+\tau\sum_{m=1}^{N}\alpha_m\vartheta_{m,1}(t)\left(\sum_{m=1}^{N}\theta_m\vartheta_{m,1}(t)\right)^2\right] + (\epsilon_2+\rho+\varphi)\mathbf{E}_0 + (\epsilon_2+\rho+\varphi)\sum_{m=1}^{N}\theta_m\vartheta_{m,1}(t)\end{array}\right\} = 0, \quad (35)$$

$$\begin{aligned}\frac{1}{\Gamma(1-\delta)}\int_0^t\sum_{m=1}^{N}\lambda_m\tilde{\psi}_m(\varpi)(t-\varpi)^{-\delta}d\varpi - \varphi\mathbf{E}_0 - \varphi\left(\sum_{m=1}^{N}\theta_m\vartheta_{m,1}(t)\right) \\ + (\lambda+\epsilon+\rho+\epsilon_3)\mathbf{I}_0 + (\lambda+\epsilon+\rho+\epsilon_3)\left(\sum_{m=1}^{N}\lambda_m\vartheta_{m,1}(t)\right) = 0,\end{aligned} \quad (36)$$

$$\begin{aligned}\frac{1}{\Gamma(1-\delta)}\int_0^t\sum_{m=1}^{N}\varpi_m\tilde{\psi}_m(\varpi)(t-\varsigma)^{-\delta}d\varpi - \epsilon_3\mathbf{I}(t)_0 - \epsilon_3\left(\sum_{m=1}^{N}\lambda_m\vartheta_{m,1}(t)\right) + \epsilon_2\mathbf{E}_0 + \epsilon_2\left(\sum_{m=1}^{N}\theta_m\vartheta_{m,1}(t)\right) \\ + \epsilon_1\mathbf{S}_0 + \epsilon_1\left(\sum_{m=1}^{N}\alpha_m\vartheta_{m,1}(t)\right) + (\rho+\sigma)\mathbf{Q}_0 + (\rho+\sigma)\left(\sum_{m=1}^{N}\varpi_m\vartheta_{m,1}(t)\right) = 0,\end{aligned} \quad (37)$$

$$\begin{aligned}\frac{1}{\Gamma(1-\delta)}\int_0^t\sum_{m=1}^{N}\zeta_m\tilde{\psi}_m(\varpi)(t-\varpi)^{-\delta}d\varpi - \eta\mathbf{S}_0 - \mathbf{S}_0\left(\sum_{m=1}^{N}\alpha_m\vartheta_{m,1}(t)\right) - \sigma\mathbf{Q}_0 \\ -\sigma\left(\sum_{m=1}^{N}\sigma_m\vartheta_{m,1}(t)\right) - \lambda\mathbf{I}_0 - \lambda\left(\sum_{m=1}^{N}\lambda_m\vartheta_{m,1}(t)\right) + \rho\mathbf{R}_0 + \rho\left(\sum_{m=1}^{N}\zeta_m\vartheta_{m,1}(t)\right) = 0.\end{aligned} \quad (38)$$

Using the method of Haar integration [52], the integral in the aforementioned system is approximately calculated as

$$\int_{\varkappa}^{\kappa}f(t)dt \approx \frac{\kappa-\varkappa}{N}\sum_{p=1}^{N}f(t_p) = \sum_{p=1}^{N}f\left(\varkappa+\frac{(\kappa-\varkappa)(p-0.5)}{N}\right) \quad (39)$$

$$\begin{cases} \frac{t}{N\Gamma(1-\delta)} \sum_{s=1}^{N} \sum_{m=1}^{N} \alpha_m \tilde{\psi}_m(\varpi_s)(t-\varpi_s)^{-\delta} - B + \theta(1+\tau \mathbf{I}_0) \times \\ \left(\mathbf{I}_0 \mathbf{S}_0 + \mathbf{I}_0 \sum_{m=1}^{N} \alpha_m \vartheta_{m,1}(t) + \mathbf{S}_0 \sum_{m=1}^{N} \theta_m \vartheta_{m,1}(t) + \sum_{m=1}^{N} \alpha_m \vartheta_{m,1}(t) \sum_{m=1}^{N} \theta_m \vartheta_{m,1}(t) \right) \\ +\theta \left[\mathbf{I}_0 \mathbf{S}_0 \tau \sum_{m=1}^{N} \theta_m \vartheta_{m,1}(t) + \mathbf{I}_0 \tau \sum_{m=1}^{N} \alpha_m \vartheta_{m,1}(t) \sum_{m=1}^{N} \theta_m \vartheta_{m,1}(t) + \mathbf{S}_0 \tau \left(\sum_{m=1}^{N} \theta_m \vartheta_{m,1}(t) \right)^2 \right. \\ \left. +\tau \sum_{m=1}^{N} \alpha_m \vartheta_{m,1}(t) \left(\sum_{m=1}^{N} \theta_m \vartheta_{m,1}(t) \right)^2 \right] + (\varepsilon_1 + \rho + \eta) \mathbf{S}_0 + (\varepsilon_1 + \rho + \eta) \sum_{m=1}^{N} \alpha_m \vartheta_{m,1}(t) \end{cases} = 0 \quad (40)$$

$$\frac{1}{N\Gamma(1-\delta)} \sum_{s=1}^{N} \sum_{m=1}^{N} \gamma_m \tilde{\psi}_m(\varpi)(t-\varpi)^{-\delta} d\varpi - \pi \left(\mathbf{S}_0 + \sum_{m=1}^{N} \alpha_m \vartheta_{m,1}(t) \right) \\ + \psi \left(\mathbf{V}_0 + \sum_{m=1}^{N} \gamma_m \vartheta_{m,1}(t) \right) + (\rho + \mathcal{K}) \left(\mathbf{V}_0 + \sum_{m=1}^{N} \gamma_m \vartheta_{m,1}(t) \right) = 0, \quad (41)$$

$$\begin{cases} \frac{t}{N\Gamma(1-\delta)} \sum_{s=1}^{N} \sum_{m=1}^{N} \theta_m \tilde{\psi}_m(\varpi_s)(t-\varpi_s)^{-\delta} - \theta(1+\tau \mathbf{I}_0) \times \\ \left(\mathbf{I}_0 \mathbf{S}_0 + \mathbf{I}_0 \sum_{m=1}^{N} \alpha_m \vartheta_{m,1}(t) + \mathbf{S}_0 \sum_{m=1}^{N} \theta_m \vartheta_{m,1}(t) + \sum_{m=1}^{N} \alpha_m \vartheta_{m,1}(t) \sum_{m=1}^{N} \theta_m \vartheta_{m,1}(t) \right) \\ +\theta \left[\mathbf{I}_0 \mathbf{S}_0 \tau \sum_{m=1}^{N} \theta_m \vartheta_{m,1}(t) + \mathbf{I}_0 \tau \sum_{m=1}^{N} \alpha_m \vartheta_{m,1}(t) \sum_{m=1}^{N} \theta_m \vartheta_{m,1}(t) + \mathbf{S}_0 \tau \left(\sum_{m=1}^{N} \theta_m \vartheta_{m,1}(t) \right)^2 \right. \\ \left. +\tau \sum_{m=1}^{N} \alpha_m \vartheta_{m,1}(t) \left(\sum_{m=1}^{N} \theta_m \vartheta_{m,1}(t) \right)^2 \right] + (\varepsilon_2 + \rho + \varphi) \mathbf{E}_0 + (\varepsilon_2 + \rho + \varphi) \sum_{m=1}^{N} \theta_m \vartheta_{m,1}(t) \end{cases} = 0, \quad (42)$$

$$\frac{t}{N\Gamma(1-\delta)} \sum_{s=1}^{N} \sum_{m=1}^{N} \lambda_m \tilde{\psi}_m(\varpi_s)(t-\varpi_s)^{-\delta} - \varphi \mathbf{E}_0 - \varphi \left(\sum_{m=1}^{N} \theta_m \vartheta_{m,1}(t) \right) \\ + (\lambda + \epsilon + \rho + \varepsilon_3) \mathbf{I}_0 + (\lambda + \epsilon + \rho + \varepsilon_3) \left(\sum_{m=1}^{N} \lambda_m \vartheta_{m,1}(t) \right) = 0, \quad (43)$$

$$\frac{t}{N\Gamma(1-\delta)} \sum_{s=1}^{N} \sum_{m=1}^{N} \varpi_m \tilde{\psi}_m(\varpi_s)(t-\varpi_s)^{-\delta} - \varepsilon_3 \mathbf{I}_0 - \varepsilon_3 \left(\sum_{m=1}^{N} \lambda_m \vartheta_{m,1}(t) \right) + \varepsilon_2 \mathbf{E}(t)_0 + \varepsilon_2 \left(\sum_{m=1}^{N} \theta_m \vartheta_{m,1}(t) \right) \\ + \varepsilon_1 \mathbf{S}_0 + \varepsilon_1 \left(\sum_{m=1}^{N} \alpha_m \vartheta_{m,1}(t) \right) + (\rho + \sigma) \mathbf{Q}_0 + (\rho + \sigma) \left(\sum_{m=1}^{N} \varpi_m \vartheta_{m,1}(t) \right) = 0, \quad (44)$$

$$\frac{t}{N\Gamma(1-\delta)} \sum_{s=1}^{N} \sum_{m=1}^{N} \zeta_m \tilde{\psi}_m(\varpi_s)(t-\varpi_s)^{-\delta} - \eta \mathbf{S}_0 - \mathbf{S}_0 \left(\sum_{m=1}^{N} \alpha_m \vartheta_{m,1}(t) \right) - \sigma \mathbf{Q}_0 \\ - \sigma \left(\sum_{m=1}^{N} \sigma_m \vartheta_{m,1}(t) \right) - \lambda \mathbf{I}_0 - \lambda \left(\sum_{m=1}^{N} \lambda_m \vartheta_{m,1}(t) \right) + \rho \mathbf{R}_0 + \rho \left(\sum_{m=1}^{N} \zeta_m \vartheta_{m,1}(t) \right) = 0. \quad (45)$$

Let

$$\Phi_{1,j} = \begin{cases} \frac{t}{N\Gamma(1-\delta)} \sum_{s=1}^{N} \sum_{m=1}^{N} \alpha_m \tilde{\psi}_m(\varpi_s)(t-\varpi_s)^{-\delta} - B + \theta(1+\tau \mathbf{I}_0) \times \\ \left(\mathbf{I}_0 \mathbf{S}_0 + \mathbf{I}_0 \sum_{m=1}^{N} \alpha_m \vartheta_{m,1}(t) + \mathbf{S}_0 \sum_{m=1}^{N} \theta_m \vartheta_{m,1}(t) + \sum_{m=1}^{N} \alpha_m \vartheta_{m,1}(t) \sum_{m=1}^{N} \theta_m \vartheta_{m,1}(t) \right) \\ +\theta \left[\mathbf{I}_0 \mathbf{S}_0 \tau \sum_{m=1}^{N} \theta_m \vartheta_{m,1}(t) + \mathbf{I}_0 \tau \sum_{m=1}^{N} \alpha_m \vartheta_{m,1}(t) \sum_{m=1}^{N} \theta_m \vartheta_{m,1}(t) + \mathbf{S}_0 \tau \left(\sum_{m=1}^{N} \theta_m \vartheta_{m,1}(t) \right)^2 \right. \\ \left. +\tau \sum_{m=1}^{N} \alpha_m \vartheta_{m,1}(t) \left(\sum_{m=1}^{N} \theta_m \vartheta_{m,1}(t) \right)^2 \right] + (\varepsilon_1 + \rho + \eta) \mathbf{S}_0 + (\varepsilon_1 + \rho + \eta) \sum_{m=1}^{N} \alpha_m \vartheta_{m,1}(t). \end{cases} \quad (46)$$

Let

$$\Phi_{2,j} = \frac{1}{N\Gamma(1-\delta)} \sum_{s=1}^{N} \sum_{m=1}^{N} \gamma_m \tilde{\psi}_m(\varpi)(t-\varpi)^{-\delta} d\varpi - \pi \left(\mathbf{S}_0 + \sum_{m=1}^{N} \alpha_m \vartheta_{m,1}(t) \right) \\ + \psi \left(\mathbf{V}_0 + \sum_{m=1}^{N} \gamma_m \vartheta_{m,1}(t) \right) + (\rho + \mathcal{K}) \left(\mathbf{V}_0 + \sum_{m=1}^{N} \gamma_m \vartheta_{m,1}(t) \right). \tag{47}$$

Let

$$\Phi_{3,j} = \begin{cases} \frac{t}{N\Gamma(1-\delta)} \sum_{s=1}^{N} \sum_{m=1}^{N} \theta_m \tilde{\psi}_m(\varpi_s)(t-\varpi_s)^{-\delta} - \theta(1+\tau \mathbf{I}_0) \times \\ \left(\mathbf{I}_0 \mathbf{S}_0 + \mathbf{I}_0 \sum_{m=1}^{N} \alpha_m \vartheta_{m,1}(t) + \mathbf{S}_0 \sum_{m=1}^{N} \theta_m \vartheta_{m,1}(t) + \sum_{m=1}^{N} \alpha_m \vartheta_{m,1}(t) \sum_{m=1}^{N} \theta_m \vartheta_{m,1}(t) \right) \\ +\theta \left[\mathbf{I}_0 \mathbf{S}_0 \tau \sum_{m=1}^{N} \theta_m \vartheta_{m,1}(t) + \mathbf{I}_0 \tau \sum_{m=1}^{N} \alpha_m \vartheta_{m,1}(t) \sum_{m=1}^{N} \theta_m \vartheta_{m,1}(t) + \mathbf{S}_0 \tau \left(\sum_{m=1}^{N} \theta_m \vartheta_{m,1}(t) \right)^2 \right. \\ \left. + \tau \sum_{m=1}^{N} \alpha_m \vartheta_{m,1}(t) \left(\sum_{m=1}^{N} \theta_m \vartheta_{m,1}(t) \right)^2 \right] + (\varepsilon_2 + \rho + \varphi) \mathbf{E}_0 + (\varepsilon_2 + \rho + \varphi) \sum_{m=1}^{N} \theta_m \vartheta_{m,1}(t). \end{cases} \tag{48}$$

Let

$$\Phi_{4,j} = \frac{t}{N\Gamma(1-\delta)} \sum_{s=1}^{N} \sum_{m=1}^{N} \lambda_m \tilde{\psi}_m(\varpi_s)(t-\varpi_s)^{-\delta} - \varphi \mathbf{E}_0 - \varphi \left(\sum_{m=1}^{N} \theta_m \vartheta_{m,1}(t) \right) \\ + (\lambda + \epsilon + \rho + \varepsilon_3) \mathbf{I}_0 + (\lambda + \epsilon + \rho + \varepsilon_3) \left(\sum_{m=1}^{N} \lambda_m \vartheta_{m,1}(t) \right). \tag{49}$$

Let

$$\Phi_{5,j} = \frac{t}{N\Gamma(1-\delta)} \sum_{s=1}^{N} \sum_{m=1}^{N} \varpi_m \tilde{\psi}_m(\varpi_s)(t-\varpi_s)^{-\delta} - \varepsilon_3 \mathbf{I}_0 - \varepsilon_3 \left(\sum_{m=1}^{N} \lambda_m \vartheta_{m,1}(t) \right) + \varepsilon_2 \mathbf{E}(t)_0 + \varepsilon_2 \left(\sum_{m=1}^{N} \theta_m \vartheta_{m,1}(t) \right) \\ + \varepsilon_1 \mathbf{S}_0 + \varepsilon_1 \left(\sum_{m=1}^{N} \alpha_m \vartheta_{m,1}(t) \right) + (\rho + \sigma) \mathbf{Q}_0 + (\rho + \sigma) \left(\sum_{m=1}^{N} \varpi_m \vartheta_{m,1}(t) \right). \tag{50}$$

Let

$$\Phi_{6,j} = \frac{t}{N\Gamma(1-\delta)} \sum_{s=1}^{N} \sum_{m=1}^{N} \zeta_m \tilde{\psi}_m(\varpi_s)(t-\varpi_s)^{-\delta} - \eta \mathbf{S}_0 - \mathbf{S}_0 \left(\sum_{m=1}^{N} \alpha_m \vartheta_{m,1}(t) \right) - \sigma \mathbf{Q}_0 \\ - \sigma \left(\sum_{m=1}^{N} \sigma_m \vartheta_{m,1}(t) \right) - \lambda \mathbf{I}_0 - \lambda \left(\sum_{m=1}^{N} \lambda_m \vartheta_{m,1}(t) \right) + \rho \mathbf{R}_0 + \rho \left(\sum_{m=1}^{N} \zeta_m \vartheta_{m,1}(t) \right). \tag{51}$$

The nodal points are placed to create the system of nonlinear algebraic equations shown below:

$$\Phi_{1,j} = \begin{cases} \frac{t_m}{N\Gamma(1-\delta)} \sum_{s=1}^{N} \sum_{m=1}^{N} \alpha_m \tilde{\psi}_m(\varpi_s)(t_m - \varpi_s)^{-\delta} - B + \theta(1+\tau\mathbf{I}_0) \times \\ \left(\mathbf{I}_0 \mathbf{S}_0 + \mathbf{I}_0 \sum_{m=1}^{N} \alpha_m \vartheta_{m,1}(t_m) + \mathbf{S}_0 \sum_{m=1}^{N} \lambda_m \vartheta_{m,1}(t_m) + \sum_{m=1}^{N} \alpha_m \vartheta_{m,1}(t_m) \sum_{m=1}^{N} \lambda_m \vartheta_{m,1}(t_m) \right) \\ +\theta\left[\mathbf{I}_0 \mathbf{S}_0 \tau \sum_{m=1}^{N} \lambda_m \vartheta_{m,1}(t_m) + \mathbf{I}_0 \tau \sum_{m=1}^{N} \alpha_m \vartheta_{m,1}(t_m) \sum_{m=1}^{N} \lambda_m \vartheta_{m,1}(t_m) \right. \\ \left. + \mathbf{S}_0 \tau \left(\sum_{m=1}^{N} \lambda_m \vartheta_{m,1}(t_m)\right)^2 + \tau \sum_{m=1}^{N} \alpha_m \vartheta_{m,1}(t_m) \left(\sum_{m=1}^{N} \lambda_m \vartheta_{m,1}(t_m)\right)^2 \right] \\ +(\varepsilon_1 + \rho + \eta)\mathbf{S}_0 + (\varepsilon_1 + \rho + \eta) \sum_{m=1}^{N} \alpha_m \vartheta_{m,1}(t_m) \end{cases}, \quad (52)$$

$$\Phi_{2,j} = \frac{1}{N\Gamma(1-\delta)} \sum_{s=1}^{N} \sum_{m=1}^{N} \gamma_m \tilde{\psi}_m(\varpi)(t_m - \varpi)^{-\delta} d\varpi - \pi \left(\mathbf{S}_0 + \sum_{m=1}^{N} \alpha_m \vartheta_{m,1}(t_m)\right)$$
$$+ \psi\left(\mathbf{V}_0 + \sum_{m=1}^{N} \gamma_m \vartheta_{m,1}(t_m)\right) + (\rho + \mathcal{K})\left(\mathbf{V}_0 + \sum_{m=1}^{N} \gamma_m \vartheta_{m,1}(t_m)\right). \quad (53)$$

$$\Phi_{3,j} = \begin{cases} \frac{t_m}{N\Gamma(1-\delta)} \sum_{s=1}^{N} \sum_{m=1}^{N} \theta_m \tilde{\psi}_m(\varpi_s)(t_m - \varpi_s)^{-\delta} - B + \theta(1+\tau\mathbf{I}_0) \times \\ \left(\mathbf{I}_0 \mathbf{S}_0 + \mathbf{I}_0 \sum_{m=1}^{N} \alpha_m \vartheta_{m,1}(t_m) + \mathbf{S}_0 \sum_{m=1}^{N} \lambda_m \vartheta_{m,1}(t_m) + \sum_{m=1}^{N} \alpha_m \vartheta_{m,1}(t_m) \sum_{m=1}^{N} \lambda_m \vartheta_{m,1}(t_m) \right) \\ +\theta\left[\mathbf{I}_0 \mathbf{S}_0 \tau \sum_{m=1}^{N} \lambda_m \vartheta_{m,1}(t_m) + \mathbf{I}_0 \tau \sum_{m=1}^{N} \alpha_m \vartheta_{m,1}(t_m) \sum_{m=1}^{N} \lambda_m \vartheta_{m,1}(t_m) \right. \\ \left. + \mathbf{S}_0 \tau \left(\sum_{m=1}^{N} \lambda_m \vartheta_{m,1}(t_m)\right)^2 + \tau \sum_{m=1}^{N} \alpha_m \vartheta_{m,1}(t_m) \left(\sum_{m=1}^{N} \lambda_m \vartheta_{m,1}(t_m)\right)^2 \right] \\ +\psi\left(\mathbf{V}_0 + \sum_{m=1}^{N} \gamma_m \vartheta_{m,1}(t_m)\right) + (\varepsilon_2 + \rho + \varphi) \sum_{m=1}^{N} \theta_m \vartheta_{m,1}(t_m) \end{cases}, \quad (54)$$

$$\Phi_{4,j} = \frac{t_m}{N\Gamma(1-\delta)} \sum_{s=1}^{N} \sum_{m=1}^{N} \lambda_m \tilde{\psi}_m(\varpi_s)(t_m - \varpi_s)^{-\delta} - \varphi \mathbf{E}_0 - \varphi\left(\sum_{m=1}^{N} \theta_m \vartheta_{m,1}(t_m)\right)$$
$$+ (\lambda + \epsilon + \rho + \varepsilon_3)\mathbf{I}_0 + (\lambda + \epsilon + \rho + \varepsilon_3)\left(\sum_{m=1}^{N} \lambda_m \vartheta_{m,1}(t_m)\right). \quad (55)$$

$$\Phi_{5,j} = \frac{t_m}{N\Gamma(1-\delta)} \sum_{s=1}^{N} \sum_{m=1}^{N} \varpi_m \tilde{\psi}_m(\varpi_s)(t_m - \varpi_s)^{-\delta} - \varepsilon_3 \mathbf{I}_0 - \varepsilon_3 \left(\sum_{m=1}^{N} \lambda_m \vartheta_{m,1}(t_m)\right)$$
$$+ \varepsilon_2 \mathbf{E}_0 + \varepsilon_2 \left(\sum_{m=1}^{N} \theta_m \vartheta_{m,1}(t_m)\right) + \varepsilon_1 \mathbf{S}_0 + \varepsilon_1 \left(\sum_{m=1}^{N} \alpha_m \vartheta_{m,1}(t_m)\right) \quad (56)$$
$$+ (\rho + \sigma)\mathbf{Q}_0 + (\rho + \sigma)\left(\sum_{m=1}^{N} \varpi_m \vartheta_{m,1}(t_m)\right).$$

Let

$$\Phi_{6,j} = \frac{t_m}{N\Gamma(1-\delta)} \sum_{s=1}^{N} \sum_{m=1}^{N} \zeta_m \tilde{\psi}_m(\varpi_s)(t_m - \varpi_s)^{-\delta} - \eta \mathbf{S}_0 - \mathbf{S}_0 \left(\sum_{m=1}^{N} \alpha_m \vartheta_{m,1}(t_m)\right) - \sigma \mathbf{Q}_0$$
$$- \sigma\left(\sum_{m=1}^{N} \sigma_m \vartheta_{m,1}(t_m)\right) - \lambda \mathbf{I}_0 - \lambda\left(\sum_{m=1}^{N} \lambda_m \vartheta_{m,1}(t_m)\right) + \rho \mathbf{R}_0 \quad (57)$$
$$+ \rho\left(\sum_{m=1}^{N} \zeta_m \vartheta_{m,1}(t_m)\right).$$

135

Using Broyden's approach, this system is resolved. The Jacobian is given by

$$\mathbf{J} = [J_{jp}]_{6N \times 6N} \qquad (58)$$

where

$$\begin{cases} \dfrac{\partial \Phi_{1,j}}{\partial \alpha_m} = \begin{cases} \dfrac{t_m}{N\Gamma(1-\delta)} \sum_{s=1}^{N} \sum_{m=1}^{N} \tilde{\psi}_m(\varpi_s)(t_m - \varpi_s)^{-\delta} + \theta(1+\tau \mathbf{I}_0) \times \\ \left(+\mathbf{I}_0 \sum_{m=1}^{N} \vartheta_{m,1}(t_m) + \sum_{m=1}^{N} \vartheta_{m,1}(t_m) \sum_{m=1}^{N} \theta_m \vartheta_{m,1}(t_m) \right) \\ +\theta \left[+\mathbf{I}_0 \tau \sum_{m=1}^{N} \vartheta_{m,1}(t_m) \sum_{m=1}^{N} \lambda_m \vartheta_{m,1}(t_m) + \tau \sum_{m=1}^{N} \vartheta_{m,1}(t_m) \left(\sum_{m=1}^{N} \lambda_m \vartheta_{m,1}(t_m) \right)^2 \right] \\ +(\varepsilon_1 + \rho + \eta) \sum_{m=1}^{N} \vartheta_{m,1}(t_m) \end{cases} \\ \dfrac{\partial \Phi_{1,j}}{\partial \gamma_m} = \sum_{m=1}^{N} \vartheta_{m,1}(t_m), \quad \dfrac{\partial \Phi_{1,j}}{\partial \theta_m} = 0, \quad \dfrac{\partial \Phi_{1,j}}{\partial \varpi_m} = 0, \quad \dfrac{\partial \Phi_{1,j}}{\partial \sigma_m} = 0, \\ \dfrac{\partial \Phi_{1,j}}{\partial \lambda_m} = \begin{cases} \theta(1+\tau \mathbf{I}_0) \times \left(\mathbf{S}_0 \sum_{m=1}^{N} \vartheta_{m,1}(t_m) + \sum_{m=1}^{N} \alpha_m \vartheta_{m,1}(t_m) \sum_{m=1}^{N} \vartheta_{m,1}(t_m) \right) \\ +\theta \left[\mathbf{I}_0 \mathbf{S}_0 \tau \sum_{m=1}^{N} \vartheta_{m,1}(t_m) + \mathbf{I}_0 \tau \sum_{m=1}^{N} \alpha_m \vartheta_{m,1}(t_m) \sum_{m=1}^{N} \vartheta_{m,1}(t_m) \right. \\ \left. +2\lambda_m \mathbf{S}_0 \tau \left(\sum_{m=1}^{N} \vartheta_{m,1}(t_m) \right)^2 + \tau \sum_{m=1}^{N} \alpha_m \vartheta_{m,1}(t_m) \left(\sum_{m=1}^{N} \vartheta_{m,1}(t_m) \right)^2 \right] \end{cases} \end{cases} \qquad (59)$$

$$\dfrac{\partial \Phi_{2,j}}{\partial \alpha_m} = -\pi \left(\sum_{m=1}^{N} \alpha_m \vartheta_{m,1}(t_m) \right), \quad \dfrac{\partial \Phi_{2,j}}{\partial \theta_m} = 0, \quad \dfrac{\partial \Phi_{2,j}}{\partial \lambda_m} = 0, \quad \dfrac{\partial \Phi_{1,j}}{\partial \varpi_m} = 0, \quad \dfrac{\partial \Phi_{1,j}}{\partial \sigma_m} = 0,$$

$$\dfrac{\partial \Phi_{2,j}}{\partial \gamma_m} = \dfrac{1}{N\Gamma(1-\delta)} \sum_{s=1}^{N} \sum_{m=1}^{N} \gamma_m \tilde{\psi}_m(\varpi)(t_m - \varpi)^{-\delta} d\varpi + \psi \left(\mathbf{V}_0 + \sum_{m=1}^{N} \vartheta_{m,1}(t_m) \right) + (\rho + \mathcal{K}) \left(\mathbf{V}_0 + \sum_{m=1}^{N} \vartheta_{m,1}(t_m) \right).$$

$$(60)$$

$$\begin{cases} \dfrac{\partial \Phi_{3,j}}{\partial \alpha_m} = \begin{cases} \theta(1+\tau \mathbf{I}_0) \times \left(+\mathbf{I}_0 \sum_{m=1}^{N} \vartheta_{m,1}(t_m) + \sum_{m=1}^{N} \vartheta_{m,1}(t_m) \sum_{m=1}^{N} \theta_m \vartheta_{m,1}(t_m) \right) \\ +\theta \left[+\mathbf{I}_0 \tau \sum_{m=1}^{N} \vartheta_{m,1}(t_m) \sum_{m=1}^{N} \lambda_m \vartheta_{m,1}(t_m) + \tau \sum_{m=1}^{N} \vartheta_{m,1}(t_m) \left(\sum_{m=1}^{N} \lambda_m \vartheta_{m,1}(t_m) \right)^2 \right] \end{cases} \\ \dfrac{\partial \Phi_{3,j}}{\partial \gamma_m} = \sum_{m=1}^{N} \vartheta_{m,1}(t_m), \quad \dfrac{\partial \Phi_{3,j}}{\partial \theta_m} = \dfrac{t_m}{N\Gamma(1-\delta)} \sum_{s=1}^{N} \sum_{m=1}^{N} \tilde{\psi}_m(\varpi_s)(t_m - \varpi_s)^{-\delta} + (\varepsilon_2 + \rho + \varphi) \sum_{m=1}^{N} \vartheta_{m,1}(t_m), \\ \dfrac{\partial \Phi_{3,j}}{\partial \varpi_m} = 0, \quad \dfrac{\partial \Phi_{3,j}}{\partial \sigma_m} = 0, \\ \dfrac{\partial \Phi_{3,j}}{\partial \lambda_m} = \begin{cases} \theta(1+\tau \mathbf{I}_0) \times \left(\mathbf{S}_0 \sum_{m=1}^{N} \vartheta_{m,1}(t_m) + \sum_{m=1}^{N} \alpha_m \vartheta_{m,1}(t_m) \sum_{m=1}^{N} \vartheta_{m,1}(t_m) \right) \\ +\theta \left[\mathbf{I}_0 \mathbf{S}_0 \tau \sum_{m=1}^{N} \vartheta_{m,1}(t_m) + \mathbf{I}_0 \tau \sum_{m=1}^{N} \alpha_m \vartheta_{m,1}(t_m) \sum_{m=1}^{N} \vartheta_{m,1}(t_m) \right. \\ \left. +2\lambda_m \mathbf{S}_0 \tau \left(\sum_{m=1}^{N} \vartheta_{m,1}(t_m) \right)^2 + \tau \sum_{m=1}^{N} \alpha_m \vartheta_{m,1}(t_m) \left(\sum_{m=1}^{N} \vartheta_{m,1}(t_m) \right)^2 \right] \end{cases} \end{cases} \qquad (61)$$

$$\dfrac{\partial \Phi_{4,j}}{\partial \alpha_m} = 0, \quad \dfrac{\partial \Phi_{4,j}}{\partial \gamma_m} = 0, \quad \dfrac{\partial \Phi_{4,j}}{\partial \varpi_m} = 0, \quad \dfrac{\partial \Phi_{4,j}}{\partial \sigma_m} = 0, \quad \dfrac{\partial \Phi_{4,j}}{\partial \theta_m} = -\varphi \left(\sum_{m=1}^{N} \theta_m \vartheta_{m,1}(t_m) \right),$$

$$\dfrac{\partial \Phi_{4,j}}{\partial \lambda_m} = \dfrac{t_m}{N\Gamma(1-\delta)} \sum_{s=1}^{N} \sum_{m=1}^{N} \tilde{\psi}_m(\varpi_s)(t_m - \varpi_s)^{-\delta} (\lambda + \epsilon + \rho + \varepsilon_3) \left(\sum_{m=1}^{N} \vartheta_{m,1}(t_m) \right).$$

$$(62)$$

$$\dfrac{\partial \Phi_{5,j}}{\partial \alpha_m} = \varepsilon_1 \left(\sum_{m=1}^{N} \vartheta_{m,1}(t_m) \right), \quad \dfrac{\partial \Phi_{5,j}}{\partial \gamma_m} = 0, \quad \dfrac{\partial \Phi_{5,j}}{\partial \varpi_m} = \dfrac{t_m}{N\Gamma(1-\delta)} \sum_{s=1}^{N} \sum_{m=1}^{N} \varpi_m \tilde{\psi}_m(\varpi_s)(t_m - \varpi_s)^{-\delta},$$

$$+ (\rho + \sigma) \left(\sum_{m=1}^{N} \varpi_m \vartheta_{m,1}(t_m) \right) \quad \dfrac{\partial \Phi_{5,j}}{\partial \sigma_m} = 0, \quad \dfrac{\partial \Phi_{5,j}}{\partial \theta_m} = \varepsilon_2 \left(\sum_{m=1}^{N} \theta_m \vartheta_{m,1}(t_m) \right),$$

$$\dfrac{\partial \Phi_{5,j}}{\partial \lambda_m} = -\varepsilon_3 \left(\sum_{m=1}^{N} \vartheta_{m,1}(t_m) \right).$$

$$(63)$$

$$\frac{\partial \Phi_{6,j}}{\partial \alpha_m} = \mathbf{S}_0 \left(\sum_{m=1}^{N} \vartheta_{m,1}(t_m) \right), \quad \frac{\partial \Phi_{6,j}}{\partial \gamma_m} = 0, \quad \frac{\partial \Phi_{6,j}}{\partial \varpi_m} = \frac{t_m}{N\Gamma(1-\delta)} \sum_{s=1}^{N} \sum_{m=1}^{N} \tilde{\psi}_m(\varpi_s)(t_m - \varpi_s)^{-\delta},$$

$$+ (\rho + \sigma) \left(\sum_{m=1}^{N} \vartheta_{m,1}(t_m) \right) \quad \frac{\partial \Phi_{6,j}}{\partial \theta_m} = 0, \frac{\partial \Phi_{6,j}}{\partial \lambda_m} = -\lambda \left(\sum_{m=1}^{N} \lambda_m \vartheta_{m,1}(t_m) \right). \quad (64)$$

$$\frac{\partial \Phi_{6,j}}{\partial \sigma_m} = \frac{t_m}{N\Gamma(1-\delta)} \sum_{s=1}^{N} \sum_{m=1}^{N} \tilde{\psi}_m(\varpi_s)(t_m - \varpi_s)^{-\delta} - \sigma \left(\sum_{m=1}^{N} \sigma_m \vartheta_{m,1}(t_m) \right) + \rho \left(\sum_{m=1}^{N} \vartheta_{m,1}(t_m) \right).$$

This system's solution yields the values of α_m's, γ_m's, λ_m's, ϖ_m's, and σ_m's unknown coefficients. By entering α_m's, γ_m's, λ_m's, ϖ_m's, and σ_m's unknown coefficients into Equation (29), it is possible to calculate the necessary solutions $\mathbf{S}(t), \mathbf{V}(t), \mathbf{E}(t), \mathbf{I}(t), \mathbf{Q}(t)$, and $\mathbf{R}(t)$ at nodal locations. The experimental rate of convergence, denoted by the formula $r_\varrho(N)$ [53], can be calculated as follows:

$$\mathbf{r}_\varrho(N) = \frac{1}{\log 2} \log \left[\frac{\text{Maximum absolute error at } \frac{N}{2}}{\text{Maximum absolute error at } N} \right]. \quad (65)$$

Graphical Results

This section presents graphical results for the fractional-order model (14). The fractional model was numerically solved based on the method outlined in [47,54] and using the information from Table 1. The resulting figures, namely Figures 4–7, illustrate the dynamics of the susceptible $\mathbf{S}(t)$, vaccinated $\mathbf{V}(t)$, exposed $\mathbf{E}(t)$, infected $\mathbf{I}(t)$, quarantined $\mathbf{Q}(t)$, and recovered $\mathbf{R}(t)$ individuals. Figure 5a shows that susceptible individuals are characterized by fractional-order derivatives ranging between 0.55 and 0.95, and as time progresses, the number of susceptible individuals decreases due to exposure to the virus. This behavior is expected and observed in other epidemiological models. The population of vaccinated individuals is shown in Figure 5b, and it grows both steadily and quickly as the fractional-order derivative approaches its classical counterpart. Similarly, the population of exposed individuals is shown in Figure 5c, and it greatly decreases in the first 15 days while both steadily and quickly increasing as the fractional-order derivative approaches its classical counterpart. This increase is brought on by more susceptible people becoming infected during the first few weeks of the outbreak and joining the exposed class. An increased risk of transmission during the early stages of the epidemic may be indicated by the rise in exposed individuals. Figure 5c displays the number of infected people, which rises as the fractional order gets closer to one. The fractional order, which becomes more sensitive as it gets closer to one, is what is causing this increase. Within a few weeks of exposure, the majority of people in the quarantined and infectious stages of the infection leave the exposed class. Figure 5d illustrates the dynamics of the quarantined individuals, who exhibit a similar behavior as the exposed population. The population of exposed individuals increases as the fractional-order derivative approaches the integer order. Figure 5e shows how the fractional order affects the number of recovered individuals, which grows steadily as the fractional-order derivative approaches the classical value. This is due to the infected individuals recovering, which aids in disease containment. Raising the fractional order can cause the population of the recovered class to grow considerably more quickly. These results suggest that the exposed population is similarly affected by the fractional order regardless of the strain of infection.

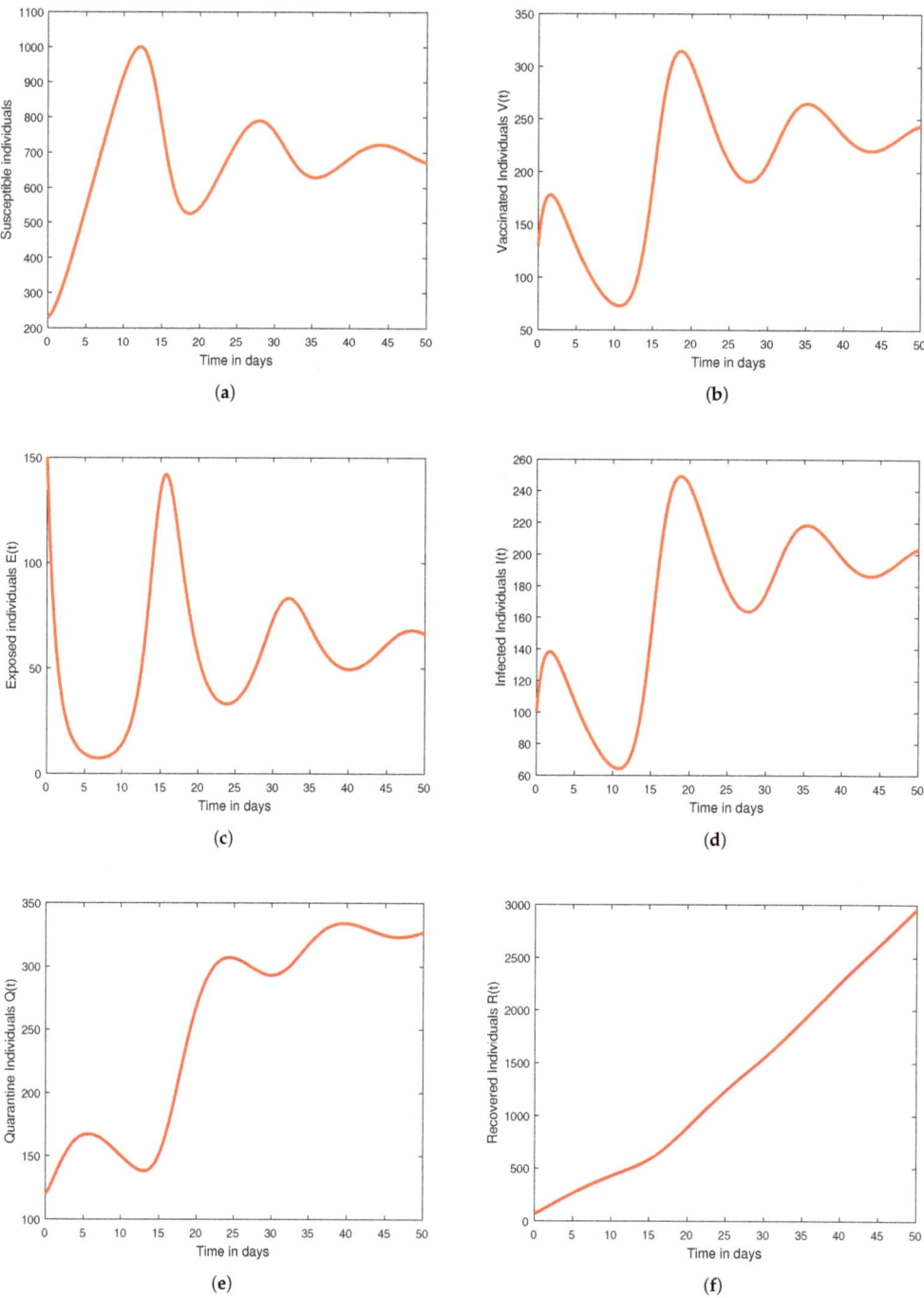

Figure 4. The caputo version of the fractional model's behavior for each state variable is depicted in the figure at $\delta = 0.95$. Furthermore, the figures (**a**–**f**) represents susceptible people $S(t)$, vaccinated people $V(t)$, exposed people $E(t)$, infectious people $I(t)$, quarantined people $Q(t)$, and recovered people $R(t)$.

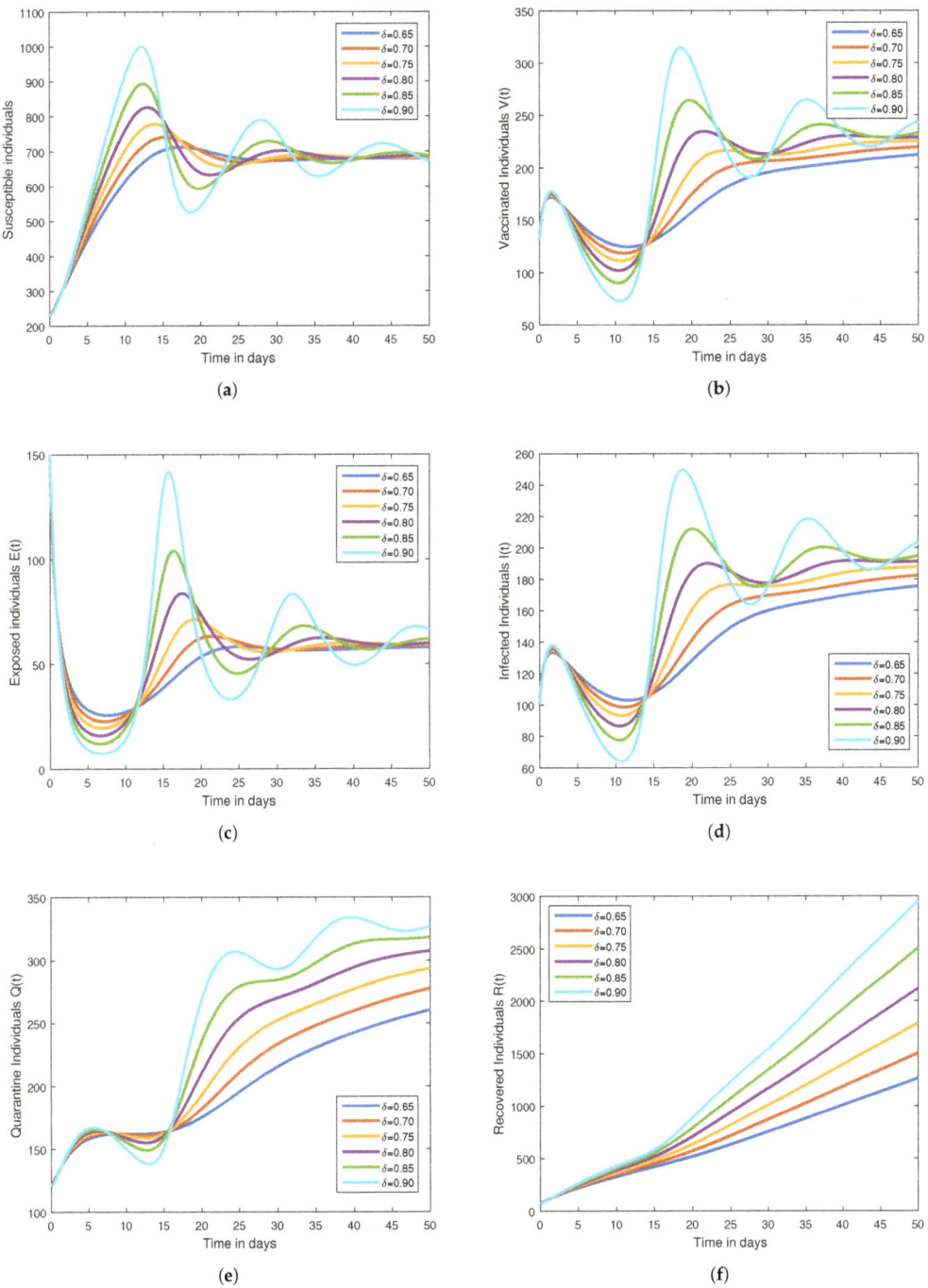

Figure 5. The caputo version of the fractional model's behavior for each state variable is depicted using the parameter values in the figure. Furthermore, the figures (**a**–**f**) represents susceptible people $S(t)$, vaccinated people $V(t)$, exposed people $E(t)$, infectious people $I(t)$, quarantined people $Q(t)$, and recovered people $R(t)$.

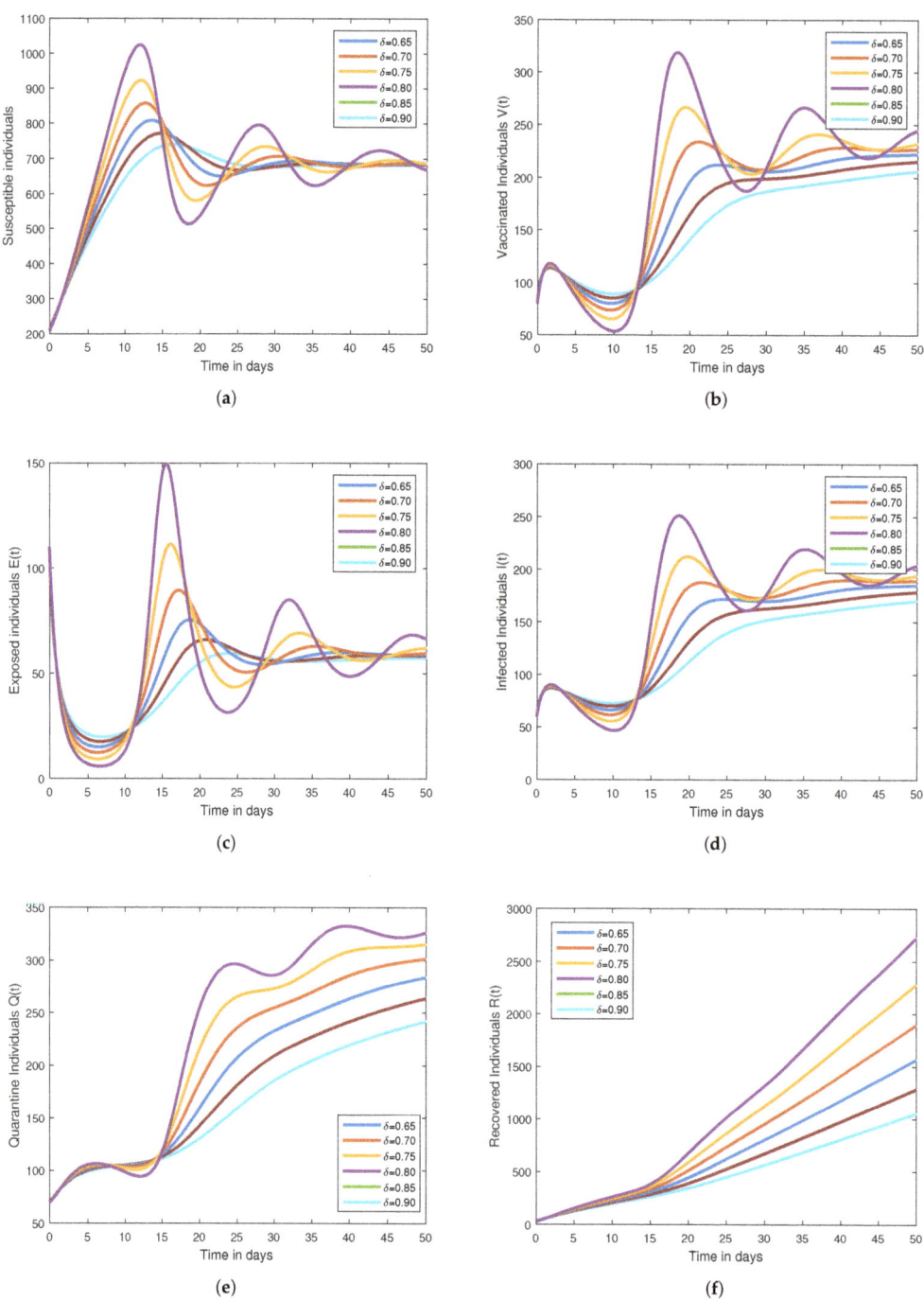

Figure 6. The behavior of each state variable for the Caputo fractional model utilizing a different set of initial conditions is depicted in the figure. Furthermore, the figures (**a**–**f**) represents susceptible people $S(t)$, vaccinated people $V(t)$, exposed people $E(t)$, infectious people $I(t)$, quarantined people $Q(t)$, and recovered people $R(t)$.

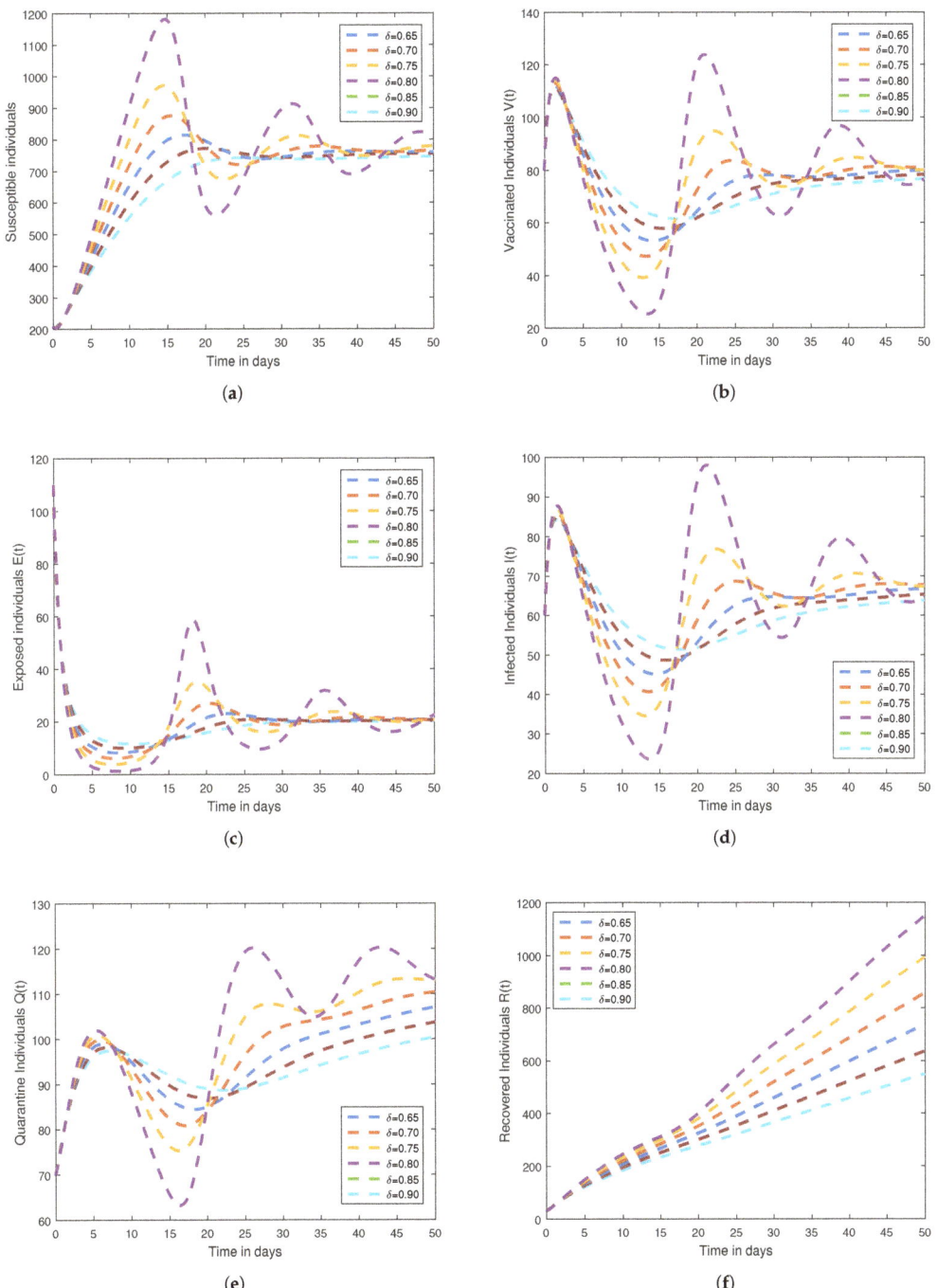

Figure 7. The behavior of each state variable is depicted in the figure for $\theta = 0.01815$. Furthermore, the figures (**a**–**f**) represents susceptible people $\mathbf{S}(t)$, vaccinated people $\mathbf{V}(t)$, exposed people $\mathbf{E}(t)$, infectious people $\mathbf{I}(t)$, quarantined people $\mathbf{Q}(t)$, and recovered people $\mathbf{R}(t)$.

8. Conclusions

The $SVEIQR$ COVID-19 epidemic model under fractional derivatives in the Caputo sense has been studied in this article. Some findings from fixed point theory are taken into consideration when determining the existence and uniqueness of the proposed model. This study simulates the spread of the COVID-19 Omicron variant in Pakistan using fractional-order models and the Haar wavelet collocation method. The study makes use of actual data from Pakistan, which improves the model's precision and applicability. To determine the factors that most influence the spread of the virus, a sensitivity analysis was also carried out. This study sheds light on the dynamics of the pandemic and emphasizes the significance of taking fractional order into account when modeling infectious diseases. In terms of future directions, this study can be extended to incorporate additional elements such as vaccination rates, travel restrictions, and socio-economic conditions. To evaluate the efficacy of various intervention strategies and compare the effects of various COVID-19 variants, the model can also be applied to other regions or nations. Additionally, this study can be expanded to assess the effects of various compliance levels with public health measures, such as mask use and social distancing. To determine the most effective and precise method, this study can be expanded to compare the performance of various numerical solution techniques for fractional-order models.

Author Contributions: Conceptualization, R.Z.; Software, Z.R.; Formal analysis, R.Z.; Investigation, Z.R.; Writing—original draft, R.Z.; Supervision, Z.R.; Project administration, Z.R. All authors have read and agreed to the published version of the manuscript.

Funding: The authors extend their appreciation to the Ministry of Education in KSA for funding this research work through project number KKU-IFP2-H-9.

Data Availability Statement: Data sharing is not applicable to this article, as no data sets were generated during the current study.

Acknowledgments: The acknowledgement to the Ministry of Education of the Kingdom of Saudi Arabia for the funding of this research work by project number KKU-IFP2-H-9.

Conflicts of Interest: The authors declare no conflict of interest.

References

1. Omrani, A.S.; Al-Tawfiq, J.A.; Memish, Z.A. Middle East respiratory syndrome coronavirus (MERS-CoV): Animal to human interaction. *Pathog. Glob. Health* **2015**, *109*, 354–362. [CrossRef]
2. World Health Organization. Weekly Epidemiological Update on COVID-19—11 January 2023. Available online: https://www.who.int/publications/m/item/weekly-epidemiological-update-on-covid-19---11-january-2023 (accessed on 11 January 2023).
3. Tay, M.Z.; Poh, C.M.; Rénia, L.; MacAry, P.A.; Ng, L.F. The trinity of COVID-19: Immunity, inflammation and intervention. *Nat. Rev. Immunol.* **2020**, *20*, 363–374. [CrossRef] [PubMed]
4. Backer, J.A.; Klinkenberg, D.; Wallinga, J. Incubation period of 2019 novel coronavirus (2019-nCoV) infections among travellers from Wuhan, China, 20–28 January 2020. *Eurosurveillance* **2020**, *25*, 2000062. [CrossRef] [PubMed]
5. Kronbichler, A.; Kresse, D.; Yoon, S.; Lee, K.H.; Effenberger, M.; Shin, J.I. Asymptomatic patients as a source of COVID-19 infections: A systematic review and meta-analysis. *Int. J. Infect. Dis.* **2020**, *98*, 180–186. [CrossRef]
6. Khan, A.; Zarin, R.; Hussain, G.; Ahmad, N.A.; Mohd, M.H.; Yusuf, A. Stability analysis and optimal control of covid-19 with convex incidence rate in Khyber Pakhtunkhawa (Pakistan). *Results Phys.* **2021**, *20*, 103703. [CrossRef]
7. Khan, A.; Zarin, R.; Akgul, A.; Saeed, A.; Gul, T. Fractional optimal control of COVID-19 pandemic model with generalized Mittag-Leffler function. *Adv. Differ. Equ.* **2021**, *2021*, 387. [CrossRef]
8. Alqarni, M.S.; Alghamdi, M.; Muhammad, T.; Alshomrani, A.S.; Khan, M.A. Mathematical modeling for novel coronavirus (COVID-19) and control. *Numer. Methods Partial. Differ. Equ.* **2022**, *38*, 760–776. [CrossRef] [PubMed]
9. Khan, A.; Zarin, R.; Khan, S.; Saeed, A.; Gul, T.; Humphries, U.W. Fractional dynamics and stability analysis of COVID-19 pandemic model under the harmonic mean type incidence rate. *Comput. Methods Biomech. Biomed. Eng.* **2022**, *25*, 619–640 [CrossRef]
10. Kucharski, A.J.; Russell, T.W.; Diamond, C.; Liu, Y.; Edmunds, J.; Funk, S.; Eggo, R.M. Early dynamics of transmission and control of COVID-19: A mathematical modelling study. *Lancet Infect. Dis.* **2020**, *20*, 553–558. [CrossRef]

11. Ferguson, N.M.; Laydon, D.; Nedjati-Gilani, G.; Imai, N.; Ainslie, K.; Baguelin, M.; Bhatia, S.; Boonyasiri, A.; Cucunuba, Z.; Cuomo-Dannenburg, G.; et al. Impact of Non-Pharmaceutical Interventions (NPIs) to Reduce COVID-19 Mortality and Healthcare Demand. Imperial College COVID-19 Response Team. 2020. Available online: https://www.imperial.ac.uk/media/imperial-college/medicine/sph/ide/gida-fellowships/Imperial-College-COVID19-NPI-modelling-16-03-2020.pdf (accessed on 16 March 2020).
12. Marathe, A.; Lewis, B.; Chen, J. COVID-19: Understanding the spread of infectious diseases. *Nat. Rev. Phys.* **2020**, *2*, 447–456. [CrossRef]
13. Hethcote, H.W.; Shuai, Z.; Van den Driessche, P. COVID-19 transmission dynamics in the United States: A mathematical model with a realistic age structure. *Math. Biosci. Eng.* **2021**, *18*, 2672–2690. [CrossRef]
14. Yopadhyay, A.; Nabar, N.R.; Salathé, M. A review of data-driven epidemiological models of infectious diseases. In *Global Dynamics of Infectious Diseases: Impact of Social Heterogeneity*; Springer: Berlin, Germany, 2021; pp. 19–41.
15. Liu, P.; Huang, X.; Zarin, R.; Cui, T.; Din, A. Modeling and numerical analysis of a fractional order model for dual variants of SARS-CoV-2. *Alex. Eng. J.* **2023**, *65*, 427–442. [CrossRef]
16. Daniloski, Z.; Guo, X.; Sanjana, N.E. The D614G mutation in SARS-CoV-2 spike increases transduction of multiple human cell types. *Nat. Commun.* **2021**, *12*, 1–9. [CrossRef]
17. Saberi, M.; Moshksayan, K.; Barati, M.; Soleymani, F.; Eftekhari, P. Modeling and analysis of COVID-19 infection dynamics with fractional-order derivatives. *Chaos Solitons Fractals* **2021**, *146*, 110844. [CrossRef]
18. Goyal, M.; Baskonus, H.M.; Prakash, A. An efficient technique for a time fractional model of lassa hemorrhagic fever spreading in pregnant women. *Eur. Phys. J. Plus* **2019**, *134*, 482. [CrossRef]
19. Gao, W.; Veeresha, P.; Prakasha, D.G.; Baskonus, H.M.; Yel, G. New approach for the model describing the deathly disease in pregnant women using Mittag-Leffler function. *Chaos Solitons Fractals* **2020**, *134*, 109696. [CrossRef]
20. Alqahtani, R.T.; Ahmad, S.; Akgül, A. Dynamical analysis of bio-ethanol production model under generalized nonlocal operator in Caputo sense. *Mathematics* **2021**, *9*, 2370. [CrossRef]
21. Agarwal, P.; Singh, R. Modelling of transmission dynamics of Nipah virus (Niv): A fractional order approach. *Phys. A Stat. Mech. Its Appl.* **2020**, *547*, 124243. [CrossRef]
22. Zarin, R.; Khan, A.; Yusuf, A.; Abdel-Khalek, S.; Mustafa Inc. Analysis of fractional COVID-19 epidemic model under Caputo operator. *Math. Methods Appl. Sci.* **2021**, *6*, 115–122. [CrossRef]
23. Zarin, R.; Khan, A.; Kumar, P. Fractional-order dynamics of Chagas-HIV epidemic model with different fractional operators. *AIMS Math.* **2022**, *7*, 18897–18924. [CrossRef]
24. Baleanu, D.; Fernez, A.; Akgül, A. On a fractional operator combining proportional and classical differintegrals. *Mathematics* **2020**, *8*, 360. [CrossRef]
25. Caputo, M.; Fabrizio, M. A new definition of fractional derivative without singular kernel. *Prog. Fract. Differ. Appl.* **2015**, *1*, 73–85.
26. Atangana, A.; Baleanu, D. New fractional derivatives with nonlocal and non-singular kernel: Theory and application to heat transfer model. *arXiv* **2016**, arXiv:1602.03408.
27. Andrew, O.; Abbas, M.; Abdel-Aty, A.-H. Assessing the impact of SARS-CoV-2 infection on the dynamics of dengue and HIV via fractional derivatives. *Chaos Solitons Fractals* **2022**, *162*, 112427.
28. Omame, A.; Abbas, M.; Onyenegecha, C.P. Backward bifurcation and optimal control in a co-infection model for SARS-CoV-2 and ZIKV. *Results Phys.* **2022**, *37*, 105481. [CrossRef]
29. Agarwal, P.; Choi, J.; Paris, R.B. Extended Riemann-Liouville fractional derivative operator and its applications. *J. Nonlinear Sci. Appl. (JNSA)* **2015**, *8*, 451–466. [CrossRef]
30. Zarin, R.; Khan, A.; Inc, M.; Humphries, U.W.; Karite, T. Dynamics of five grade leishmania epidemic model using fractional operator with Mittag–Leffler kernel. *Chaos Solitons Fractals* **2021**, *147*, 110985. [CrossRef]
31. Agarwal, P.; Choi, J. Fractional calculus operators and their image formulas. *J. Korean Math. Soc.* **2016**, *53*, 1183–1210. [CrossRef]
32. Atangana, A. Non validity of index law in fractional calculus: A fractional differential operator with Markovian and non-Markovian properties. *Phys. A Stat. Mech. Its Appl.* **2018**, *505*, 688–706. [CrossRef]
33. Zarin, R. Modeling and numerical analysis of fractional order hepatitis B virus model with harmonic mean type incidence rate. *Comput. Methods Biomech. Biomed. Eng.* **2022**, 1–16. [CrossRef] [PubMed]
34. Zarin, R.; Ahmed, I.; Kumam, P.; Zeb, A.; Din, A. Fractional modeling and optimal control analysis of rabies virus under the convex incidence rate. *Results Phys.* **2021**, *28*, 104665. [CrossRef]
35. Ahmad, S.; Ullah, A.; Akgül, A.; Baleanu, D. Analysis of the fractional tumour-immune-vitamins model with Mittag–Leffler kernel. *Results Phys.* **2020**, *19*, 103559. [CrossRef]
36. Atangana, A. A novel Covid-19 model with fractional differential operators with singular and non-singular kernels: Analysis and numerical scheme based on Newton polynomial. *Alexandria Eng. J.* **2021**, *60*, 3781–3806. [CrossRef]
37. Bansal, K.; Arora, S.; Pritam, K.S.; Mathur, T.; Agarwal, S. Dynamics of Crime Transmission Using Fractional-Order Differential Equations. *Fractals* **2022**, *30*, 2250012. [CrossRef]
38. Pritam, K.S.; Mathur, T.; Agarwal, S. Underlying dynamics of crime transmission with memory. *Chaos Solitons Fractals* **2021**, *146*, 110838. [CrossRef]
39. Rahman, M.U.; Ahmad, S.; Arfan, M.; Akgül, A.; Jarad, F. Fractional Order Mathematical Model of Serial Killing with Different Choices of Control Strategy. *Fractal Fract.* **2022**, *6*, 162. [CrossRef]

40. Zhi, S.; Deng, L.-Y.; Qing, J.C. Numerical Solution of Differential Equations by Using Haar Wavelets. In Proceedings of the International Conference on Wavelet Analysis and Pattern Recognition, Beijing, China, 2–4 November 2007; pp. 1039–1044.
41. Shah, K.; Khan, Z.A.; Ali, A.; Amin, R.; Khan, H.; Khan, A. Haar wavelet collocation approach for the solution of fractional order COVID-19 model using Caputo derivative. *Alex. Eng. J.* **2020**, *59*, 3221–3231. [CrossRef]
42. Prakash, B.; Setia, A.; Alapatt, D. Numerical solution of nonlinear fractional SEIR epidemic model by using Haar wavelets. *J. Comput. Sci.* **2017**, *22*, 109–118. [CrossRef]
43. Kumar, D.; Singh, J.; Baleanu, D. A new analysis of the Fornberg-Whitham equation pertaining to a fractional derivative with Mittag-Leffler-type kernel. *Eur. Phys. J. Plus* **2018**, *133*, 70. [CrossRef]
44. Kumar, D.; Singh, J.; Purohit, S.D.; Swroop, R. A hybrid analytical algorithm for nonlinear fractional wave-like equations. *Math. Model. Nat. Phenom.* **2019**, *14*, 304. [CrossRef]
45. Caputo, M.; Mainardi, F. A new dissipation model based on memory mechanism. *Pure Appl. Geophys.* **1971**, *91*, 134–147. [CrossRef]
46. Chen, Y.; Yi, M.; Yu, C. Error analysis for numerical solution of fractional differential equation by Haar wavelets method. *J. Comput. Sci.* **2012**, *3*, 367–373. [CrossRef]
47. Lepik, Ü.; Hein, H. Haar wavelets. In *Haar Wavelets*; Springer: Cham, Switzerland, 2014; pp. 7–20.
48. Van den Driessche, P.; Watmough, J. Reproduction number and sub-threshold endemic equilbria for compartmental models of disease transmission. *Math. Biosci.* **2002**, *180*, 29–38. [CrossRef]
49. Taylor, A.E.; Lay, D.C. *Introduction to Functional Analysis*; Wiley: New York, NY, USA, 1958; Volume 1.
50. Available online: https://www.who.int/countries/pak/ (accessed on 23 August 2022).
51. Mandal, M.; Jana, S.; Nandi, S.K.; Khatua, A.; Adak, S.; Kar, T.K. A model based study on the dynamics of COVID-19: Prediction and control. *Chaos Solitons Fractals* **2020**, *136*, 109889. [CrossRef]
52. Li, Y.; Zhao, W. Haar wavelet operational matrix of fractional order integration and its applications in solving the fractional order differential equations. *Appl. Math. Comp.* **2010**, *216*, 2276–2285. [CrossRef]
53. Majak, J.; Shvartsman, B.; Karjust, K.; Mikola, M.; Haavajõe, A.; Pohlak, M. On the accuracy of the Haar wavelet discretization method. *Compos. Part B Eng.* **2015**, *80*, 321–327. [CrossRef]
54. Zarin, R.; Khaliq, H.; Khan, A.; Ahmed, I.; Humphries, U.W. A Numerical Study Based on Haar Wavelet Collocation Methods of Fractional-Order Antidotal Computer Virus Model. *Symmetry* **2023**, *15*, 621. [CrossRef]

Disclaimer/Publisher's Note: The statements, opinions and data contained in all publications are solely those of the individual author(s) and contributor(s) and not of MDPI and/or the editor(s). MDPI and/or the editor(s) disclaim responsibility for any injury to people or property resulting from any ideas, methods, instructions or products referred to in the content.

Article

Differential and Time-Discrete SEIRS Models with Vaccination: Local Stability, Validation and Sensitivity Analysis Using Bulgarian COVID-19 Data

Svetozar Margenov [1], Nedyu Popivanov [1,2,*], Iva Ugrinova [3] and Tsvetan Hristov [2]

1. Institute of Information and Communication Technologies, Bulgarian Academy of Sciences, 1113 Sofia, Bulgaria; margenov@parallel.bas.bg
2. Faculty of Mathematics and Informatics, Sofia University "St. Kliment Ohridski", 1164 Sofia, Bulgaria; tsvetan@fmi.uni-sofia.bg
3. Institute of Molecular Biology, Bulgarian Academy of Sciences, 1113 Sofia, Bulgaria; ugryiva@gmail.com
* Correspondence: nedyu@parallel.bas.bg

Abstract: Bulgaria has the lowest COVID-19 vaccination rate in the European Union and the second-highest COVID-19 mortality rate in the world. That is why we think it is important better to understand the reason for this situation and to analyse the development of the disease over time. In this paper, an extended time-dependent SEIRS model SEIRS-VB is used to investigate the long-term behaviour of the COVID-19 epidemic. This model includes vaccination and vital dynamics. To apply the SEIRS-VB model some numerical simulation tools have been developed and for this reason a family of time-discrete variants are introduced. Suitable inverse problems for the identification of parameters in discrete models are solved. A methodology is proposed for selecting a discrete model from the constructed family, which has the closest parameter values to these in the differential SEIRS-VB model. To validate the studied models, Bulgarian COVID-19 data are used. To obtain all these results for the discrete models a mathematical analysis is carried out to illustrate some biological properties of the differential model SEIRS-VB, such as the non-negativity, boundedness, existence, and uniqueness. Using the next-generation method, the basic reproduction number associated with the model in the autonomous case is defined. The local stability of the disease-free equilibrium point is studied. Finally, a sensitivity analysis of the basic reproduction number is performed.

Keywords: COVID-19 epidemic; Cauchy problem; non-linear ordinary differential equations; time-discrete models; basic reproduction number; stability analysis; sensitivity analysis

MSC: 34A34; 34C60; 34D23; 65L05; 92C60

1. Introduction

The first mathematical epidemiological model was created by Daniel Bernoulli modelling smallpox in 1766 (see [1]). However, it was not until 1927 that Kermack and McKendrick [2] introduced the SIR model for a more precise analysis of epidemic diseases. Nowadays, almost a hundred years later, SIR-type models are the most commonly used deterministic models to describe the COVID-19 epidemic, caused by severe acute respiratory syndrome coronavirus 2 (SARS-CoV-2). Actually, there are two different types of models: statistical and mathematical, and both approaches depend upon prior estimates as well as reliable data. Historical remarks and comparisons between both methods can be found in [3]. In the present paper we only use the mathematical deterministic approach, based on systems of non-linear differential equations which can be solved analytically or numerically. In these models, the host population is divided into different categories according to infection, vaccination, hospitalization, quarantine, etc. The dynamics of the infection in the categories are modelled by a Cauchy problem for a system of non-linear ordinary differential equations. Different epidemiological parameters are involved in the models as coefficients in the differential equations. If the values of these parameters are known, one could numerically solve the corresponding Cauchy problem. Unfortunately, when it comes to new viruses, as in the case of SARS-CoV-2, these parameters are unknown and their change over time cannot be determined. Of course, after a long enough period of time, based on the statistics, one can guess the values of some of the parameters. It is often assumed that some parameters do not change during the dominance of a particular variant of the virus or for a short period of time. However, this is not sufficient in general cases for the successful application of these models. At the same time, available databases contain information on the number of individuals in some of the categories in the models, for example, infected, vaccinated, recovered, deceased cases. In this way, various inverse problems arise that must be solved to determine all the parameters in the relevant model (see [4–9]) . Only then, analysing the behaviour of the parameters over time, would it be possible to make an assumption about their change in the future and make realistic forecasts for the development of the epidemic.

A key role for the invasion and persistence of an infection in a new host population [10], i.e., for the duration of the epidemic, is played by the so-called basic reproduction number $\Re_0$. In deterministic models $\Re_0$ relates to the stability of the equilibrium points (in the more general case of models with non-linear incidence we refer to [11,12]). The classical SIR/SEIR and SEIRS (susceptible-exposed-infectious-recovered-susceptible) models with constant parameters, from a mathematical perspective, have an infinite number of equilibrium points. More precise models involving, for example, vital dynamics usually have a finite number of equilibrium points—the disease-free equilibrium and the endemic equilibriums. Then, the typical situation from a mathematical perspective is the following (see [13–17]). When $\Re_0 < 1$, the disease-free equilibrium is globally asymptotically stable. Therefore, the number of infected individuals tends to zero as time $t \to \infty$ and the epidemic subsides. If $\Re_0 > 1$, the disease-free equilibrium is unstable, but there exists a unique endemic equilibrium which is globally asymptotically stable. This means that the epidemic does not end, but the number of infected individuals tends to a non-zero constant as $t \to \infty$. In other words, the number of infected people will be approximately constant after a certain point in time and the outbreak will stop being a global emergency. A similar scenario is being observed with the current COVID-19 epidemic, although it is a much more complex process and its long-term behaviour cannot be described by a model with constant coefficients. SIR-based models with time-dependent coefficients are more suitable for this purpose. Official datasets contain information for the daily variation in the number of individuals in some categories. Based on these, various time-discrete analogues of the differential models have been introduced (see [18–24]). The time-discrete models with step size $h = 1$ day provide two interpretations [25]. First, each quantity in the model $z_k = z(t_k)$ is expressed in days $^{-1}$ or alternatively, $z_k = z(t_k)h$ is dimensionless, where t_k is a fixed day of the time frame under consideration.

Recently, there has been a lot of interest in the construction of various integer- or fractional-order time-discrete variants of different deterministic models. To model the memory effects of the disease, different time-fractional models have been used (see, for example, refs. [26–28]). By default, the use of such non-local models requires larger refined datasets. They are also quite computationally expensive, which is why well-specialized numerical algorithms are required [29]. Currently, integer-order explicit or implicit time-discrete models have been used to the model COVID-19 epidemics. For example, time-continuous and time-discrete versions of the classical SIR model are discussed in [25] Stability analysis, parameter identification and validation with Brazilian and UK COVID-19 data have been conducted. Some time-continuous and explicit time-discrete SEIR models with eventual linear feedback vaccination and partial re-susceptibility are studied in [30] The stability of both the disease-free and the endemic equilibrium points is discussed. The proposed model is tested with Italian COVID-19 data. In [31], an implicit time-discrete SIR models is used for estimation of the basic reproduction number for the second wave of

COVID-19 in Fiji. An implicit time-discrete SIR model is established in [32]. It showed that many of the desired properties of the time-continuous case are still valid in the time-discrete implicit case. An upper error estimate was also derived. The developed time-discrete SIR model was applied to COVID-19 data in Germany and Iran.

Replacing differential models with time-discrete ones raises several questions. What is the relationship between the values of the parameters in the time-discrete models with step size $h = 1$ and the daily values of parameters in the differential model? More precisely, whether the parameter values found in the discrete model with $h = 1$ by solving an appropriate inverse problem are close enough to the daily values of the parameters in the corresponding differential problem? The values of which parameters in the models can be assumed to be known based on the known statistics? The answer to the last question affects the answers to the previous two questions. This answer, on the one hand, depends on the correctness of the official data, and on the other hand, on the results of the sensitivity analysis of the input parameters. The sensitivity analysis could also provide an answer to another interesting question: which parameters have the greatest influence on the basic reproduction number? This, in turn, is extremely useful for developing a good vaccination strategy and adequate measures to limit the spread of the virus. For similar or different approaches to designing optimal vaccination strategies see [33] or [34], respectively, and the references therein.

In this article, we seek answers to the questions formulated above in the case of the SEIRS-VB model introduced in our previous work [4]. This is an extended SEIRS model with additional categories: V for susceptible vaccinated individuals and B for individuals with vaccine-induced immunity. The rest of the paper is organized as follows:

In Section 2, we briefly describe the differential SEIRS-VB model with time-dependent coefficients (see [4] for more details). It includes vaccination and vital dynamics. In Section 3, we consider a SEIRS-VB model with constant coefficients that can be used for short periods of time. By studying analytic properties we determine its positively invariant region. In Section 4, we find the disease-free equilibrium point (DFEP) of the same model. After this, using the proposed next-generation method we define the basic reproduction number $\Re_0$. The impact of $\Re_0$ on the local stability of the DFEP is studied in Section 5. In Section 6.1, to numerically solve the differential model SEIRS-VB, we construct a family of time-discrete SEIRS-VB models (difference schemes) and find sufficient conditions for the step size and parameters under which the model preserves the positivity property. Furthermore, in Section 6.2, bearing in mind the available data on the spread of SARS-CoV-2, we formulate an appropriate inverse problem. Then, we propose an algorithm to solve this problem and find the parameters in the suggested time-discrete models with a step size of 1. This allows us to compute the daily values of $\Re_0$. Validation of the SEIRS-VB models is presented in Section 7. We use reported COVID-19 data for Bulgaria to show that the suggested parameter identification method leads to biologically reasonable results. Then, we find that the time-discrete model form family (with a step size of 1) for which the computed parameter values are closest to the parameter values in the differential model with time-dependent coefficients. On the other hand, the obtained results show that the parameters values in the differential SEIRS-VB model and in the time-discrete model, realized through an explicit (forward) Euler method (with a step size of 1), are rather different. This observation deserves more serious attention. Let us note that there are some authors who use time-discrete models based on the explicit Euler method, omitting that in general this method is conditionally stable. Parameter sensitivity analysis of the basic reproduction number is conducted in Section 8. A discussion and conclusions are outlined in Sections 9 and 10, respectively.

2. The Differential SEIR-VB Model

This paper deals with the mathematical modelling of COVID-19 transmission using the SEIRS-VB model, introduced in our previous work [4]. The model is described by the following Cauchy problem for a system of non-linear ordinary differential equations

$$\begin{cases} \dfrac{dS}{dt} = \Lambda(t)N(t) - (\alpha(t)+\theta(t))S(t) - \dfrac{\beta(t)}{N(t)}S(t)I(t) + \lambda(t)R(t) + \nu(t)B(t), \\[4pt] \dfrac{dE}{dt} = \dfrac{\beta(t)}{N(t)}S(t)I(t) + \dfrac{\beta(t)}{N(t)}V(t)I(t) - (\omega(t)+\theta(t))E(t), \\[4pt] \dfrac{dI}{dt} = \omega(t)E(t) - (\gamma(t)+\tau(t)+\theta(t))I(t), \\[4pt] \dfrac{dR}{dt} = \gamma(t)I(t) - (\lambda(t)+\theta(t))R(t), \\[4pt] \dfrac{dV}{dt} = \alpha(t)S(t) - (\mu(t)+\theta(t))V(t) - \dfrac{\beta(t)}{N(t)}I(t)V(t), \\[4pt] \dfrac{dB}{dt} = \mu(t)V(t) - (\nu(t)+\theta(t))B(t) \end{cases} \qquad (1)$$

with non-negative initial conditions

$$S(t_0) = S_0,\ E(t_0) = E_0,\ I(t_0) = I_0,\ R(t_0) = R_0,\ V(t_0) = V_0,\ B(t_0) = B_0, \qquad (2)$$

where $t_0 \geq 0$ is an integer number.

Here, the total population size is the number of all living individuals

$$N(t) := S(t) + E(t) + I(t) + R(t) + B(t) + V(t).$$

A full description of the SEIRS-VB model's categories and parameters is given in [4]. Here, we will only note them briefly and give the model diagram in Figure 1:

Categories of the SEIRS-VB model:

- $S(t)$—susceptible individuals (unvaccinated, not fully vaccinated, vaccinated people for whom the vaccine is ineffective, fully vaccinated or recovered individuals who have lost their immunity).
- $E(t)$—exposed individuals. These are virus carriers in the latent stage, during which they are not virus spreaders. They usually have no symptoms.
- $I(t)$—infectious individuals. These are virus carriers and virus spreaders of extremely high infectivity. The former are likely to transmit the virus in case of contact.
- $R(t)$—Recovered individuals with disease-acquired immunity. These individuals have disease-acquired immunity. They have recovered, and thus are protected from the disease.
- $V(t)$—vaccinated susceptible individuals. These are fully vaccinated individuals for whom the vaccine is effective. However, they have not developed antibodies. They can do so after a certain period of time or otherwise they will become exposed individuals before that. It is worth pointing out that, due to vaccine imperfections, some of the vaccinated individuals cannot develop antibodies, and they cannot pass from group $S(t)$ to group $V(t)$.
- $B(t)$—vaccinated individuals with vaccination-acquired immunity. These are vaccinated individuals who are well protected from future infection because they have antibodies.

Parameters of the SEIRS-VB model:

- $\Lambda(t)$—birth rate;
- $\alpha(t)$—vaccination rate;

- σ—vaccine effectiveness;
- $\alpha(t) = \sigma a(t)$—vaccination parameter;
- $\beta(t)$—transmission rate;
- $\gamma(t)$—recovery rate;
- $\omega(t)$—latency rate;
- $\theta(t)$—natural mortality rate;
- $\tau(t)$—mortality rate of infectious people;
- $\lambda(t)$—reinfection rate of recovered individuals;
- $\nu(t)$–reinfection rate of vaccinated individuals;
- $\mu(t)$–antibody rate.

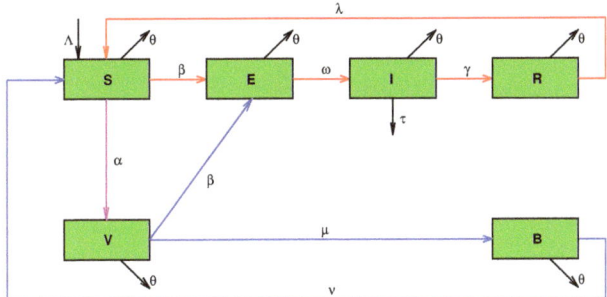

Figure 1. The SEIRS-VB model diagram.

Available datasets contain daily numbers of individuals in some category in the SEIRS-BV model. In [4], using Bulgarian COVID-19 data we find daily values of the parameters in the model (1). In other words we assume that the parameters are step functions in time and for every single day we consider the problem (1) with constant coefficients. That is why, for simplicity, in the next three sections we study a SEIRS-VB model with time-independent coefficients, and define the basic reproduction number associated with it.

3. SEIRS-VB Model with Time-Independent Coefficients

To further introduce the basic reproduction number, in this section we begin by studying the simplified model with constant coefficients. To do this we analyse the disease-free equilibrium points (see explanation below) of the corresponding system of differential equations. Let us rewrite (1) using different notation

$$\begin{cases} \dfrac{dS}{dt} = \Omega - (\alpha + \theta)S(t) - bS(t)I(t) + \lambda R(t) + \nu B(t), \\[4pt] \dfrac{dE}{dt} = bS(t)I(t) + bV(t)I(t) - (\omega + \theta)E(t), \\[4pt] \dfrac{dI}{dt} = \omega E(t) - (\gamma + \tau + \theta)I(t), \\[4pt] \dfrac{dR}{dt} = \gamma I(t) - (\lambda + \theta)R(t), \\[4pt] \dfrac{dV}{dt} = \alpha S(t) - (\mu + \theta)V(t) - bI(t)V(t), \\[4pt] \dfrac{dB}{dt} = \mu V(t) - (\nu + \theta)B(t), \end{cases} \qquad (3)$$

where the new parameters are

$$\Omega := \Lambda N > 0, \ b := \frac{\beta}{N} > 0. \quad (4)$$

We suppose that all parameters $\Omega, b, \omega > 0, \theta > 0, \alpha \geq 0, \gamma \geq 0, \tau \geq 0, \nu \geq 0, \mu \geq 0$ are constants.

Remark 1. *Let us introduce the vector notations* $X(t) = (S(t), E(t), I(t), R(t), V(t), B(t))$, $X_0 = (S_0, E_0, I_0, R_0, V_0, B_0)$, $N_0 = S_0 + E_0 + I_0 + R_0 + V_0 + B_0$, $p = (\alpha, \beta, \gamma, \tau, \Lambda, \theta, \omega, \lambda, \mu, \nu)$. *Here and in the following we call a vector non-negative/positive if all its components are non-negative/positive.*

Since the system (3) is a partial case of the SEIRS-VB model for the Cauchy problem (1), (2), we can apply the following theorem, proven in [4]:

Theorem 1. *(non-negativity, existence, uniqueness) [4] Let* $X_0 \geq 0$, $S_0 > 0$, $I_0 > 0$, $p \in C([t_0, T])$ *and* $p(t) \geq 0$, $\Lambda(t) > 0$, $\beta(t) > 0$, $\omega(t) > 0$ *for* $t_0 \leq t \leq T$. *Then, there exists a solution* $X(t)$ *of the Cauchy problem (1), (2), which is defined for* $t \in [t_0, T]$ *and* $X(t) \geq 0$, $S(t) > 0$, $I(t) > 0$ *for* $t \in [t_0, T]$. *The solution with such properties is unique.*

We returned (see [10,35,36]) to the concept of the **disease-free equilibrium point** which means the equilibrium point $(S^*, E^*, I^*, R^*, V^*, B^*)$ of system (3), in which $E^* = 0$ and $I^* = 0$. We will use this equilibrium point to define the basic reproduction number $\Re_0$ associated with the model (3). We then find the daily $\Re_0$ values which are very important because they measure the transmission potential of a disease. Actually, the non-linear autonomous system (3) also has other equilibrium points, which are called endemic. However, the study of these points is not the subject of the present work, because they are only related to an autonomous system, not to a system with time-variable coefficients (1) used to model the long-term development of the COVID-19 epidemic.

Next, the following theorem gives a feasible region of the SEIRS-VB model (3).

Theorem 2. *(non-negativity, boundedness) A positively invariant set of the system (3) is*

$$\Sigma := \Psi \cup \left\{ X \geq 0 : \ S > 0, \ I > 0, \ 0 < N \leq \frac{\Omega}{\theta} \right\}, \quad (5)$$

where Ψ *is the set of all disease-free equilibrium points of the system (3) with non-negative components.*

Proof. Suppose $X_0 \in \Sigma$:

Case 1. Let $X_0 \in \Psi$, i.e., X_0 is a disease-free equilibrium point of the system (3) with non-negative components. If $X(t)$ is a solution to (3) with $X(t_0) = X_0$, we have $X(t) = X_0 \in \Sigma$ for all $t \geq t_0$.

Case 2. Let $X_0 \in \Sigma \setminus \Psi$, then for each arbitrary, but with fixed $T > 0$, Theorem 1 implies that the unique solution $X(t)$ to the Cauchy problems (3) and (2) are defined in $[t_0, T]$ and $X(t) \geq 0$, $S(t) > 0$, $I(t) > 0$ and as a result $N(t) > 0$ for $t \in [t_0, T]$.

Adding the equations of the system (3) and initial conditions (2), we obtain

$$\left| \begin{array}{l} \dfrac{dN(t)}{dt} = \Omega - \theta N(t) - \tau I(t), \\ N(t_0) = N_0. \end{array} \right. \quad (6)$$

The unique solution of the Cauchy problem (6) is

$$N(t) = \frac{\Omega}{\theta} + \left(N_0 - \frac{\Omega}{\theta} \right) e^{\theta(t_0 - t)} - \tau e^{-\theta t} \int_{t_0}^{t} e^{\theta s} I(s) \, ds. \quad (7)$$

Since $I(t) > 0$ for $t \in [t_0, T]$ and $N_0 \leq \Omega/\theta$, we obtain

$$N(t) \leq \frac{\Omega}{\theta}.$$

Therefore, $X(t) \in \Sigma$ for $t \in [t_0, T]$.

The coefficients in the system (3) are constant. The number $T > 0$ is arbitrary and therefore there exists a unique solution $X(t)$ of the system (3), defined as belonging to Σ for all $t \geq t_0$. The proof is complete. □

4. The Disease-Free Equilibrium Point and Basic Reproduction Number

To find the disease-free equilibrium points of the autonomous system (3), we search for solutions $(S, E, I, R, V$ and $B)$ with $E = 0$ and $I = 0$ of the system

$$\begin{vmatrix} \Omega - (\alpha + \theta)S - bSI + \lambda R + \nu B = 0, \\ bSI + bVI - (\omega + \theta)E = 0, \\ \omega E - (\gamma + \tau + \theta)I = 0, \\ \gamma I - (\lambda + \theta)R = 0, \\ \alpha S - (\mu + \theta)V - bIV = 0, \\ \mu V - (\nu + \theta)B = 0. \end{vmatrix} \quad (8)$$

In this case, the system (3) becomes linear and has a unique disease-free equilibrium point

$$X^* := (S^*, E^*, I^*, R^*, V^*, B^*) = \left(1, 0, 0, 0, \frac{\alpha}{\mu + \theta}, \frac{\alpha \mu}{(\mu + \theta)(\nu + \theta)}\right) S^*, \quad (9)$$

where

$$S^* := \frac{\Omega(\nu + \theta)(\mu + \theta)}{\theta(\alpha \nu + (\mu + \theta)(\alpha + \nu + \theta))}.$$

Remark 2. *We find that* $\Psi = X^*$ *in (5) and* $N^* = S^* + E^* + I^* + R^* + V^* + B^* = \Omega/\theta$. *Therefore, for all elements of* Σ *we have the boundedness estimate* $N \leq \Omega/\theta$ *(see Theorem 2).*

The Jacobian method (stability of the first approximation) is a widely used approach to derive a parameter that reflects the stability of the disease-free equilibrium. This parameter is used to establish the herd immunity threshold, but it sometimes does not yield the true value of the basic reproduction number. It is called the basic reproduction number if it has the same biological interpretation as the spectral radius of the next-generation matrix (see [35,36] and references therein). We find the basic reproduction number using the next-generation method and then using the Jacobian method to show that it is the stability parameter of the disease-free equilibrium.

Following [14,35], since we are concerned with the populations that spread the infection we only need to model the exposed E and infectious I categories. The dynamics in these two categories are modelled by the following equations

$$\begin{vmatrix} \dfrac{dE}{dt} = bS(t)I(t) + bV(t)I(t) - (\omega + \theta)E(t), \\ \dfrac{dI}{dt} = \omega E(t) - (\gamma + \tau + \theta)I(t). \end{vmatrix} \quad (10)$$

Let
$$F(X) := (b(S+V)I, \, 0)^T$$
be the rate of appearance of new infections in each category of (10),
$$W^+(X) := (0, \, \omega E)^T$$
is the rate of transfer of individuals into each category of (10) by all other means and
$$W^-(X) := ((\omega + \theta)E, \, (\gamma + \tau + \theta)I)^T$$
is the rate of transfer of individuals from one category of (10).

Let us denote by $Y = (E, I)$ the group with infected individuals and reformulate the system (10) in the form
$$\frac{dY}{dt} = F - W, \tag{11}$$
where $W(X) = W^-(X) - W^+(X)$.

Calculating the Jacobian matrix for F and W at the disease-free equilibrium point X^*, we obtain
$$\frac{\partial F}{\partial Y}(X^*) := \begin{pmatrix} 0 & b(S^* + V^*) \\ 0 & 0 \end{pmatrix}, \quad \frac{\partial W}{\partial Y}(X^*) := \begin{pmatrix} \omega + \theta & 0 \\ -\omega & \gamma + \tau + \theta \end{pmatrix}.$$

Reminding (see [35,36]) that the next-generation matrix (operator) is defined by
$$G := \frac{\partial F}{\partial Y}(X^*) \left(\frac{\partial W}{\partial Y}(X^*) \right)^{-1} = \begin{pmatrix} \mathfrak{R}_0 & \frac{\omega + \theta}{\omega} \mathfrak{R}_0 \\ 0 & 0 \end{pmatrix},$$
where
$$\mathfrak{R}_0 := \frac{\omega b(S^* + V^*)}{(\omega + \theta)(\gamma + \tau + \theta)} = \frac{\omega b \Omega (\nu + \theta)(\alpha + \mu + \theta)}{\theta(\omega + \theta)(\gamma + \tau + \theta)(\alpha \nu + (\mu + \theta)(\alpha + \nu + \theta))}. \tag{12}$$

The eigenvalues of the matrix G are 0 and $\mathfrak{R}_0$. The basic reproduction number is defined as the spectral radius of the matrix G (the maximum absolute values of its eigenvalues), and is $\mathfrak{R}_0$. From a biological perspective the basic reproduction number is the number of secondary cases or new infections produced by a single infectious individual in a completely susceptible host population [13,36].

Remark 3. *We use this notation and definition in the case of constant coefficients in our method in Sections 6–8 for a fixed time period of one day, assuming that all parameters are appropriate constants and consequently our reproduction number $\mathfrak{R}_0$ is also constant. Of course, the values of $\mathfrak{R}_0$ depend on t and will differ for different days, making the basic reproduction number a step function $\mathfrak{R}_0(t)$.*

5. Local Stability Analysis of the Disease-Free Equilibrium Point

Here we introduce the concept of the locally stable equilibrium point, meaning that any solution starting near that point will be near it all the time. More precisely, following Barbashin ([37], page 19), we call the equilibrium point X^* of the system (8) locally stable on Σ if for every $\varepsilon > 0$, there exists $\delta(\varepsilon) > 0$ such that for any other solution $X(t)$ of the system (8) on Σ from the inequality $||X(t_0) - X^*|| \leq \delta(\varepsilon)$, there follows $||X(t) - X^*|| \leq \varepsilon$ for $t \geq t_0$. We call X^* locally asymptotically stable on Σ if it is locally stable on Σ and there exists a positive number h, such that for $||X(t_0) - X^*|| \leq h$ we have $||X(t) - X^*|| \to 0$ as $t \to +\infty$.

Now, we use this theory in the case of the disease-free equilibrium point X^*.

Theorem 3. *Let the coefficients of the system (3) be non-negative and $\omega > 0$, $\theta > 0$, $\Omega > 0$. If $\Re_0 < 1$, then the disease-free equilibrium point X^* of the system (3) is locally asymptotically stable on Σ. If $\Re_0 > 1$, then the point X^* is locally unstable on Σ.*

Proof. The Jacobian matrix of (3) evaluated at the disease-free equilibrium point X^* is

$$J(X^*) = \begin{pmatrix} -\alpha - \theta & 0 & -bS^* & \lambda & 0 & \nu \\ 0 & -\omega - \theta & b(S^* + V^*) & 0 & 0 & 0 \\ 0 & \omega & -\gamma - \tau - \theta & 0 & 0 & 0 \\ 0 & 0 & \gamma & -\lambda - \theta & 0 & 0 \\ \alpha & 0 & -bV^* & 0 & -\mu - \theta & 0 \\ 0 & 0 & 0 & 0 & \mu & -\nu - \theta \end{pmatrix}.$$

It is easy to calculate that the characteristic polynomial $P(\zeta)$ of $J(X^*)$ has the following representation

$$P(\zeta) = P_1(\zeta) P_2(\zeta),$$

where

$$P_1(\zeta) := (\lambda + \theta + \zeta)(\theta + \zeta)\left\{(\theta + \zeta)^2 + (\nu + \mu + \alpha)(\theta + \zeta) + \alpha\mu + \alpha\nu + \mu\nu\right\}$$

and

$$P_2(\zeta) := \zeta^2 + (\omega + \gamma + \tau + 2\theta)\zeta + (\omega + \theta)(\gamma + \tau + \theta)(1 - \Re_0).$$

The polynomial $P_1(\zeta)$ has four roots $\zeta_1, \zeta_2, \zeta_3, \zeta_4$, where $\zeta_1 = -\theta < 0$ and $\zeta_2 = -\lambda - \theta < 0$. The last two roots ζ_3 and ζ_4 satisfy

$$\zeta_3 + \zeta_4 = -(\nu + \mu + \alpha) - 2\theta < 0, \quad (\zeta_3 + \theta)(\zeta_4 + \theta) = \alpha\mu + \alpha\nu + \mu\nu > 0.$$

Hence, the roots of $P_1(\zeta)$ have negative real parts.

Similar arguments for the roots ζ_5 and ζ_6 of the other polynomials $P_2(\zeta)$ give

$$\zeta_5 + \zeta_6 = -(\omega + \gamma + \tau + 2\theta) < 0, \quad \zeta_5 \zeta_6 = (\omega + \theta)(\gamma + \tau + \theta)(1 - \Re_0).$$

If $\Re_0 < 1$, then ζ_5 and ζ_6 have negative real parts. According to the stability theorem of the first approximation (Theorem 11.1 in [37], p. 45) X^* is locally asymptotically stable on Σ. If $\Re_0 > 1$, then one of the eigenvalues ζ_5 or ζ_6 is positive and according to the instability theorem of the first approximation (Theorem 11.2 in [37], p. 45) X^* is locally unstable.

The proof is complete. □

Remark 4. *It is possible to study the impact of the basic reproduction number $\Re_0$ on the global asymptotic stability of the disease-free equilibrium point X^* and the epidemic equilibrium points (with $I^*E^* \neq 0$) of the autonomous SEIRS-VB model (3). Commonly used methods to perform a global stability analysis are the method of Carlos Castillo-Chavez [14,38,39] and the direct Lyapunov method [12,15,37,40]. However, this would be of no practical importance because the long-term behaviour of a complex real-world process such as the COVID-19 epidemic cannot be described by models with constant parameters. For this reason, as already noted, we applied the autonomous system (3) to describe the daily changes in the number of individuals in different categories in the SEIRV-BD model. The results are step functions, as are the available data.*

6. Time-Discrete SEIRS-VB Model

6.1. Construction of a New Family of Semi-Implicit Difference Schemes

In this section, we introduce a new family of semi-implicit difference schemes as time-discrete analogues of the differential model SEIRS-VB (1).

We consider the time-frame $t_1, t_2, \ldots, t_K$, where $t_0 \leq t_1 < t_2 < \ldots < t_K \leq T$ and introduce the notation for the values of functions

$$X_k := (S_k, E_k, I_k, R_k, V_k, B_k) = (S(t_k), E(t_k), I(t_k), R(t_k), V(t_k), B(t_k)),$$

$$N_k := N(t_k) = S_k + E_k + I_k + R_k + V_k + B_k,$$

for $k = 1, 2, \ldots, K$ and the values of the parameters

$$p_k := (\alpha_k, \beta_k, \gamma_k, \tau_k, \Lambda_k, \theta_k, \omega_k, \lambda_k, \mu_k, \nu_k)$$
$$= (\alpha(t_k), \beta(t_k), \gamma(t_k), \tau(t_k), \Lambda(t_k), \theta(t_k), \omega(t_k), \lambda(t_k), \mu(t_k), \nu(t_k)),$$

$k = 1, 2, \ldots, K - 1$.

Starting with the given initial data $S(t_1) = S_1$, $E(t_1) = E_1$, $I(t_1) = I_1$, $R(t_1) = R_1$, $V(t_1) = V_1$, $B(t_1) = B_1$ we consider the following family of time-discrete models with weights $0 \leq \xi \leq 1$, $0 \leq \eta \leq 1$:

$$\begin{cases} \dfrac{S_k - S_{k-1}}{h} = \Lambda_{k-1} N_{k-1} - (\alpha_{k-1} + \theta_{k-1}) S_{k-1} - \dfrac{\beta_{k-1}}{N_{k-1}} S_{k-1}[(1-\xi) I_{k-1} + \xi I_k] + \lambda_{k-1} R_{k-1} + \nu_{k-1} B_{k-1}, \\[6pt] \dfrac{E_k - E_{k-1}}{h} = \dfrac{\beta_{k-1}}{N_{k-1}} (S_{k-1} + V_{k-1})[(1-\xi) I_{k-1} + \xi I_k] - (\omega_{k-1} + \theta_{k-1}) E_{k-1}, \\[6pt] \dfrac{I_k - I_{k-1}}{h} = \omega_{k-1} E_{k-1} - (\gamma_{k-1} + \tau_{k-1})[(1-\eta) I_{k-1} + \eta I_k] - \theta_{k-1} I_{k-1}, \\[6pt] \dfrac{R_k - R_{k-1}}{h} = \gamma_{k-1}[(1-\eta) I_{k-1} + \eta I_k] - (\lambda_{k-1} + \theta_{k-1}) R_{k-1}, \\[6pt] \dfrac{V_k - V_{k-1}}{h} = \alpha_{k-1} S_{k-1} - \dfrac{\beta_{k-1}}{N_{k-1}} V_{k-1}[(1-\xi) I_{k-1} + \xi I_k] - (\mu_{k-1} + \theta_{k-1}) V_{k-1}, \\[6pt] \dfrac{B_k - B_{k-1}}{h} = \mu_{k-1} V_{k-1} - (\nu_{k-1} + \theta_{k-1}) B_{k-1}, \end{cases} \quad (13)$$

where $k = 2, 3, \ldots, K$.

Remark 5. *In fact, when $\xi = \eta = 0$ the difference scheme (13) coincides the explicit Euler scheme, and when $\xi = 1, \eta = 0$ it is the semi-implicit scheme used in [4].*

Summing up all equations in system (13) we obtain

$$\frac{N_k - N_{k-1}}{h} = [\Lambda_{k-1} - \theta_{k-1}] N_{k-1} - \tau_{k-1}[(1-\eta) I_{k-1} + \eta I_k]. \quad (14)$$

Equation (14) can be considered as a discretization of differential Equation (7).

In the next theorem, we study the positivity features of the proposed family of the difference schemes. We find a dependence between the step size h and the parameters in the difference scheme (13), preserving the component-wise non-negativity of the initial vector X_1 in time for the numerical solution X_k, $k = 2, 3, \ldots, K$.

Theorem 4. *Let $\xi, \eta \in [0,1]$, $p_{k-1} \geq 0$, $\Lambda_{k-1} > 0$, $\beta_{k-1} > 0$, $\omega_{k-1} > 0$ for $k = 2, 3, \ldots, K$ and*

$$\max_{k=2,\ldots,K} q_{k-1} \leq 1/h, \quad (15)$$

where

$$q_{k-1} := \theta_{k-1} + \max\{\alpha_{k-1} + \beta_{k-1}, \mu_{k-1} + \beta_{k-1}, \gamma_{k-1} + \tau_{k-1} + \omega_{k-1}, \lambda_{k-1}, \nu_{k-1}\}.$$

If $X_1 \geq 0$ with $S_1 > 0$, $I_1 > 0$ then for the values calculated by (13), we have $X_k \geq 0$, $S_k > 0$, $I_k > 0$ for all $k = 1, 2, \ldots, K$.

Proof. For X_1, the statement of the theorem holds. We suppose that $X_{k-1} \geq 0$, $I_{k-1} > 0$, $S_{k-1} > 0$ for some $2 \leq k \leq K$. Therefore, $N_{k-1} > 0$.

Now, we rewrite (13) in the form

$$\begin{aligned}
S_k &= \left[1 - h\left(\alpha_{k-1} + \theta_{k-1} + \frac{\beta_{k-1}}{N_{k-1}}[(1-\xi)I_{k-1} + \xi I_k]\right)\right]S_{k-1} \\
&\quad + h(\Lambda_{k-1}N_{k-1} + \lambda_{k-1}R_{k-1} + \nu_{k-1}B_{k-1}), \\
E_k &= [1 - h(\omega_{k-1} + \theta_{k-1})]E_{k-1} + h\frac{\beta_{k-1}}{N_{k-1}}(S_{k-1} + V_{k-1})[(1-\xi)I_{k-1} + \xi I_k], \\
I_k &= \left[1 - \frac{h(\gamma_{k-1} + \tau_{k-1} + \theta_{k-1})}{1 + h\eta(\gamma_{k-1} + \tau_{k-1})}\right]I_{k-1} + \frac{h\omega_{k-1}}{1 + h\eta(\gamma_{k-1} + \tau_{k-1})}E_{k-1}, \\
R_k &= [1 - h(\lambda_{k-1} + \theta_{k-1})]R_{k-1} + h\gamma_{k-1}[(1-\eta)I_{k-1} + \eta I_k], \\
V_k &= \left[1 - h\left(\mu_{k-1} + \theta_{k-1} + \frac{\beta_{k-1}}{N_{k-1}}[(1-\xi)I_{k-1} + \xi I_k]\right)\right]V_{k-1} + h\alpha_{k-1}S_{k-1}, \\
B_k &= [1 - h(\nu_{k-1} + \theta_{k-1})]B_{k-1} + h\mu_{k-1}V_{k-1}.
\end{aligned} \quad (16)$$

Using (15) we perform the following steps:

1. From the I-equation in (16), it follows that $I_k \geq 0$ and

$$\begin{aligned}
I_k &\leq \left[1 - \frac{h(\gamma_{k-1} + \tau_{k-1} + \theta_{k-1})}{1 + h\eta(\gamma_{k-1} + \tau_{k-1})}\right](I_{k-1} + E_{k-1}) \\
&\leq \left[1 - \frac{h(\gamma_{k-1} + \tau_{k-1} + \theta_{k-1})}{1 + h\eta(\gamma_{k-1} + \tau_{k-1})}\right]N_{k-1} \\
&\leq N_{k-1}.
\end{aligned} \quad (17)$$

2. Since $\Lambda_{k-1} > 0$, $N_{k-1} > 0$, and using (17), from the S-equation in (16) we obtain

$$S_k \geq [1 - h(\alpha_{k-1} + \theta_{k-1} + \beta_{k-1})]S_{k-1} + h(\Lambda_{k-1}N_{k-1} + \lambda_{k-1}R_{k-1} + \nu_{k-1}B_{k-1}) > 0.$$

3. The E-equation, R-equation and B-equation in (16) give $E_k \geq 0$, $R_k \geq 0$ and $B_k \geq 0$, respectively.
4. Finally, using (17) from the V-equation in (16) we obtain

$$V_k \geq [1 - h(\mu_{k-1} + \theta_{k-1} + \beta_{k-1})]V_{k-1} + h\alpha_{k-1}S_{k-1} \geq 0.$$

The proof is complete. □

6.2. Parameter Identification Algorithm

As in [4], we divide the parameters p_k into two groups

$$p_k = (\bar{p}_k, \hat{p}_k), \ k = 1, 2, \ldots, K - 1,$$

where $\tilde{p}_k$ are the known parameter values (they can be selected from the available statistical data) and $\hat{p}_k$ are the unknown values that we have to find.

Let us introduce the notations for the available measurements for $k = 1, 2, \ldots, K$

$$\begin{aligned} m_k &:= (A_k, Rtotal_k, Dtotal_k, Vtotal_k) \\ &= (E_k + I_k, Rtotal_k, Dtotal_k, Vtotal_k) \\ &= (E(t_k) + I(t_k), Rtotal(t_k), Dtotal(t_k), Vtotal(t_k)), \\ \tilde{p}_k &:= (\Lambda_k, \theta_k, \omega_k, \lambda_k, \mu_k, \nu_k), \end{aligned} \quad (18)$$

where for each day t_k

1. $A_k := E_k + I_k$ is the number of the active cases.
2. $Rtotal_k$ is the cumulative number of the individuals recovered from the disease to time t_k. Unlike R_k, individuals who have already lost disease-acquired immunity are counted in $Rtotal_k$.
3. $Dtotal_k$ is the cumulative number of COVID-19 deaths.
4. $Vtotal_k$ is the cumulative number of the fully vaccinated individuals.

In the same manner, we denote the values of the unknown functions and parameters

$$\begin{aligned} g_k &:= (S_k, E_k, I_k, R_k, V_k, B_k), \\ \hat{p}_k &:= (\alpha_k, \beta_k, \gamma_k, \tau_k), \end{aligned} \quad (19)$$

for $k = 1, 2, \ldots, K$.

Now, we are ready to formulate the following appropriate "inverse" problem.

Inverse discrete problem (**IDP**): Using the given data g_1, $\{\tilde{p}_k\}_{k=1}^{K-1}$, $\{m_k\}_{k=1}^{K}$, find the values $\left(\{g_k\}_{k=2}^{K}, \{\hat{p}_{k-1}\}_{k=2}^{K}\right)$, such that the relations (13) hold.

Remark 6. *As in [4], we introduce two groups that do not appear explicitly in the differential SEIRS-BV model: $Rtotal(t)$, the cumulative number of individuals that have recovered from the disease; $Dtotal(t)$, the cumulative number of COVID-19 deaths to time t. It is clear that $Rtotal_k = Rtotal(t_k)$ and $Dtotal_k = Dtotal(t_k)$ for $k = 1, 2, \ldots, K$. As in SIR-type models, we have (see [4], page 13, Equations (23) and (24)):*

$$\frac{dRtotal}{dt} = \gamma(t)I(t) \quad (20)$$

and

$$\frac{dDtotal}{dt} = \tau(t)I(t). \quad (21)$$

Discretizing Equations (20) and (21) in Remark 6 in accordance with the difference scheme (13) we derive the following relations:

$$\begin{aligned} Rtotal_k - Rtotal_{k-1} &:= \gamma_{k-1}[\eta_{k-1}I_{k-1} + \eta I_k], \\ Dtotal_k - Dtotal_{k-1} &:= \tau_{k-1}[\eta_{k-1}I_{k-1} + \eta I_k], \end{aligned} \quad (22)$$

where $k = 2, 3, \ldots, K$.

Let us introduce the notations

$$\begin{aligned} I_{\beta,k-1} &:= \frac{\beta_{k-1}}{N_{k-1}}[(1-\xi)I_{k-1} + \xi I_k], \\ I_{\gamma,k-1} &:= \gamma_{k-1}[(1-\eta)I_{k-1} + \eta I_k], \\ I_{\tau,k-1} &:= \tau_{k-1}[(1-\eta)I_{k-1} + \eta I_k]. \end{aligned} \quad (23)$$

Now, we recall that $A_k = E_k + I_k$, and sum the E_k-equation and I_k-equation in scheme (16) with $h = 1$

$$\begin{aligned}
S_k &= \left[1 - \alpha_{k-1} - \theta_{k-1} - I_{\beta,k-1}\right] S_{k-1} + \Lambda_{k-1} N_{k-1} + \lambda_{k-1} R_{k-1} + \nu_{k-1} B_{k-1}, \\
A_k &= (1 - \theta_{k-1}) A_{k-1} + I_{\beta,k-1}(S_{k-1} + V_{k-1}) - I_{\gamma,k-1} - I_{\tau,k-1}, \\
I_k &= [1 - \omega_{k-1} - \theta_{k-1}] I_{k-1} + \omega_{k-1} A_{k-1} - I_{\gamma,k-1} - I_{\tau,k-1}, \\
R_k &= [1 - \lambda_{k-1} - \theta_{k-1}] R_{k-1} + h I_{\gamma,k-1}, \\
V_k &= \left[1 - \mu_{k-1} - \theta_{k-1} - I_{\beta,k-1}\right] V_{k-1} + h \alpha_{k-1} S_{k-1}, \\
B_k &= [1 - \nu_{k-1} - \theta_{k-1}] B_{k-1} + \mu_{k-1} V_{k-1}.
\end{aligned} \quad (24)$$

Algorithm to solve the **IDP** with ($h = 1$): Starting with the non-negative initial data A_1 and S_1, I_1, R_1, V_1, B_1, where $S_1 > 0$ and $A_1 \geq I_1 > 0$, we perform the following steps:

1. Since the values $\{Dtotal_k\}_{k=1}^K$ and $\{Rtotal_k\}_{k=1}^K$ are given and non-decreasing with respect to k, we find via (23) the non-negative values $\{I_{\gamma,k-1}\}_{k=2}^K$, $\{I_{\tau,k-1}\}_{k=2}^K$.

2. Since $N_1 = S_1 + A_1 + R_1 + V_1 + B_1$ is given, the relations (14), (21) and (22) imply

$$N_k = (1 + \Lambda_{k-1} - \theta_{k-1}) N_{k-1} - I_{\tau,k-1}, \quad k = 2, 3, \ldots, K. \quad (25)$$

3. Since the values $\{Vtotal_k\}_{k=1}^K$ are given, we find the values of the vaccination parameter

$$\alpha_{k-1} = \sigma \frac{Vtotal_k - Vtotal_{k-1}}{N_{k-1}}, \quad k = 2, 3, \ldots, K, \quad (26)$$

where σ is the vaccine effectiveness.

4. Now, using (24), we calculate for $k = 2, 3, \ldots, K$ the values

$$I_{\beta,k-1} = \frac{1}{S_{k-1} + V_{k-1}} [A_k + (\theta_{k-1} - 1) A_{k-1} + I_{\gamma,k-1} + I_{\tau,k-1}] \quad (27)$$

and

$$\begin{pmatrix} S_k \\ I_k \\ R_k \\ V_k \\ B_k \end{pmatrix} = Q_{k-1} \begin{pmatrix} S_{k-1} \\ I_{k-1} \\ R_{k-1} \\ V_{k-1} \\ B_{k-1} \end{pmatrix} + u_{k-1},$$

where

$$Q_k := \begin{pmatrix} 1 - \alpha_k - \theta_k - I_{\beta,k} & 0 & \lambda_k & 0 & \nu_k \\ 0 & 1 - \omega_k - \theta_k & 0 & 0 & 0 \\ 0 & 0 & 1 - \lambda_k - \theta_k & 0 & 0 \\ \alpha_k & 0 & 0 & 1 - \mu_k - \theta_k - I_{\beta,k} & 0 \\ 0 & 0 & 0 & \mu_k & 1 - \nu_k - \theta_k \end{pmatrix},$$

$$u_k := \left(\Lambda_k N_k, \omega_k A_k - I_{\gamma,k} - I_{\tau,k}, I_{\gamma,k}, 0, 0\right)^T.$$

5. Now, we are able to calculate the number of exposed individuals

$$E_k = A_k - I_k, \quad k = 1, 2, \ldots, K.$$

6. Finally, if $(1-\xi)I_{k-1} + \xi I_k \neq 0$ and $(1-\eta)I_{k-1} + \eta I_k \neq 0$, we calculate

$$\beta_{k-1}(\xi,\eta) = N_{k-1} \frac{I_{\beta,k-1}}{(1-\xi)I_{k-1} + \xi I_k},$$

$$\gamma_{k-1}(\xi,\eta) = \frac{I_{\gamma,k-1}}{(1-\eta)I_{k-1} + \eta I_k}, \qquad (28)$$

$$\tau_{k-1}(\xi,\eta) = \frac{I_{\tau,k-1}}{(1-\eta)I_{k-1} + \eta I_k}$$

for $k = 2, 3, \ldots, K$.

7. If one or more of the calculated values are negative or some of the values $S_{k-1} + V_{k-1}$, $(1-\xi)I_{k-1} + \xi I_k$, $(1-\eta)I_{k-1} + \eta I_k$ are equal to zero for some k, the algorithm must stop. Otherwise, the algorithm continues and the problem IDP can be solved.

Now we are able to compute the daily values of the basic reproduction number (12). According to (4) we have $b\Omega = \beta \Lambda$ and using the obtained parameters we obtain

$$\mathfrak{R}_{0,k}(\xi,\eta) := \frac{\omega_k \beta_k(\xi,\eta) \Lambda_k (\nu_k + \theta_k)(\alpha_k + \mu_k + \theta_k)}{\theta_k(\omega_k + \theta_k)(\gamma_k(\xi,\eta) + \tau_k(\xi,\eta) + \theta_k)(\alpha_k \nu_k + (\mu_k + \theta_k)(\alpha_k + \nu_k + \theta_k))}. \qquad (29)$$

for $k = 1, 2, \ldots, K-1$.

Remark 7. *We note that the computed values $\{g_k\}_{k=2}^K$ do not depend on the weights ξ and η (see the above algorithm to solve the IDP). However, the calculated parameters $\{\beta_{k-1}(\xi,\eta)\}_{k=2}^K$, $\{\gamma_{k-1}(\xi,\eta)\}_{k=2}^K$, $\{\tau_{k-1}(\xi,\eta)\}_{k=2}^K$ (see (28)) and $\{\mathfrak{R}_{0,k-1}(\xi,\eta)\}_{k=2}^K$ are different for different weights ξ and η, i.e., for different schemes from family (13).*

7. Validation of SEIRS-VB Models

For validation of the SEIRS-VB model we used official COVID-19 data from Bulgaria (see [41,42]). The first reported COVID-19 case in Bulgaria was on 8 March 2020. We consider the time period: 8 March 2020–12 February 2023.

7.1. Time-Discrete SEIRS-VB Model

In this section, we show that solving the inverse problem IDP with official COVID-19 data for Bulgaria leads to a solution with biologically reasonable properties.

We make several assumptions analogous to those in [4]:

1. The following parameters are constant during the time period under consideration:
   - $\Lambda_k = \Lambda$ is the average birth rate for 2015–2020;
   - $\theta_k = \theta$ is the average natural mortality rate for 2015–2020;
   - $\omega_k = 1/T_e$, where T_e is the incubation (latency) period for the dominant variant of SARS-CoV-2;
   - $\mu_k = 1/T_a$, where T_a is the average time taken for antibodies to develop;
   - $\nu_k = 1/T_b$, where T_b is the duration of the immune responses in individuals with vaccination-acquired immunity;
   - $\lambda_k = 1/T_r$, where T_r is the duration of the immune responses in recovered individuals.

2. Several vaccines were in use during the COVID-19 mass vaccination campaign in Bulgaria—Comirnaty (Pfizer/BioNTech), Comirnaty Original/Omicron BA.1, Comirnaty Original/Omicron BA.4-5, Spikevax (COVID-19 Vaccine Moderna), Spikevax Bivalent Original/Omicron BA.1, COVID-19 Vaccine Janssen, Vaxzevria (AstraZeneca) Nuvaxovid (NVX-CoV2373), and COVID-19 Vaccine (inactivated, adjuvanted) Valneva, Vidprevtyn Beta. Taking into account the number of the fully vaccinated and boosted people with these vaccines, and the product information in [43], we applied the average parameters specified in Table 1.

Table 1. The given parameter's values.

Parameter	Description	Values
Λ	birth rate	2.4095×10^{-5}
θ	natural mortality rate	4.1904×10^{-5}:
T_e	latency period	7 days [1], 6 days [2], 5 days [3], 4 days [4]
T_a	time taken for antibodies to develop	14 days
T_b	duration of vaccine-based immunity	180 days
T_r	duration of disease-based immunity	180 days
σ	vaccine effectiveness	0.85 [2], 0.70 [3], 0.45 [4]

For [1] Wuhan variant, [2] Alpha variant, [3] Delta variant, [4] Omicron variant.

The initial data for the COVID-19 pandemic in Bulgaria are

$$g_1 = (S_1, E_1, I_1, R_1, V_1, B_1) = (6941259, 0, 4, 0, 0, 0). \tag{30}$$

Let us introduce the grid

$$\begin{aligned}\Pi &:= \Pi_\xi \times \Pi_\eta, \\ \Pi_\xi &:= \{\xi_m = m/100, \ m = 0, 1, 2, \ldots 100\}, \\ \Pi_\eta &:= \{\eta_n = n/100, \ n = 0, 1, 2, \ldots 100\}.\end{aligned} \tag{31}$$

Then we solve the IDP for each point (ξ_m, η_n) from the grid Π. The algorithm described in Section 6.2 provides a unique non-negative solution

$$\{(S_k, E_k, I_k, R_k, V_k, B_k)\}_{k=1}^K, \ \{(\alpha_{k-1}, \beta_{k-1}(\xi_m, \eta_n), \gamma_{k-1}(\xi_m, \eta_n), \tau_{k-1}(\xi_m, \eta_n))\}_{k=2}^K$$

to the *IDP* and $S_k > 0$, $I_k > 0$ for $k = 1, 2, \ldots, K$.

Then, using (29) we compute the daily values $\Re_{0,k}(\xi, \eta)$, $k = 1, 2, \ldots, K-1$ of the basic reproduction number. Figure 2 give $\Re_0$ values on a weakly basis (starting from the 5th week of the epidemic), obtained with the difference schemes from (13) with $\xi = 0.83$, $\eta = 0.55$ (In the next section we explain this choice of the weights). We will note that the average value of the basic reproduction number in the first week of dominance of the Delta variant is 1.0179, while in the first week of dominance of the Omicron variant is 2.2660. The Omicron variant invasion begins with high numbers of active cases and causes the biggest COVID-19 wave in Bulgaria. We observe that the vaccination campaign and the introduction of green certificates have a similar ripple effect as lockdowns in previous variants of the SARS-CoV-2 virus. The subsequent removal of all restrictive measures and extremely low vaccination rates in Bulgaria have maintained high levels of the basic reproduction number - the mean value of $\Re_0$ for the time-period 4 September 2022–12 February 2023 is 1.8357.

With the same weights $\xi = 0.83$, $\eta = 0.55$, we also calculate the number of individuals in categories for which there is no official data. These are exposed individuals E_k, infectious individuals I_k and vaccinated susceptible individuals V_k in Figure 3, as well as individuals with antibodies B_k and individuals with disease-acquired immunity R_k in Figure 4. We see the low number of individuals with vaccine-induced immunity. At the peak of the vaccination campaign, it was less than 12% of Bulgaria's population. In our opinion, this is one of the main reasons for the high number of susceptible individuals and the high mortality rate in the country.

Remark 8. *Figures 2 and 3 clearly show that peaks in the basic reproduction number $\Re_0$ correspond to peaks in the number of exposed individuals E, followed (corresponding to the incubation period) by peaks in the number of infectious individuals I. Furthermore, in Figures 3 and 4 we see that changes in the number of vaccinated susceptible individuals V (which depends on the vaccination rate and the effectiveness of the vaccines) lead to changes in the number of individuals with vaccine-induced immunity B. The dependence is not directly proportional because the individuals from category V* may become infected and the duration of the immunity in individuals with vaccination-acquired immunity is much longer than the time taken for antibodies to develop.

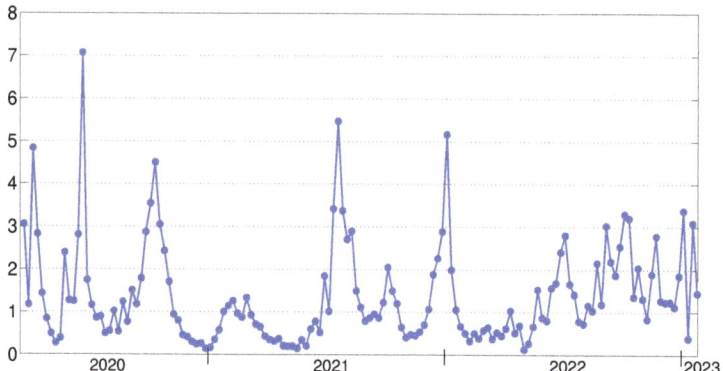

Figure 2. Basic reproduction number on a weekly basis when $\xi = 0.83$, $\eta = 0.55$.

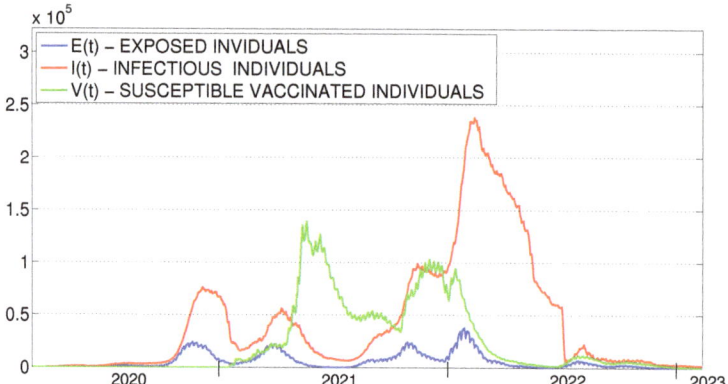

Figure 3. Model curves for exposed, infectious and vaccinated susceptible individuals.

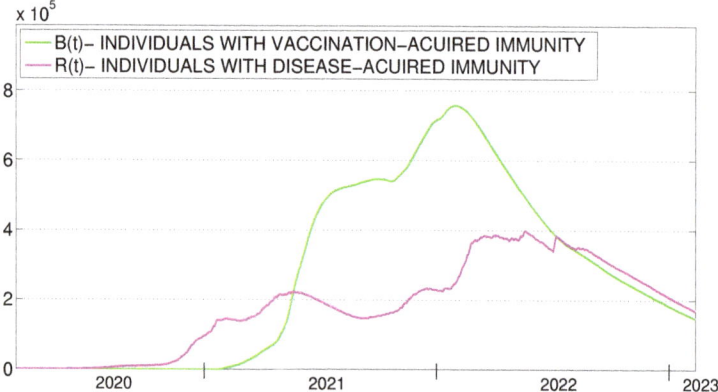

Figure 4. Model curves for individuals with disease-acquired immunity and individuals with vaccine-induced immunity.

At the end of this section, note that we do not use the SEIRS-VB model with known values of infection, recovery and diseases mortality rate, which are determined depending on the measures of social distancing, the characteristics of the virus and effectiveness and safety of available treatments for COVID-19. We do not know the exact dependence between them. We solve an inverse problem and find the daily values of these parameters and basic reproduction number that mathematically represents the complex behaviour of the official available data. This is why the obtained values are sensitive to changes in the data (see Section 8). Thus, with the obtained values, we can model the daily changes in the number of individuals in the different groups in the model with very high accuracy (see Section 7.2). If we use a constant values we lose this accuracy.

7.2. Differential SEIRS-VB Model

In this section we solve the differential problems (1) and (2) for the considered time period $[t_0, T]$ using parameters obtained by different difference schemes from (31) (with step $h = 1$). By the (reference) solution $\widetilde{X} = (\widetilde{S}, \widetilde{E}, \widetilde{I}, \widetilde{R}, \widetilde{V}, \widetilde{B})$ of the differential problem we mean a numerical solution, obtained with a Runge–Kutta 4th/5th-order method with step size 0.01. Then, we compare the obtained values of the functions $\widetilde{A}(t_k) = \widetilde{E}(t_k) + \widetilde{I}(t_k)$ with available official values of the active cases $A(t_k) = E(t_k) + I(t_k)$.

We denote by $f(\widetilde{p}, \hat{p}, \widetilde{X})$ the right-hand side function of the system (1), where parameters $\widetilde{p}$ and $\hat{p}$ are given by (18) and (19).

Now, we fix an arbitrary point (ξ, η) from the grid Π (see (31)). For each day of the considered time frame $t_1, t_2, \ldots, t_k$ we solve a Cauchy problem with constant parameters—the daily values found in Section 7.1.

Let $\widetilde{X}_k(t, \xi, \eta) := (\widetilde{S}_k(t, \xi, \eta), \widetilde{E}_k(t, \xi, \eta), \widetilde{I}_k(t, \xi, \eta), \widetilde{R}_k(t, \xi, \eta), \widetilde{V}_k(t, \xi, \eta), \widetilde{B}_k(t, \xi, \eta))$, $\widetilde{N}_k(t)$ be the numerical solution of the Cauchy problem

$$
\left|
\begin{array}{l}
\dfrac{d\widetilde{X}_k}{dt}(t, \xi, \eta) = f(\widetilde{p}_{k-1}, \hat{p}_{k-1}(\xi, \eta), \widetilde{X}_k(t, \xi, \eta)), \\[6pt]
\dfrac{d\widetilde{N}_k}{dt}(t, \xi, \eta) = (\Lambda_{k-1} - \theta_{k-1})\widetilde{N}_k(t, \xi, \eta) - \tau_{k-1}(\xi, \eta)\widetilde{I}(t, \xi, \eta), \\[6pt]
\widetilde{X}_k(t_{k-1}, \xi, \eta) = \widetilde{X}_{k-1}(t_{k-1}, \xi, \eta), \ \widetilde{N}_k(t_{k-1}) = \widetilde{N}_{k-1}(t_{k-1}, \xi, \eta)
\end{array}
\right.
\tag{32}
$$

in the interval $[t_{k-1}, t_k]$, obtained with the built-in function ode45 in Matlab. The solution to the problem at the end of the day t_{k-1} are used as initial data in the next problem for the day t_k. Of course $\widetilde{X}(t_1) = g_1$, $\widetilde{N}(t_1) = S_1 + E_1 + I_1 + R_1 + V_1 + B_1$ (see (30)). For the solution we have $\widetilde{N}_k(t, \xi, \eta) = \widetilde{S}_k(t, \xi, \eta) + \widetilde{E}_k(t, \xi, \eta) + \widetilde{I}_k(t, \xi, \eta) + \widetilde{R}_k(t, \xi, \eta) + \widetilde{V}_k(t, \xi, \eta) + \widetilde{B}_k(t, \xi, \eta)$ for $t_{k-1} \le t \le t_k$.

To find which difference scheme from (31) for $(\xi, \eta) \in \Pi$ gives the best approximation of the reported numbers of active cases $A_{pandemic} = (A_1, A_2, \ldots, A_K)$, we study the relative errors

$$Error(\ell_2, A)(\xi, \eta) := \frac{\|\widetilde{A}_{pandemic}(\xi, \eta) - A_{pandemic}\|_2}{\|A_{pandemic}\|_2} \tag{33}$$

and

$$Error(\ell_\infty, A)(\xi, \eta) := \frac{\|\widetilde{A}_{pandemic}(\xi, \eta) - A_{pandemic}\|_\infty}{\|A_{pandemic}\|_\infty} \tag{34}$$

on the grid Π. Here, $\widetilde{A}_{pandemic}(\xi, \eta) = (\widetilde{A}_1(t_1, \xi, \eta), \widetilde{A}_2(t_2, \xi, \eta), \ldots, \widetilde{A}_K(t_K, \xi, \eta))$ are the computed values by (32).

Conducted numerical experiments show that both relative errors (36) and (34) are largest when $\xi = \eta = 0$ (see Figures 5 and 6), that is, when we use an explicit Euler method. We observe that the smallest values of the two errors are reached at two adjacent grid points. In Figure 5 we see that the minimum of the $Error(l_2, A)(\xi, \eta)$ is reached at point $\xi^* = 0.83, \eta^* = 0.56$ and its value is $Error(l_2, A)(\xi^*, \eta^*) = 0.007$. At the same time, in Figure 6 we see that the minimum of the $Error(l_\infty, A)(\xi, \eta)$ is reached at point $\hat{\xi} = 0.83$, $\hat{\eta} = 0.55$ with a value of $Error(l_\infty, A)(\hat{\xi}, \hat{\eta}) = 0.0048$.

In Figure 7 we see that in the case $\xi = \hat{\xi}$, $\eta = \hat{\eta}$ the model curve and the curve of reported data for active cases almost coincide. In this case the suggested method for parameter identification in the discrete problem leads to the parameter values which are very close to the time-dependent coefficients in the differential problem. In fact, we improve (Figures 7 and 8) our result in [4] where the difference scheme with $\xi = 1, \eta = 0$ is used. On the other hand, it is noted that the explicit Euler method ($\xi = 0, \eta = 0$) cannot be recommended for parameter approximation in the differential SEIRS-VB model. Figure 9 clearly shows the large difference between the model curve and the curve of reported data in this case.

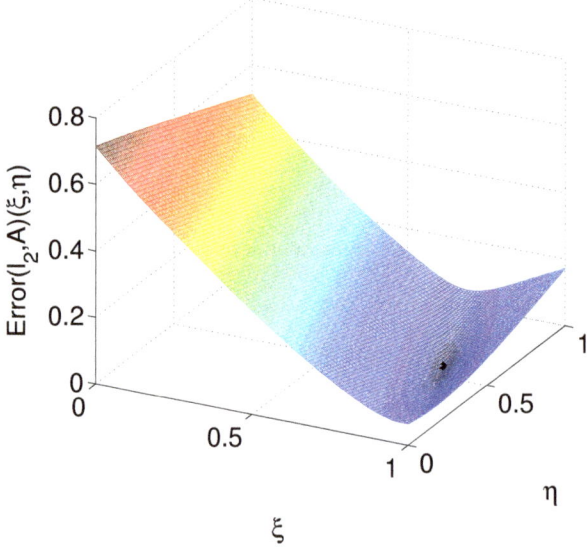

Figure 5. Values of the $Error(\ell_2, A)(\xi, \eta)$ on the grid Π.

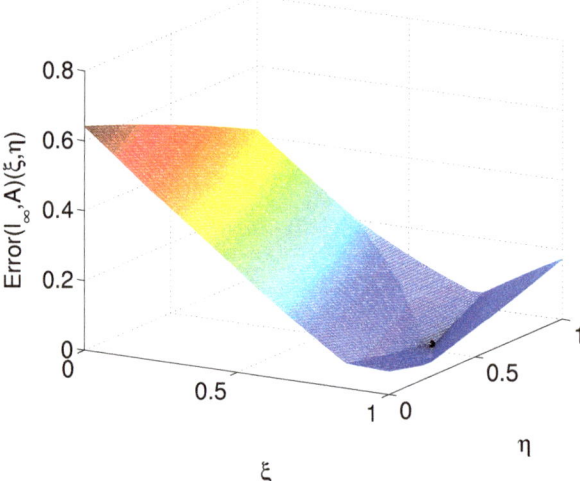

Figure 6. Values of the $Error(\ell_\infty, A)(\xi, \eta)$ on the grid Π.

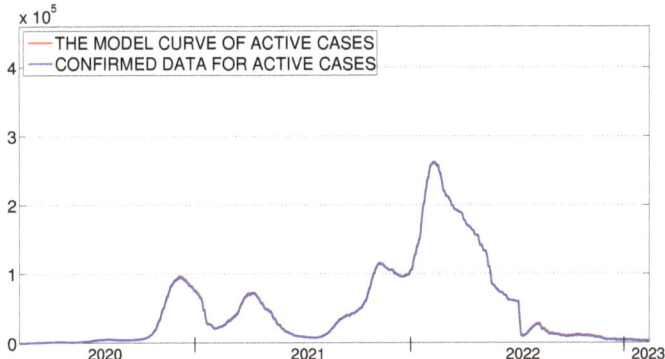

Figure 7. Confirmed data and SEIRS-VB model's curve of active cases, when $\xi = 0.83$, $\eta = 0.55$.

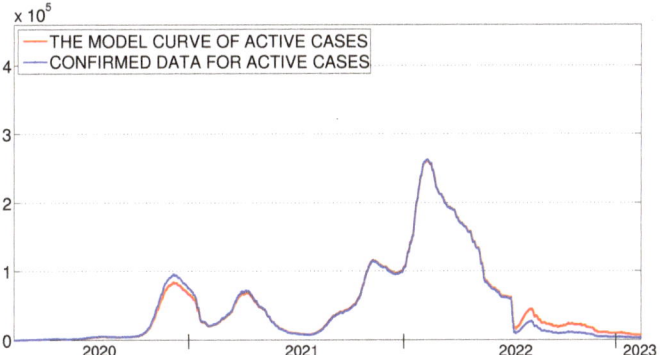

Figure 8. Confirmed data and SEIRS-VB model's curve of active cases, when $\xi = 1$, $\eta = 0$.

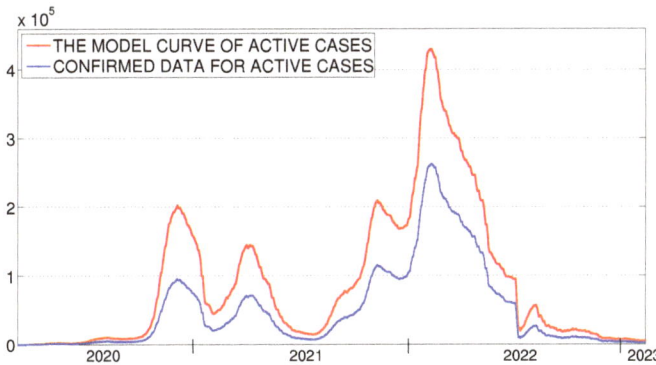

Figure 9. Confirmed data and SEIRS-VB model's curve of active cases, when $\xi = 0$, $\eta = 0$ (explicit Euler method).

Remark 9. *Euler's time-stepping methods are widely used in computational sciences. The advantage of the explicit (forward) Euler scheme is its simple implementation but it is conditionally stable. The implicit (backward) Euler scheme is unconditionally stable. Both explicit and implicit Euler schemes have first-order accuracy. It is worth noting that a weighted convex linear combination of these can provide higher accuracy while maintaining stability. This is convincingly confirmed by the results presented in this section when $\xi = 0.83$ and $\eta = 0.55$.*

8. Parameter Sensitivity Analysis

The aim of this section is to perform a sensitivity analysis of the basic reproduction number $\Re_0$. Sensitivity analysis helps to find the parameters in the SEIRS-VB model that have a dominant effect on $\Re_0$. The obtained results can be used for design strategies to reduce the base reproduction number and the number of active cases. The aim is to limit the spread of the virus and to prevent the health system from being overwhelmed.

Let us fix one arbitrary parameter

$$\varrho \in \{\alpha, \beta, \gamma, \tau, \Lambda, \theta, \omega, \lambda, \mu, \nu\}$$

in the differential model SEIRS-VB (1).

To evaluate the influence of a single input parameter on the basic reproduction number in different SEIR-type models with constant coefficients, a normalized forward sensitivity index was used in [44–47].

To apply a similar technique to the case of the SEIRS-VB model with time-varying coefficients we consider the time-dependent normalized forward sensitivity index of $\Re_0$ with respect to ϱ:

$$Y_\varrho(t) := \frac{\varrho(t)}{\Re_0(t)} \frac{\partial \Re_0}{\partial \varrho}(t). \tag{35}$$

Let us denote by $\hat{\Re}_0 := (\hat{\Re}_{0,1}, \hat{\Re}_{0,2}, \ldots, \hat{\Re}_{0,K-1})$ the daily values of $\Re_0$, found in Section 7, to solve IDPs for the difference scheme with weights $\xi = \hat{\xi} = 0.83$, $\eta = \hat{\eta} = 0.55$ and using (29). In a similar way, using the daily values of the parameters and (12) we found the values $\frac{\partial \Re_0}{\partial \varrho}(t_k)$ for each day t_k of the considered time-period of the epidemic in Bulgaria (8 March 2020–12 February 2023). Then, using (35), we calculate the daily sensitivity indices of the parameters of $\Re_0$. The mean values of the sensitivity indices are given in Table 2.

Table 2. The mean values of the sensitivity indices of $\Re_0$ with respect to the SEIRS-VB model parameters.

Parameter	Description	Sensitivity Index
$\alpha(t)$	vaccination parameter	−0.0571
$\beta(t)$	transmission rate	1
$\gamma(t)$	recovery rate	−0.9251
$\tau(t)$	mortality rate of infectious people	−0.0568
$\Lambda(t)$	birth rate	1
$\theta(t)$	natural mortality rate	−0.9818
$\omega(t)$	latency rate	2.2655×10^{-4}
$\lambda(t)$	reinfection rate of recovered individuals	0
$\mu(t)$	antibody rate	-6.4420×10^{-4}
$\nu(t)$	reinfection rate of vaccinated individuals	0.0407

From the sensitivity analysis as presented in Table 2, it follows that:

Increasing the vaccination parameter α (i.e., increasing vaccine effectiveness or increasing vaccination rate) by 1% leads to a decrease in $\Re_0$ by 0.0571%. An increase in the rate of re-infection of vaccinated people (i.e., reducing the duration of vaccine-based immunity) by 1% leads to an increase in $\Re_0$ by 0.0407%. The influence of the antibody rate (i.e., the time taken for antibodies to develop) on $\Re_0$ is significantly smaller. Therefore, to reduce the basic reproduction number it is of great importance to increase vaccination rates, vaccines inventions with greater efficiency and prolong the immunity duration.

The transmission rate, the birth rate, the natural mortality rate and the recovery rate have a big impact on $\Re_0$. Therefore, it is extremely important to take the correct measures to limit the spread of the virus and effectively treat infected individuals. The latency rate and the antibody rate have a lower impact. $\Re_0$ does not depend on the reinfection rate of recovered individuals because at the disease-free equilibrium point the recovery rate of individuals is zero.

In order to solve the inverse problem IDP in Section 7 we assume some values for parameters $\omega, \lambda, \mu, \nu$. Now, we see that they have low impact on $\mathfrak{R}_0$ if other parameters in the SEIRS-VB model do not depend on them. In fact, Table 2 provides information on how the basic reproduction number changes when each parameter in the model is changed independently of the other parameters. In fact, the situation is more complicated. When we solve the IDP, we use daily values of the functions $A, Rtotal, Dtotal, Vtotal$ and parameters $\Lambda, \theta, \omega, \lambda, \nu, \mu$ to find the daily values of parameters $\alpha, \beta, \gamma, \tau$. Then we find the basic reproduction number. To study the influence of each input quantity in the IDP on the $\mathfrak{R}_0$ we fix an arbitrary quantity

$$z \in \{A, Rtotal, Dtotal, Vtotal, \Lambda, \theta, \omega, \lambda, \nu, \mu\}$$

and we denote by $\check{z}$ the value of z, selected in Section 7, when solving the IDP and $\check{\mathfrak{R}}_0$.

Now, we introduce the relative errors

$$Error(\ell_2, \mathfrak{R}_0; z, w) := \frac{\|\check{\mathfrak{R}}_0 - \mathfrak{R}_{0,z}^w(\hat{\xi}, \hat{\eta})\|_2}{\|\check{\mathfrak{R}}_0\|_2}, \qquad (36)$$

$$Error(\ell_\infty, \mathfrak{R}_0; z, w) := \frac{\|\check{\mathfrak{R}}_0 - \mathfrak{R}_{0,z}^w(\hat{\xi}, \hat{\eta})\|_\infty}{\|\check{\mathfrak{R}}_0\|_\infty}, \qquad (37)$$

where $\mathfrak{R}_{0,z}^w(\hat{\xi}, \hat{\eta})$ is the vector whose components are the daily values of $\mathfrak{R}_0(\hat{\xi}, \hat{\eta})$, calculated by solving the IDP when the daily values of the input quantity $\check{z}$ are multiplied by weight w and the daily values of the other input quantities are unchanged.

In Tables 3 and 4 we give the values of $Error(\ell_2, \mathfrak{R}_0; z, w)$ and $Error(\ell_\infty, \mathfrak{R}_0; z, w)$ calculated for all input quantities z of the IDP with weights $w = 1 \pm k/100$, i.e., $(\pm k\%)$ for $k = 1, 2, 3, 4$. We see that the influence of the parameters λ, μ and ν on the considered relative errors is very small. This once again confirms their choice as known parameters when solving the IDP. The influence of the latent rate is greater and it must be carefully chosen for each SARS-CoV-2 variant. Since the vaccination parameter α is a product of the vaccination rate and effectiveness, the influence of $Vtotal$ and σ is the same. The values of Λ and θ have a big impact on $\mathfrak{R}_0$, but we selected them from the available statistical data from the information system INFOSTAT of the National Statistical Institute of the Republic of Bulgaria.

Table 3. The values of $Error(\ell_2, \mathfrak{R}_0; z, w)$, calculated for all input values in the IDP with different rates of gauge.

IV [1]/RG [2]	−4%	−3%	−2%	−1%	+1%	+2%	+3%	+4%
A	2.8757×10^{-3}	2.1369×10^{-3}	1.4116×10^{-3}	6.9948×10^{-4}	6.8726×10^{-4}	1.3627×10^{-3}	2.0267×10^{-3}	2.6797×10^{-3}
$Dtotal$	7.6732×10^{-4}	5.7465×10^{-4}	3.8256×10^{-4}	1.9101×10^{-4}	1.9051×10^{-4}	3.8052×10^{-4}	5.7007×10^{-4}	7.5917×10^{-4}
$Rtotal$	2.1109×10^{-3}	1.5800×10^{-3}	1.0513×10^{-3}	5.2469×10^{-4}	5.2282×10^{-4}	1.0439×10^{-3}	1.5632×10^{-3}	2.0810×10^{-3}
$Vtotal$	1.7604×10^{-3}	1.3192×10^{-3}	8.7878×10^{-4}	4.3904×10^{-4}	4.3834×10^{-4}	8.7598×10^{-4}	1.3129×10^{-3}	1.7492×10^{-3}
σ	1.7604×10^{-3}	1.3192×10^{-3}	8.7878×10^{-4}	4.3904×10^{-4}	4.3834×10^{-4}	8.7598×10^{-4}	1.3129×10^{-3}	1.7492×10^{-3}
Λ	4.0000×10^{-2}	3.0000×10^{-2}	2.0000×10^{-2}	1.0000×10^{-2}	1.0000×10^{-2}	2.0000×10^{-2}	3.0000×10^{-2}	4.0000×10^{-2}
θ	8.5025×10^{-2}	6.2780×10^{-2}	4.1212×10^{-2}	2.0294×10^{-2}	1.9694×10^{-2}	3.8811×10^{-2}	5.7375×10^{-2}	7.5406×10^{-2}
ω	2.1812×10^{-2}	1.6168×10^{-2}	1.0655×10^{-2}	5.2667×10^{-3}	5.1491×10^{-3}	1.0184×10^{-2}	1.5109×10^{-2}	1.9927×10^{-2}
λ	4.1466×10^{-5}	3.0837×10^{-5}	2.0385×10^{-5}	1.0108×10^{-5}	9.9416×10^{-6}	1.9721×10^{-5}	2.9341×10^{-5}	3.8806×10^{-5}
μ	3.4809×10^{-6}	2.5785×10^{-6}	1.6980×10^{-6}	8.3874×10^{-7}	8.1893×10^{-7}	1.6187×10^{-6}	2.4000×10^{-6}	3.1633×10^{-6}
ν	6.6911×10^{-5}	4.9657×10^{-5}	3.2759×10^{-5}	1.6210×10^{-5}	1.5880×10^{-5}	3.1438×10^{-5}	4.6683×10^{-5}	6.1622×10^{-5}

[1] Input value, [2] Rate of gauge of input value.

Table 4. The values of $Error(\ell_\infty, \Re_0; z, w)$, calculated for all input values in IDP problem with different rates of gange.

IV [1]/RG [2]	−4%	−3%	−2%	−1%	+1%	+2%	+3%	+4%
A	2.6254×10^{-3}	1.9474×10^{-3}	1.2842×10^{-3}	6.3518×10^{-4}	6.2181×10^{-4}	1.2307×10^{-3}	1.8270×10^{-3}	2.4111×10^{-3}
$Dtotal$	6.8446×10^{-4}	5.1343×10^{-4}	3.4235×10^{-4}	1.7120×10^{-4}	1.7126×10^{-4}	3.4258×10^{-4}	5.1396×10^{-4}	6.8539×10^{-4}
$Rtotal$	1.8274×10^{-3}	1.3712×10^{-3}	9.1456×10^{-4}	4.5749×10^{-4}	4.5791×10^{-4}	9.1624×10^{-4}	1.3750×10^{-3}	1.8342×10^{-3}
$Vtotal$	1.3908×10^{-3}	1.0432×10^{-3}	6.9554×10^{-4}	3.4781×10^{-4}	3.4789×10^{-4}	6.9586×10^{-4}	1.0439×10^{-3}	1.3921×10^{-3}
σ	1.3908×10^{-3}	1.0432×10^{-3}	6.9554×10^{-4}	3.4781×10^{-4}	3.4789×10^{-4}	6.9586×10^{-4}	1.0439×10^{-3}	1.3921×10^{-3}
Λ	4.0000×10^{-2}	3.0000×10^{-2}	2.0000×10^{-2}	1.0000×10^{-2}	1.0000×10^{-2}	2.0000×10^{-2}	3.0000×10^{-2}	4.0000×10^{-2}
θ	8.5060×10^{-2}	6.2805×10^{-2}	4.1228×10^{-2}	2.0302×10^{-2}	1.9702×10^{-2}	3.8827×10^{-2}	5.7398×10^{-2}	7.5436×10^{-2}
ω	2.1824×10^{-2}	1.6177×10^{-2}	1.0661×10^{-2}	5.2696×10^{-3}	5.1519×10^{-3}	1.0190×10^{-2}	1.5118×10^{-2}	1.9938×10^{-2}
λ	1.3093×10^{-5}	9.7225×10^{-6}	6.4176×10^{-6}	3.1773×10^{-6}	3.1159×10^{-6}	6.1718×10^{-6}	9.1691×10^{-6}	1.2109×10^{-5}
μ	1.1444×10^{-6}	8.4707×10^{-7}	5.5739×10^{-7}	2.7511×10^{-7}	2.6817×10^{-7}	5.2962×10^{-7}	7.8456×10^{-7}	1.0332×10^{-6}
ν	1.5961×10^{-5}	1.1823×10^{-5}	7.7856×10^{-6}	3.8454×10^{-6}	3.7530×10^{-6}	7.4162×10^{-6}	1.0992×10^{-5}	1.4483×10^{-5}

[1] Input value, [2] Rate of gaugeof input value.

9. Discussion

Since the emergence of SARS-CoV-2 in December 2019, the number of confirmed COVID-19 infections worldwide has surpassed 682 million, resulting in nearly 7 million deaths. To date, nine COVID-19 vaccine candidates based on the original Wuhan-Hu-1 strain have been developed, all demonstrating efficacies over 50% against symptomatic COVID-19 disease. These include NVX-CoV2373 (∼96%), BNT162b2 (∼95%), mRNA-1273 (∼94%), Sputnik V (∼92%), AZD1222 (∼81%), BBIBP-CorV (∼79%), Covaxin (∼78%), Ad26.CoV.S (∼66%), and CoronaVac (∼51%) [48–51]. However, the rapid emergence and spread of SARS-CoV-2 variants of concern (VOCs) could jeopardize the efficacy of these vaccines by evading neutralizing antibodies and/or cell-mediated immunity. Moreover, rare adverse events have been reported shortly after administration of viral vector and mRNA vaccines.

The effectiveness of COVID-19 vaccines can vary depending on the specific vaccine, the variant of the virus, and other factors such as the age and health of the person receiving the vaccine. Generally speaking, however, all vaccines currently authorized for emergency use by various regulatory agencies around the world have been shown to be highly effective in preventing severe illness, hospitalization, and death from COVID-19.

For example, the Pfizer–BioNTech vaccine has been shown to be approximately 95% effective in preventing symptomatic COVID-19 infections, while the Moderna vaccine has been shown to be approximately 94.1% effective. The Johnson & Johnson vaccine has been shown to be approximately 66.3% effective at preventing moderate to severe COVID-19 infections. It is important to note that these numbers are based on clinical trials and real-world studies, and they can change over time as more data become available However, the general consensus among health experts is that the COVID-19 vaccines are highly effective at preventing severe illness, hospitalization, and death, and they are a crucial tool in ending the COVID-19 pandemic.

No one can predict the future with certainty, but one can say that as of our knowledge cutoff in September 2021, the COVID-19 pandemic was still ongoing and evolving, while vaccines have been developed and authorized for emergency use in many countries, new variants of the virus have emerged, and there have been ongoing efforts to distribute and administer vaccines globally.

By the end of 2023, the COVID-19 pandemic may have been largely contained through a combination of vaccination, public health measures, and other interventions. However it is also possible that new variants of the virus may emerge or that there may be other challenges in controlling the spread of the disease.

Ultimately, the trajectory of the pandemic will depend on a variety of factors, including the effectiveness of vaccines, the uptake of vaccines, the emergence of new variants, and the ongoing efforts of individuals and governments to contain the spread of the virus.

In this aspect, the development and application of mathematical epidemiological models could provide valuable tools to understand the spread of COVID-19 and develop strategies to control its spread. However, it is essential to remember that these models are based on simplifying assumptions and may not fully capture the complex dynamics of the disease in real-world settings.

10. Conclusions

In conclusion, since the emergence of SARS-CoV-2 in December 2019, the COVID-19 pandemic has had a significant global impact, with millions of confirmed infections and deaths worldwide. Several COVID-19 vaccine candidates have been developed, demonstrating varying degrees of efficacy against symptomatic COVID-19 disease caused by different strains of the virus. However, the emergence of SARS-CoV-2 variants of concern and the occurrence of rare adverse events following vaccination pose challenges to the effectiveness of these vaccines. Nevertheless, authorised vaccines have shown high effectiveness in preventing severe illness, hospitalisation, and death from COVID-19. Vaccines such as Pfizer–BioNTech, Moderna, and Johnson & Johnson have demonstrated substantial efficacy rates in clinical trials and real-world studies. It is important to note that these effectiveness figures can evolve as more data becomes available and new variants emerge. While the ongoing vaccination efforts and public health measures offer hope for containing the pandemic, the future trajectory of the COVID-19 pandemic remains uncertain. Factors such as vaccine effectiveness, vaccine uptake, the emergence of new variants, and continued efforts to control the spread of the virus will shape the course of the pandemic. Mathematical epidemiological models provide valuable tools to understand the dynamics of the disease and develop strategies to mitigate its spread, although they may not fully capture the complexities of real-world scenarios.

In this paper, a time-continuous category model SEIRS-VB (introduced in [4]) was used to model the long-term behaviour of the COVID-19 pandemic. The basic reproduction number $\Re_0$ associated with this model in the autonomous case was defined. It is shown that the unique disease-free equilibrium point is locally asymptotically stable when $\Re_0 < 1$ and locally unstable if $\Re_0 > 1$. Furthermore, a family of time-discrete models with weights was proposed that preserves the biological properties of the differential model. The weights wee then chosen to minimize the relative error of the solution to the time-discrete SEIRS-VB model with respect to the exact (reference) solution to the time-continuous SEIRS-VB model. The constructed weighted time-discrete model may be different for a different dataset or a different country. In this paper, the efficiency of the method was verified on COVID-19 data from Bulgaria. In this situation we found the discrete model from a weighted family, giving a better approximation of the continuous model compared to the more standard discrete model used in [4]. The sensitivity analysis presented showed the impact of the applied measures to limit the spread of the virus and the treatment protocols used on the basic reproduction number $\Re_0$. The sensitivity analysis also confirmed the importance of the vaccination rates and efficacy of the vaccines in use.

We note that in our previous work [52] we proposed a multiple vaccine model which takes into account the different parameters of each vaccine—effectiveness and vaccination rate. A similar approach could be applied to the SEIRS-VB model. However, it will become more complicated for analytical study. However, in this case, more complex numerical experiments could be performed, generating improved accuracy. This is a very promising topic for future work.

The topic of short-term forecasting of the COVID-19 epidemic in Bulgaria using the SEIRS model with vaccination was discussed in our previous work [4]. The new results presented here create new opportunities to improve the accuracy of the forecasting of the COVID-19 epidemic, including the analysis of the longer-term evolution of the underlying processes.

It is crucial to stay informed through reliable sources such as health authorities and follow recommended guidelines and precautions to protect oneself and others from COVID-19.

Author Contributions: Methodology, S.M., N.P., I.U. and T.H.; investigation, S.M., N.P. and T.H.; validation, T.H.; visualization, T.H.; writing—original draft, N.P. and T.H.; writing—review and editing, S.M. and I.U. All authors have read and agreed to the published version of the manuscript.

Funding: The work of N. Popivanov and Ts. Hristov was partially supported by the Bulgarian NSF under grant KP-06-H52/4-2021 and by the Sofia University under grant 80-10-81/2023. The work of I. Ugrinova was partially supported by the Bulgarian Ministry of Education under Grant D 01-397/18.12.2020.

Data Availability Statement: The codes used for this study are available upon request from the corresponding author.

Conflicts of Interest: The authors declare no conflict of interest.

References

1. Dietz, K.; Heesterbeek, J.A.P. Daniel Bernoulli's epidemiological model revisited. *Math. Biosci.* **2002**, *180*, 1–21. [CrossRef] [PubMed]
2. Kermack, W.; McKendrick, A. A Contribution to the Mathematical Theory of Pandemics. *Proc. R. Soc. Lond. Ser. A* **1927**, *115*, 700–721. [CrossRef]
3. Khan, A.H. *Modeling the Spread of COVID-19 Pandemic in Morocco. Challenges in Modeling of an Outbreak's Prediction, Forecasting and Decision Making for Policy Makers, Infosys Science Foundation Series in Mathematical Sciences*; Springer: Singapore, 2021; pp. 377–408. [CrossRef]
4. Margenov, S.; Popivanov, N.; Ugrinova, I.; Hristov, T. Mathematical Modeling and Short-Term Forecasting of the COVID-19 Epidemic in Bulgaria: SEIRS Model with Vaccination. *Mathematics* **2022**, *10*, 2570. [CrossRef]
5. Kabanikhin, S.; Bektemessov, I.; Krivorotko, O.; Bektemessov, Z. Determination of the coefficients of nonlinear ordinary differential equations systems using additional statistical information. *Int. J. Math. Phys.* **2019**, *10*, 36–42. [CrossRef]
6. Krivorotko, O.; Kabanikhin, S.; Sosnovskaya, M.; Andornaya, D. Sensitivity and Identifiability Analysis of COVID-19 Pandemic Models. *Vavilov J. Genet. Breed.* **2021**, *25*, 82–91. [CrossRef] [PubMed]
7. Marinov, T.; Marinova, R. Dynamics of COVID-19 using inverse problem for coefficient identification in SIR epidemic models. *Chaos Solitons Fractals X* **2020**, *5*, 100041. [CrossRef]
8. Marinov, T.; Marinova, R. Inverse problem for adaptive SIR model: Application to COVID-19 in Latin America. *Infect. Dis. Model.* **2022**, *5*, 134–148. [CrossRef]
9. Leonov, A.S.; Nagornov, O.V.; Tyuflin, S.A. Inverse problem for coefficients of equations describing propagation of COVID-19 epidemic. *J. Phys. Conf. Ser.* **2021**, *2036*, 012028. [CrossRef]
10. Hethcote, H. The Mathematics of Infectious Diseases. *Siam Rev.* **2000**, *42*, 599–653. Available online: https://www.jstor.org/stable/2653135 (accessed on 15 February 2023). [CrossRef]
11. Li, M.; Muldowney, J. Global stability for the SEIR model in epidemiology. *Math. Biosci.* **1995**, *125*, 155–164. [CrossRef]
12. Korobeinikov, A.; Maini, P. A Lyapunov function and global properties for SIR and SEIR epidemiological models with nonlinear incidence. *Math. Biosci. Eng.* **2004**, *1*, 57–60. [CrossRef]
13. Ghostine, R.; Gharamti, M.; Hassrouny, S.; Hoteit, I. An Extended SEIR Model with Vaccination for Forecasting the COVID-19 Pandemic in Saudi Arabia Using an Ensemble Kalman Filter. *Mathematics* **2021**, *9*, 636. [CrossRef]
14. Al-Shbeil, I.; Djenina, N.; Jaradat, A.; Al-Husban, A.; Ouannas, A.; Grassi, G. A New COVID-19 Pandemic Model Including the Compartment of Vaccinated Individuals: Global Stability of the Disease-Free Fixed Point. *Mathematics* **2023**, *11*, 576. [CrossRef]
15. Xu, D.-G.; Xu, X.-Y.; Yang, C.-H.; Gui, W.-H. Global Stability of a Variation Epidemic Spreading Model on Complex Networks. *Math. Probl. Eng.* **2015**, *2015*, 365049. [CrossRef]
16. Wangari, I. Condition for Global Stability for a SEIR Model Incorporating Exogenous Reinfection and Primary Infection Mechanisms. *Comput. Math. Methods Med.* **2020**, *2020*, 9435819. [CrossRef] [PubMed]
17. Li, M.Y.; Wang, L. Global Stability in Some Seir Epidemic Models. In *Mathematical Approaches for Emerging and Reemerging Infectious Diseases: Models, Methods, and Theory. The IMA Volumes in Mathematics and its Applications 126*; Springer: New York, NY, USA, 2002. [CrossRef]
18. Lobatog, F.; Plattg, M.; Libottea, B.; Silva Neto, A. Formulation and Solution of an Inverse Reliability Problem to Simulate the Dynamic Behavior of COVID-19 Pandemic. *Trends Comput. Appl. Math.* **2021**, *22*, 91–107. [CrossRef]
19. Georgiev, S.; Vulkov, L. Coefficient Identification for SEIR Model and Economic Forecasting in the Propagation of COVID-19. In *Advanced Computing in Industrial Mathematics, Studies in Computational Intelligence*; Springer: Berlin/Heidelberg, Germany, 2023; Volume 1076, pp. 34–44. Available online: https://link.springer.com/content/pdf/10.1007/978-3-031-20951-2_4 (accessed on 16 February 2023).
20. Ibeas, A.; De la Sen, M.; Alonso-Quesada, S.; Zamani, I.; Shafiee, M. Observer design for SEIR discrete-time epidemic models. In Proceedings of the 13th International Conference on Control Automation Robotics & Vision (ICARCV), Singapore, 10–12 December 2014; pp. 1321–1326. [CrossRef]

21. Leonov, A.; Nagornov, O.; Tyuflin, S. Modeling of Mechanisms of Wave Formation for COVID-19 Epidemic. *Mathematics* **2023**, *11*, 167. [CrossRef]
22. Li, B.; Eskandari, Z.; Avazzadeh, Z. Dynamical Behaviors of an SIR Epidemic Model with Discrete Time. *Fractal Fract.* **2022**, *6*, 659. [CrossRef]
23. Carcione, J.; Santos, J.; Bagaini, C.; Ba, J. A Simulation of a COVID-19 Epidemic Based on a Deterministic SEIR Model. *Front. Public Health* **2020**, *8*, 230. [CrossRef]
24. Khalsaraei, M.; Shokri, A.; Ramos, H.; Yao, S.-W.; Molayi, M. Efficient Numerical Solutions to a SIR Epidemic Model. *Mathematics* **2022**, *10*, 3299. [CrossRef]
25. Costa, J.A.; Martinez, A.C.; Geromel, J.C. On the Continuous-time and Discrete-Time Versions of an Alternative Epidemic Model of the SIR Class. *J. Control Electr. Syst.* **2022**, *33*, 38–48. [CrossRef]
26. Qin, H.; Chen, X.; Zhou, B. A Family of Transformed Difference Schemes for Nonlinear Time-Fractional Equations. *Fractal Fract.* **2023**, *7*, 96. [CrossRef]
27. Alharbi, W.; Shater, A.; Ebaid, A.; Cattani, C.; Areshi, M.; Jalal, M.; Alharbi, M.; Communicable disease model in view of fractional calculus. *AIMS Math.* **2023**, *8*, 10033–10048. [CrossRef]
28. He, Z.-Y.; Abbes, A.; Jahanshahi, H.; Alotaibi, N.D.; Wang, Y. Fractional-Order Discrete-Time SIR Epidemic Model with Vaccination: Chaos and Complexity. *Mathematics* **2022**, *10*, 165. [CrossRef]
29. Islam, M.R.; Peace, A.; Medina, D.; Oraby, T. Integer Versus Fractional Order SEIR Deterministic and Stochastic Models of Measles. *Int. J. Environ. Res. Public Health* **2020**, *17*, 2014. [CrossRef] [PubMed]
30. De la Sen, M.; Alonso-Quesada, S.; Ibeas, A. On a Discrete SEIR Epidemic Model with Exposed Infectivity, Feedback Vaccination and Partial Delayed Re-Susceptibility. *Mathematics* **2021**, *9*, 520. [CrossRef]
31. Singh, R.A.; Lal, R.; Kotti, R.R. Time-discrete SIR model for COVID-19 in Fiji. *Epidemiol. Infect.* **2022**, *150*, e75. [CrossRef]
32. Wacker, B.; Schlüter, J. Time-continuous and time-discrete SIR models revisited: Theory and applications. *Adv. Differ. Eqs.* **2020**, *2020*, 556. [CrossRef]
33. Zhao, Z.; Niu, Y.; Luo, L.; Hu, Q.; Yang, T.; Chu, M.; Chen, Q.; Lei, Z.; Rui, J.; Song, C.; et al. The optimal vaccination strategy to control COVID-19: A modeling study in Wuhan City, China. *Infect. Dis. Poverty* **2021**, *10*, 140. [CrossRef]
34. Angelov, G.; Kovacevic, R.; Stilianakis, N.I.; Veliov, V. Optimal vaccination strategies using a distributed model applied to COVID-19. *Cent. Eur. J. Oper. Res.* **2023**, *31*, 499–521. [CrossRef]
35. Heffernan, J.; Smith, R.; Wahl, L. Perspectives on the basic reproductive ratio. *J. R. Soc. Interface* **2005**, *2*, 281–293. [CrossRef] [PubMed]
36. Van Den Driessche, P. Reproduction numbers of infectious disease models. *Infect. Dis. Model.* **2017**, *2*, 288–303. [CrossRef] [PubMed]
37. Barbashin, E. *Introduction to the Theory of Stability*; Wolters-Noordhoff Publishing: Groningen, The Netherlands, 1970; p. 223.
38. Tiwari, S.; Porwal, P.; Barve, T. Transmission Dynamics of Coronavirus and the Effect of Vaccination Using SEIR Model. *Serdica Math. J.* **2021**, *47*, 161–178.
39. Castillo-Chavez, C.; Feng, Z.; Huang, W. Mathematical approaches for emerging and reemerging infectious diseases: An introduction. In *The IMA Volumes in Mathematics and Its Applications*; Springer: Berlin/Heidelberg, Germany; New York, NY, USA, 2002; pp. 229–250.
40. Hartman, P. *Ordinary Differential Equations*, 2nd ed.; Society for Industrial and Applied Mathematics: Philadelphia, PA, USA, 2002; p. 624.
41. The Open Data Portal of the Republic of Bulgaria. Available online: https://data.egov.bg (accessed on 16 February 2023).
42. The Official Bulgarian Unified Information Portal. Available online: https://coronavirus.bg/ (accessed on 16 February 2023).
43. European Medicines Agency. Vaccines Authorised in the European Union (EU) to Prevent COVID-19. Available online: https://www.ema.europa.eu/en/human-regulatory/overview/public-health-threats/coronavirus-disease-COVID-19/treatments-vaccines/vaccines-COVID-19/COVID-19-vaccines-authorised (accessed on 16 February 2023).
44. Deressa, C.; Mussa, Y.; Duressa, G. Optimal control and sensitivity analysis for transmission dynamics of Coronavirus. *Results Phys.* **2020**, *19*, 103642. [CrossRef]
45. Wachira, C.; Lawi, G.; Omondi, L. Sensitivity and Optimal Control Analysis of an Extended SEIR COVID-19 Mathematical Model. *J. Math.* **2022**, *2022*, 1476607. [CrossRef]
46. Ma, C.; Li, X.; Zhao, Z.; Liu, F.; Zhang, K.; Wu, A.; Nie, X. Understanding Dynamics of Pandemic Models to Support Predictions of COVID-19 Transmission: Parameter Sensitivity Analysis of SIR-Type Models. *IEEE J. Biomed. Health Inform.* **2022**, *26*, 2458–2468. [CrossRef]
47. Zine, H.; Lotfi, E.M.; Mahrouf, M.; Boukhouima, A.; Aqachmar, Y.; Hattaf, K.; Torres, D.; Yousfi, N. Modeling the Spread of COVID-19 Pandemic in Morocco. In *Analysis of Infectious Disease Problems (COVID-19) and Their Global Impact, Infosys Science Foundation Series in Mathematical Sciences*; Springer: Singapore, 2021; pp. 599–616. [CrossRef]
48. Polack, F.P.; Thomas, S.J.; Kitchin, N.; Absalon, J.; Gurtman, A.; Lockhart, S.; Perez, J.L.; Pérez Marc, G.; Moreira, E.D.; Zerbini, C.; et al. Safety and efficacy of the BNT162b2 mRNA COVID-19 vaccine. *N. Engl. J. Med.* **2020**, *383*, 2603–2615. [CrossRef] [PubMed]

49. Voysey, M.; Clemens, S.A.C.; Madhi, S.A.; Weckx, L.Y.; Folegatti, P.M.; Aley, P.K.; Angus, B.; Baillie, V.L.; Barnabas, S.L.; Bhorat, Q.E.; et al. Safety and efficacy of the ChAdOx1 nCoV-19 vaccine (AZD1222) against SARS-CoV-2: An interim analysis of four randomised controlled trials in Brazil, South Africa, and the UK. *Lancet* **2021**, *397*, 99–111. [CrossRef]
50. Sadoff, J.; Le Gars, M.; Shukarev, G.; Heerwegh, D.; Truyers, C.; de Groot, A.M.; Stoop, J.; Tete, S.; Van Damme, W.; Leroux-Roels, I.; et al. Interim results of a phase 1-2a trial of Ad26.COV2.S COVID-19 vaccine. *N. Engl. J. Med.* **2021**, *384*, 1824–1835. [CrossRef]
51. Zhang, Y.J.; Zeng, G.; Pan, H.X.; Li, C.; Hu, Y.; Chu, K.; Han, W.; Chen, Z.; Tang, R.; Yin, W.; et al. Safety, tolerability, and immunogenicity of an inactivated SARS-CoV-2 vaccine in healthy adults aged 18–59 years: A randomised, double-blind, placebo-controlled, phase 1/2 clinical trial. *Lancet Infect. Dis.* **2021**, *21*, 181–192. [CrossRef] [PubMed]
52. Margenov, S.; Popivanov, N.; Ugrinova, I.; Harizanov, S.; Hristov, T. Parameters Identification and Forecasting of COVID-19 Transmission Dynamics in Bulgaria with Mass Vaccination Strategy. *AIP Conf. Proc.* **2022**, *2505*, 080010. [CrossRef] [PubMed]

Disclaimer/Publisher's Note: The statements, opinions and data contained in all publications are solely those of the individual author(s) and contributor(s) and not of MDPI and/or the editor(s). MDPI and/or the editor(s) disclaim responsibility for any injury to people or property resulting from any ideas, methods, instructions or products referred to in the content.

Article

Cumulative Incidence Functions for Competing Risks Survival Data from Subjects with COVID-19

Mohammad Anamul Haque *,† and Giuliana Cortese

Department of Statistical Sciences, University of Padova, 35121 Padova, Italy; giuliana.cortese@unipd.it
* Correspondence: haque-sta@sust.edu
† Current address: Department of Statistics, Shahjalal University of Science & Technology, Sylhet 3114, Bangladesh.

Abstract: Competing risks survival analysis is used to answer questions about the time to occurrence of events with the extension of multiple causes of failure. Studies that investigate how clinical features and risk factors of COVID-19 are associated with the survival of patients in the presence of competing risks (CRs) are limited. The main objective of this paper is, under a CRs setting, to estimate the Cumulative Incidence Function (CIF) of COVID-19 death, the CIF of other-causes death, and the probability of being cured in subjects with COVID-19, who have been under observation from the date of symptoms to the date of death or exit from the study because they are cured. In particular, we compared the non-parametric estimator of the CIF based on the naive technique of Kaplan–Meier (K–M) with the Aalen–Johansen estimator based on the cause-specific approach. Moreover, we compared two of the most popular regression approaches for CRs data: the cause-specific hazard (CSH) and the sub-distribution hazard (SDH) approaches. A clear overestimation of the CIF function over time was observed under the K–M estimation technique. Moreover, exposure to asthma, diabetes, obesity, older age, male sex, black and indigenous races, absence of flu vaccine, admission to the ICU, and the presence of other risk factors, such as immunosuppression and chronic kidney, neurological, liver, and lung diseases, significantly increased the probability of COVID-19 death. The highest hazard ratio of 2.03 was observed for subjects with an age greater than 70 years compared with subjects aged 50–60 years. The SDH approach showed slightly higher survival probabilities compared with the CSH approach. An important foundation for producing precise individualized predictions was provided by the competing risks regression models discussed in this paper. This foundation allowed us, in general, to more realistically model complex data, such as the COVID-19 data, and can be used, for instance, by many modern statistical learning and personalized medicine techniques to obtain more accurate conclusions.

Keywords: competing risks; COVID-19; risk factors; cause-specific hazard; sub-distribution hazard

MSC: 62Nxx

1. Introduction

A novel coronavirus had been discovered by the end of 2019 as the source of a cluster of pneumonia cases in Wuhan, China's Hubei Province. It rapidly spread, resulting in an epidemic throughout China, followed by several outbreaks in other countries worldwide [1]. In February 2020, as the situation worsened, the World Health Organization named the disease COVID-19, caused by Severe Acute Respiratory Syndrome Coronavirus 2 (SARS-CoV-2). Later, on 11 March, COVID-19 was classified as a global pandemic. In COVID-19 subjects, the interval between exposure and the onset of symptoms is expected to be around 5 days, but it might be as long as 14 days. The median number of days between the onset of symptoms and death among those who died from the condition is 14, ranging from 6 to 41 days [2].

Data analysis from COVID-19 subjects is required to study clinical prognostic exposures, generate possible treatment drugs, and design intervention strategies. Many studies have investigated COVID-19 data [3–6] to identify important exposures for the occurrence of death or cure. However, statistical models were presented by either ignoring the competing events or using inappropriate regression-based statistical methods. Thus, one of the objectives of this paper is to consider the competing risks (CRs) settings to estimate the likelihood of the event of interest among the numerous potential outcomes over time using the Cumulative Incidence Function (CIF). The quantity CIF estimates the marginal likelihood of patients who actually developed the event of interest, no matter if a patient was censored or failed in other competing events. The graphical representation of CIF curves is always appealing and, thus, is popular in medical research. CR extends the conventional survival techniques of Kaplan–Meier (K–M) estimate, the log-rank test, and the Cox regression to handle data that have multiple event types. However, in the presence of CR data, the K–M method for the estimation of CIF, the log-rank test for comparison of CIF curves, and the conventional Cox model for assessing exposures lead to incorrect and biased results [7]. This bias arises because the aforementioned conventional techniques assume that all events are independent, which means they censor events other than the event of interest. Moreover, with the CR settings, the log-rank test and Cox regression do not automatically lead to a correct analysis of the CIF, although they can be adapted with minimal effort to make inferences about the CSH function [8]. If one wants to apply the K–M estimator using the CIF, then obtaining the correct estimation is possible when there is only one event of interest (which equals the complementary survival function). As an alternative, regression approaches can be employed. In this context, this paper aims to compare a frequently used conventional technique and two regression approaches to estimate the CIF in the presence of competing events. These are Kaplan–Meier (K–M), cause-specific hazard (CSH), and sub-distribution hazard (SDH), proposed by [9–11], respectively. The SDH approach is also known as the Fine–Gray method [11], where the CIF can be modeled for one particular event of interest. Alternatively, the CIF can be computed by modeling the CSHs, which models the CSHs of all causes. The Fine–Gray method provides an important contribution to modeling the CIF. With the SDH approach, the CIF can be modeled by its direct relation with the SDH rate (λ_k^*) under the assumption that only one event is possible at a given time t. Furthermore, the CSH and SDH approaches differ in the definition of the risk set: in the CSH approach, the risk set decreases when an event of the competing cause or censoring is observed, whereas under the SDH approach, patients who failed from an event other than the one of interest before t remain in the risk set. The SDH approach is similar to a Cox proportional regression model, but it also takes into account cumulative incidence and the SDH rate. In particular, the effect of exposure on the CSH function may be quite different from the effect on the CIF. This implies that exposure may have a strong influence on the CSH function, but have no effect on the CIF [11]. Therefore, the SDH approach takes into account the informative censoring nature of the CR events, while the CSH approach views CR events as non-informative censorship [12].

The competing risks regression models that are discussed in this study offer a crucial basis for obtaining precise individualized predictions. In particular, using a competing risk strategy, this study will help to prioritize patients for vaccination and/or guide clinical decisions either for close monitoring or admission to the ICU, or approval for new intervention. In addition to that, for any other applicable disciplines, the development of precise regression models, for instance, under competing risks data with the CSH and SDH regression approaches has the potential to be of significant importance.

This paper is organized as follows. In Sections 2.1 and 2.3, the non-parametric (without covariate) estimation technique of the CSH and SDH approaches are discussed, respectively. Semi-parametric and parametric (with covariate) estimation techniques are discussed in Sections 2.2 and 2.4, respectively. Section 3 reports the results from COVID-19 data. In Section 3.4, regression analyses to estimate the parameters are compared and in Section 3.5, model prediction between the CSH and SDH approaches is compared. Finally, a discussion is reported in Section 4.

2. Materials and Methods

Consider a CR setting with an event (i.e., cause) of interest (type 1; $k = 1$) and a competing event (type 2; $k = 2$). Here, the indicator variable is $\varepsilon \in \{1,2\}$. Then, assume that T_1 and T_2 are the potential unobservable event times of type $k = 1$ and $k = 2$, respectively. For the CR data, $T = \min(T_1, T_2)$ is observed, and the indicators of the type of event are $\varepsilon = 1$ if $T = T_1$ and $\varepsilon = 2$ if $T = T_2$.

Denote the observed data on the i-th individual by (T_i, C_i), $i = 1, \ldots, n$, respectively. Right-censored CR data, $T_i^* = \min(T_i, C_i)$, for each patient are observed. The event is $\delta_i = 1(T_i \leq C_i)$, where $1(.)$ is an indicator function, $\delta_i = 1$ if $\{T_i \leq C_i\}$ and $\delta_i = 0$ if $\{C_i < T_i\}$, and $k_i \in \{1,2\}$, for the causes of event types 1 and 2. The CIF for event type 1 is the probability that an event of type 1 occurs at or before time t, i.e., $CIF_1(t) = P(T \leq t, k = 1)$. In this context, the CIF in clinical trial settings can be defined as follows: assume that a is the patient's accrued time and $\tilde{f}$ is the follow-up time. Then, the probability of a patient who has the event of death within the time interval $[t, a + \tilde{f}]$ can be estimated, given that he/she entered the study at time t. This is a conditional CIF that can be rewritten as $CIF_1(a + \tilde{f} - t) = P(\tilde{T} \leq a + \tilde{f} - t, k = 1)$, where $\tilde{T} = T - t$ is the survival time given that the patient enters at time t without having an event before t.

2.1. Non-Parametric Estimation Technique: CSH Approach

It is convenient to model survival times through the hazard function because of censoring [13]. The joint distribution of event time and event cause may be completely specified through the CSHs. The advantage of presenting the non-parametric estimator of the CSH approach in this subsection is that it provides a template for predicting the CIF in CSH regression models. In the regression approach, the Nelson–Aalen estimator is replaced with its model-based counterparts [14]. The CSH function of event type k is defined as follows:

$$\lambda_k(t) = \lim_{\Delta t \downarrow 0} \frac{P(t \leq T < t + \Delta t, \varepsilon = k | T \geq t)}{\Delta t}$$

For simplicity, event type 1 (main event of interest) and event type 2 (competing event) are considered in this paper. The CIF for type 1 is then determined by also accounting for the competing event type 2, and it is:

$$CIF_1(t) = \int_0^t \lambda_1(u) e^{-\{\Lambda_1(u) + \Lambda_2(u)\}} du$$

where $\Lambda_k(u) = \int_0^u \lambda_k(v) dv$ is the cumulative CSH function for event k and $k = 1, 2$. It is clear that $CIF_1(t)$ involves not only the hazard function, but also all the competing CSH functions when $k > 1$. When $k = 1$, the sub-distribution function degenerates to $CIF_1(t) = 1 - \exp(-\Lambda_1(t))$ and becomes a function of only $\lambda_1(t)$.

To estimate $CIF_1(t)$ non-parametrically, let us assume D distinct event time-points, $0 = t_0 < t_1 < \ldots < t_D$. Then, at a particular event time t_i, let d_1 and d_2 be the number of patients who experienced event types 1 and 2, respectively, and assume that $\mathcal{R}(t_i)$ denotes the risk set at event time t_i and includes individuals who did not fail due to any causes or are not censored just before t_i. Here, it should be noted that under the CSH approach, a patient is no longer at risk for having the event of interest if he/she experiences a competing event and thus leaves the risk set. Therefore, the CSH rate λ_k is estimated by counting the number of events of type k, divided by the observed number at risk:

$$\widehat{\lambda_k}(t_i) = \frac{d_k(t_i)}{\mathcal{R}(t_i)}.$$

The overall survival function for T can be obtained by using the Kaplan–Meier estimate [9]:

$$\hat{S}(t) = \prod_{t_i \leq t}\left(1 - \frac{d(t_i)}{\mathcal{R}(t_i)}\right)$$

where $d(t_i) = d_1(t_i) + d_2(t_i)$. Alternatively, $S(t)$ can be obtained through $\hat{S}(t) = \exp[-(\hat{\Lambda}_1(t) + \hat{\Lambda}_2(t))]$. Here, $\hat{\Lambda}_k(t)$ is the Nelson–Aalen estimator for the cumulative CSH function for the event type k.

Finally, the CIF function for event type k can be obtained from the CSHs through $CIF_k(t) = \int_0^t \lambda_k(u)S(u)du$, and a natural non-parametric estimate of $CIF_k(t)$ is

$$\widehat{CIF}_k(t) = \int_0^t \hat{\lambda}_k(u)\hat{S}(u)du = \sum_{t_i \leq t} \frac{d_k(t_i)}{\mathcal{R}(t_i)}\hat{S}(t_i^-) \quad \text{for, } k = 1, 2.$$

A step function is returned with jumps at time points of observed events of type k, and constant values at times where no events or a competing event is observed [15]. That estimator for the CIF in a CR setting is a special case of the Aalen–Johansen estimator for transition probabilities in multi-state models [16]. The Aalen–Johansen estimator can be obtained as the product-integral of the Nelson–Aalen estimators for the cumulative transition intensities [17].

2.2. Semi-Parametric Regression Models for the CSH Approach

The difference in the cumulative incidence curves between treatment groups is identified either indirectly using a Cox proportional hazard (PH) model for the main event of interest (considering other CRs as censored), or with the direct regression model with the effect of covariates on the CIF.

The PH model assumes that hazards are proportional in the follow-up period, and a separate model can be fit for each event type. However, the analysis is more powerful when all competing events are combined. Specifically, the literature in [18,19] considered the following Cox proportional hazards models for all causes:

$$\lambda_k(t|X) = \lambda_{k0}(t)\exp\left\{\beta_k^T x\right\} \tag{1}$$

where $\lambda_{k0}(t)$ is the baseline hazard function for cause k, x is a vector of covariates that is assumed to be equal among events, and β_k is the vector of regression coefficients.

The regression coefficients β_k can be estimated for cause k by maximizing the Cox partial likelihood and log partial likelihood

$$\ln[L(\beta_k)] = \ln\left[\prod_{i=1}^n \left(\frac{\exp(\beta_k^T x_i)}{\sum_{j \in \mathcal{R}(t_i)} \exp(\beta_k^T x_j)}\right)^{\delta_i}\right] = \sum_{i=1}^n \delta_i\left[\beta_k^T x_i - \ln\left(\sum_{j \in \mathcal{R}(t_i)} \exp(\beta_k^T x_j)\right)\right],$$

The score function is

$$s(\beta_k) = \frac{\partial \ln L(\beta_k)}{\partial \beta_k} = \sum_{i=1}^n \delta_i\left[x_i - \frac{\sum_{j \in \mathcal{R}(t_i)} x_j \exp(\beta_k^T x_j)}{\sum_{j \in \mathcal{R}(t_i)} \exp(\beta_k^T x_j)}\right].$$

Asymptotically, the maximum likelihood estimate $\hat{\beta}_k$ is normally distributed, as $\sqrt{n}\left(\hat{\beta}_k - \beta_k\right) \simeq \mathcal{N}(0, \mathbf{V})$, where $\mathbf{V} = \mathbf{I}_{\beta_k}^{-1}$ is the asymptotic variance–covariance matrix of the $\sqrt{n}\hat{\beta}_k$ and $\mathbf{I}_{\beta_k}$ is the Fisher Information matrix. In practice, the asymptotic variance of the estimator of each single coefficient β_{kj} is obtained from the diagonal elements V_{jj} of the Information matrix. Then, the frequently used Wald test can be applied, where the test statistic under the null hypothesis $H_0 : \beta_{kj} = \beta_0$ is $Z_{kj} = \sqrt{n}\left(\hat{\beta}_{kj} - \beta_0\right)\sqrt{\hat{V}_{jj}}$ and asymptotically follows a standard Normal distribution. However, in practice, the variance is evaluated under the alternative hypothesis $H_1 : \beta_{kj} \neq \beta_0$. Let V^* be the variance of $\sqrt{n}\hat{\beta}_k$ under the alternative. Theoretically, it is proved by Slutsky's theorem that the distributions of Z_{kj} and $\sqrt{n}(\hat{\beta}_{kj} - \beta_0)/\sqrt{V_{jj}^*}$ are equivalent for large n [20–22].

Furthermore, predicting the CIF is not straightforward when using the CSH approach. To do so for a particular event type, the fitted cause-specific Cox model has to be used for each event type. Here, if we assume that the goal is to fit separate models to each of the k events for the given covariates $\mathbf{x}$, then the cause-specific Cox model leads to

$$\hat{\Lambda}_k(t|\mathbf{x}^*) = \exp\left(\hat{\beta}_k^T \mathbf{x}^*\right) \hat{\Lambda}_{k0}(t),$$

where $\hat{\beta}_k$ is the maximum partial likelihood estimate, $\hat{\Lambda}_{k0}(t)$ is the estimate from the Breslow estimator of the baseline cumulative CSH function, and $\mathbf{x}^*$ is the specific-subject covariate values for which we are interested in obtaining predictions.

Then, the predicted CIF is

$$\widehat{CIF}_k(t|\mathbf{x}^*) = \int_0^t \hat{S}(s^-|\mathbf{x}^*) d\hat{\Lambda}_k(s|\mathbf{x}^*)$$

where the predicted survival function is

$$\hat{S}(t|\mathbf{x}^*) = \prod_{t_i : t_i \leq t} \left[1 - \hat{\Lambda}(t_i|\mathbf{x}^*)\right]$$

with $\hat{\Lambda}(t|\mathbf{x}^*) = \sum_{k=1}^{2} \hat{\Lambda}_k(t|\mathbf{x}^*)$ being the predicted cumulative function estimated for a patient with covariates $\mathbf{x}^*$.

2.3. Non-Parametric Estimation Technique: The SDH Approach

Contrary to the CSH approach, patients who experienced an earlier competing event remain included in the risk set. Thus, in the SDH, the risk set at time t is

$$\mathcal{R}^*(t_i) = \{i : (t \leq T_i) \cup (t \geq T_i \cap \varepsilon_i \neq 1), i = 1, \ldots, n\}.$$

A patient who has not experienced failure or death due to the event of interest by time t is at risk. Those who are included in this risk set can be divided into two distinct categories: patients who never failed due to any cause and patients who have previously failed due to competing causes. Here, the SDH can be interpreted as the likelihood of observing an event that is of main interest in the next time interval, with the condition that either the main event of interest did not occur until that time or that the CR event had occurred previously. A sub-distribution function is one in which the value does not increase from 0 to 1 as time progresses due to competing events which can prevent the event from occurring. Literature [23] describes SDH for cause 1 as:

$$\lambda_1^*(t) = \lim_{\Delta t \to 0} \frac{P(t \leq T \leq t + \Delta t, \varepsilon = 1 | T \geq t \cup \{T \leq t \cap \varepsilon \neq 1\})}{\Delta t} = \frac{-\partial \log\{1 - CIF_1(t)\}}{\partial t}.$$

The cumulative SDH for cause 1 is defined as $\Lambda_1(t) = \int_0^t \lambda_1^*(s) ds$. Moreover, for SDH approach, a direct relationship exists between the Cumulative Incidence Function $CIF_1(t)$ and the SDH rate λ_1^*) [11]:

$$CIF_1(t) = 1 - S_1^*(t) = 1 - e^{-\Lambda_1^*(t)} = 1 - e^{-\int_0^t \lambda_1^*(u) du}.$$

This implies

$$\int_0^t \lambda_1^*(u) du = -\log(1 - CIF_1(t)) = g(CIF_1(t)) \qquad (2)$$

where $g(\cdot)$ is, here, the complementary-log link function. In terms of estimation, this means that the occurrence of a competing event is ignored and such patients remain in the risk set until the time at which they are censored for a reason other than the competing event. This suggests the following estimator: $\widehat{\lambda}_k^*(t_i) = \frac{d_k(t_i)}{\mathcal{R}^*(t_i)}$, where $\mathcal{R}^*(t_i)$ is never smaller than $\mathcal{R}(t_i)$. Therefore, the classical K–M is always at least as steep as the estimator of the cause-specific cumulative incidence, due to overestimation.

2.4. Semi-Parametric Regression Models for the SDH Approach

Under the SDH approach, a frequently used regression model is the so-called Fine–Gray model [11]. The likelihood function differs from that of the CSH approach in terms of the definition of risk set. Although the risk set is unconventional, it leads to a proper partial likelihood [11], which can be expressed as:

$$\tilde{L}(\beta_k) = \prod_{i=1}^n \left[\frac{\exp(\beta_k^T x_i)}{\sum_{j \in \mathcal{R}_i^*} \exp(\beta_k^T x_j)} \right]^{\delta_i}.$$

After fitting the CR models, one can use these models to make predictions about CIFs. For the Fine and Gray model, predicting them for the event of interest is a straightforward task because the sub-distribution hazard is modeled directly, and the CIF is only one transformation away. The Cox-type proportional sub-distributional hazard model for cause k can be written as [11]:

$$-\log\{1 - CIF_k(t \mid \mathbf{x})\} = \int_0^t \lambda_{k0}^*(u) \exp(\beta_k^T \mathbf{x}) du = \exp(\beta_k^T \mathbf{x}) \int_0^t \lambda_{k0}^*(u) du,$$

where $\lambda_{k0}^*(t)$ is the baseline sub-distribution hazard for cause k. Then, the predicted CIF with time-invariant covariates $\mathbf{x}$ can be estimated by

$$\widehat{CIF}_k(t|\mathbf{x}) = 1 - \exp\left[-\hat{\Lambda}_{k0}^*(t) \exp\left(\hat{\beta}_k^T \mathbf{x}\right)\right]$$

where $\hat{\Lambda}_{k0}^*(t)$ is the baseline cumulative sub-distribution hazard function for cause k.

3. Application to COVID-19 Data

We applied the competing risk survival analyses described above for estimating the CIF of dying from COVID-19 and the CIF of dying from other causes in Brazilian subjects with COVID-19.

3.1. Data Sources and Variables

We analyzed data from subjects who had COVID-19 symptoms and were under observation from the date of symptoms to the date of death or exit from the study because they were cured or no longer in danger. Our time-to-event data were obtained from the Brazilian Ministry of Health for all COVID-19 patients from 1 January 2020 to 30 April 2021. Figure 1 summarises the main outcomes that we analysed on these data.

The exposures that we considered as risk factors were some patient characteristics and types of COVID-19 symptoms. We considered the binary variable risk factor (does the subject present some risk factor? 1: yes, 2: no) to categorize patients in two groups: those who did not have any risk factors and those who had one or more risk factors prior to COVID-19 symptoms. The considered risk factors were: *asthma* (1: yes, 2: no), *cardio.dis* (chronic cardiovascular disease 1: yes, 2: no), *diabetes* (1: yes, 2: no), *hepatic.dis* (chronic liver disease 1: yes, 2: no), *immuno* (immunosuppression which is decreased from immunological system function 1: yes, 2: no), kidney (chronic kidney disease 1: yes, 2 no), *neuro* (neurological diseases 1: yes, 2: no), *obesity* (1: yes, 2: no), *pneumo* (lung chronic disease 1: yes, 2: no), *pneumo.dis* (other chronic pneumatopathy 1: yes, 2: no), *other.risk* (other risk factors 1: yes, 2: no). In addition, the considered COVID-19 symptoms were *loss.smell* (1: yes, 2: no), *loss.taste* (1: yes, 2: no), *cough* (1: yes, 2: no), *diarrhea* (1: yes, 2: no), *dyspnea* (1: yes, 2: no), *fatigue* (1: yes, 2: no), *fever* (1: yes, 2: no), *resp.disc* (respiratory discomfort 1: yes, 2: no), *sore.throat* (1: yes, 2: no), *vomit* (1: yes, 2: no), *saturation* (oxygen saturation < 95%? 1: yes, 2: no), *abdom.pain* (abdominal pain 1: yes, 2: no), *other.symp* (1: yes, 2: no). The patient characteristics were: *flu.vaccine* (flu vaccine last campaign 1: yes, 2: no), *race* (1: White; 2: Black; 3: East Asian; 4: Brown; 5: Indigenous), *age* (age in years at first symptoms), *sex* (male = 1; female = 2), *ICU* (admitted to Intensive Care Unit 1: yes, 2: no), *parto* (has the subject given birth less than 45 days from the first symptoms? 1: yes, 2: no).

It was found in the recent literature (see, e.g., [2]) that about 80% of COVID-19 deaths were in those over 60 years of age, and 75% had pre-existing health problems. Thus, it was meaningful to study the effect of COVID-19 outcomes on different age groups. The variable *age* has been categorized as follows: less than 40 years ("Young"); between 40 and 50 years ("Young-Old"); between 50 and 60 years ("Medium-Old"), between 60 and 70 years ("Old") and, finally, age greater than 70 years ("Old-old").

In our preliminary analysis (Sections 3.2 and 3.3), the time to become cured was considered as a cause of interest and investigated on its own (see Figure 1 for a data summary of the outcome). However, in the competing risks regression analysis, this was considered as a censored time, since the main focus was on the causes of death. Moreover, we did not possess the exact dates of exit from the hospital due to being cured, but only the date formally registered by the Brazilian Health Ministry, which could also be considerably later than the actual date. The latter violates the assumption of non-informative censoring, i.e., the cured patients are not representative of the whole population of those who were admitted to the hospital in terms of their risk of dying, because they are associated with a lower risk. However, when the regression approach is based on the Inverse of the Probability of Censoring Weights (IPCW) technique, as in the Fine–Gray model, this setting is particularly relevant because regression models can also account for dependent censoring.

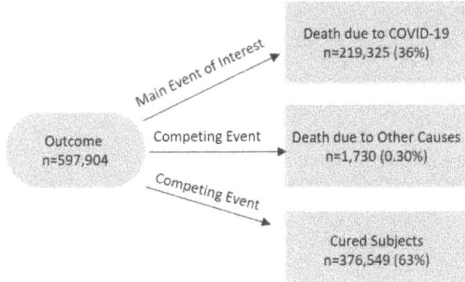

Figure 1. Outcome variable for one main event of interest and two competing events.

3.2. Results of Non-Parametric Estimation of the CIF

Figure 2 shows the CIFs for cured subjects, death due to COVID-19, and death due to other causes, which were estimated non-parametrically using the Aalen–Johansen estimator. Here, the estimated probability of COVID-19 death was 24% after the first 20 days from the day of symptoms and became 35% after 30 days. Meanwhile, the likelihood of becoming cured was 50% after the first 20 days and around 60% after 40 days. Death due to other causes was found to be negligible, as the probability of death over time was slightly over zero. This is because, when compared to COVID-19 death and cured subjects, there were very few patients who experienced death due to other causes (only 0.30% death due to other causes, whereas $n = 219,325$ for COVID-19 death and $n = 376,549$ for cured events, see Figure 1).

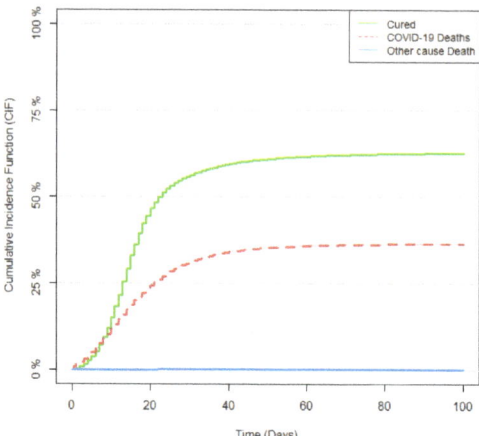

Figure 2. Cumulative Incidence Functions (CIFs) for subjects who were cured from COVID-19, death due to COVID-19, and other causes of death in the whole population.

3.3. Comparison between the Kaplan–Meier and CSH Approaches

The objective here was to compute the CIFs based on the conventional technique (K–M) and then compare the results with the competing risks CSH approach. The K–M plot for COVID-19 death estimates the survival probability of subjects who did not experience COVID-19 death. The CIF can be obtained by plotting the complimentary function (1-KM), which estimates the cumulative risk of dying from COVID-19 over time, in the absence of the competing events (here, we treated all of them as right-censored times). Overall, Figures 3–12 satisfied the proportional hazards assumption since risk curves do not cross during the analyzed period. A clear overestimation of the CIFs over time was observed under the K–M estimation technique compared with the CSH approach. The overestimation gap between the (1-KM) and CSH approaches was severe, mainly for COVID-19 death. For these reasons, the following results are shown for both the K–M and CSH approaches, but their interpretation is provided only under the CSH approach.

The subjects who developed the exposures of chronic liver disease (hepatic.dis), other symptoms, respiratory discomfort (resp.disc), oxygen saturation level, and ICU admission had a lower probability of survival after 20 days of hospitalization than subjects who did not experience these characteristics. In particular, the most severe exposure group was the one who entered the ICU, and they had about a 40% less probability of survival after 20 days than those subjects who were not admitted to this unit (Figure 10). Furthermore, the cumulative risk for COVID-19 death was slightly higher among subjects who had a fever, as compared with subjects who did not experience fever (Figure 4). Moreover, the CIFs for subjects who had been vaccinated for the flu and those who had not received the flu vaccine were almost indistinguishable (Figure 5). Additionally, the probability of COVID-19 death for male subjects was higher as compared with female subjects (Figure 3). However, under the CSH approach, flu vaccine and sex exposures were both found to be statistically significant with a hazard ratio of 0.94 and 1.06, respectively (Table 1). The probability of COVID-19 death increased with older age; in particular, it was more severe for the group with age greater than 70 years, being 50% after 30 days ('Old-old', blue lines in Figure 11). Black subjects were found to have a higher probability of dying from COVID-19 and White subjects were associated with lower risk, as compared with the other races (Figure 12).

Note that, as expected, the results of cured subjects in Figures 3–12 show an inverse situation on the CIFs with respect to the curves for COVID-19 death.

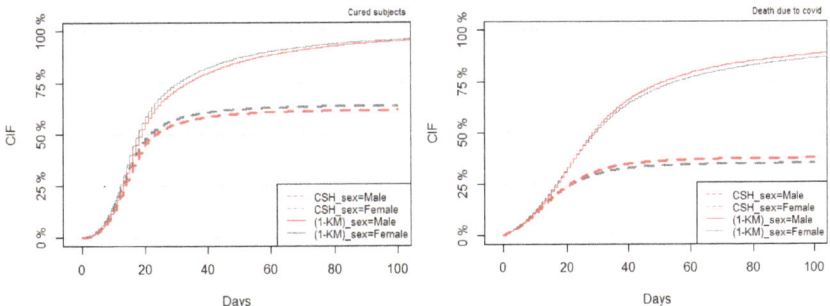

Figure 3. Cumulative Incidence Function (CIF) curves for exposure sex.

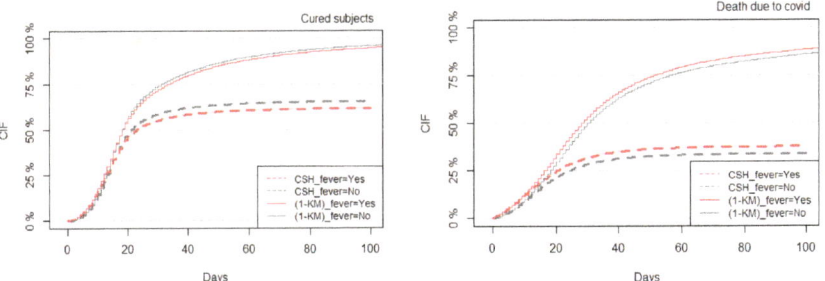

Figure 4. Cumulative Incidence Function (CIF) curves for exposure fever.

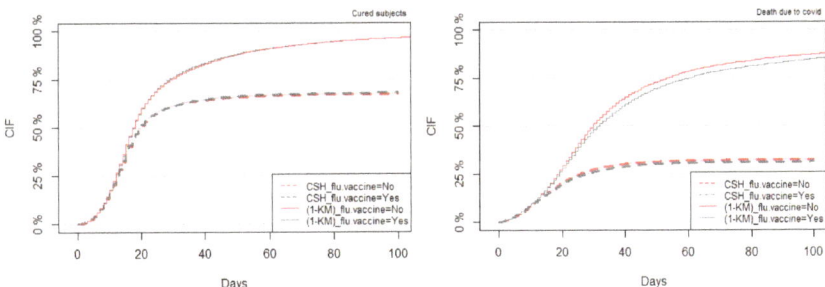

Figure 5. Cumulative Incidence Function (CIF) curves for exposure flu vaccine.

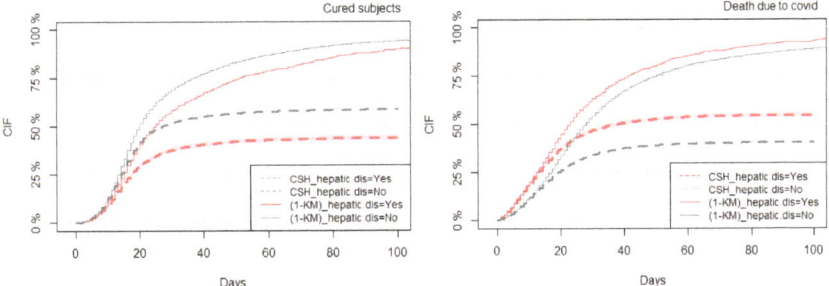

Figure 6. Cumulative Incidence Function (CIF) curves for exposure hepatic.dis (chronic liver disease).

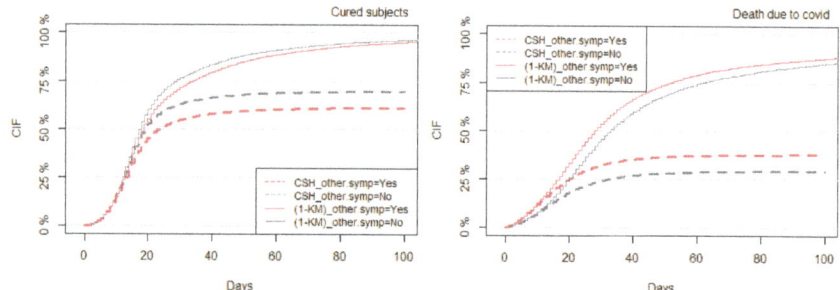

Figure 7. Cumulative Incidence Function (CIF) curves for exposure other.symp (other symptoms).

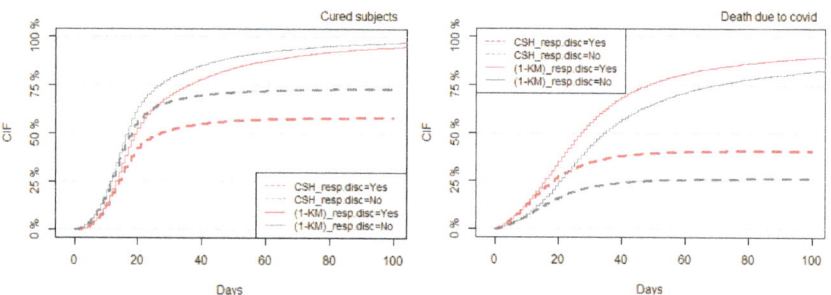

Figure 8. Cumulative Incidence Function (CIF) curves for exposure resp.disc (respiratory discomfort).

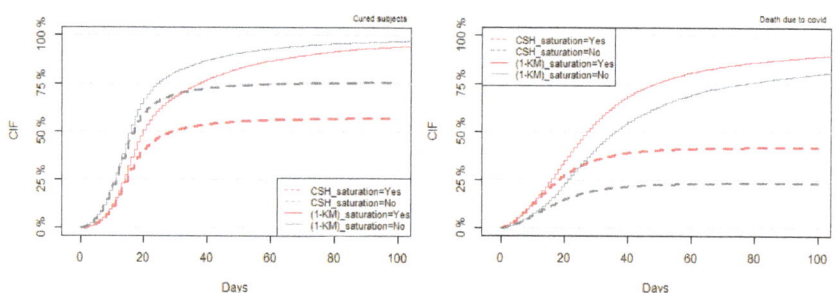

Figure 9. Cumulative Incidence Function (CIF) curves for exposure saturation (oxygen saturation).

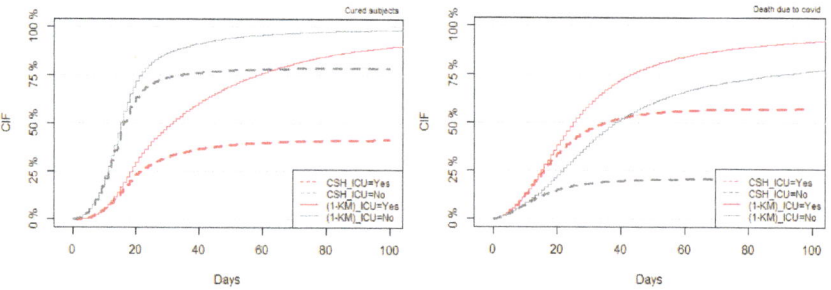

Figure 10. Cumulative Incidence Function (CIF) curves for exposure ICU.

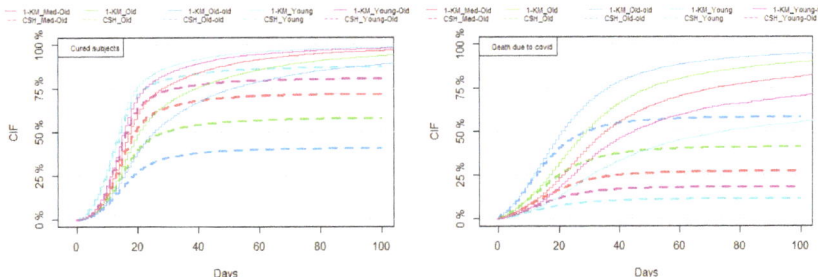

Figure 11. Cumulative Incidence Function (CIF) curves for exposure age.

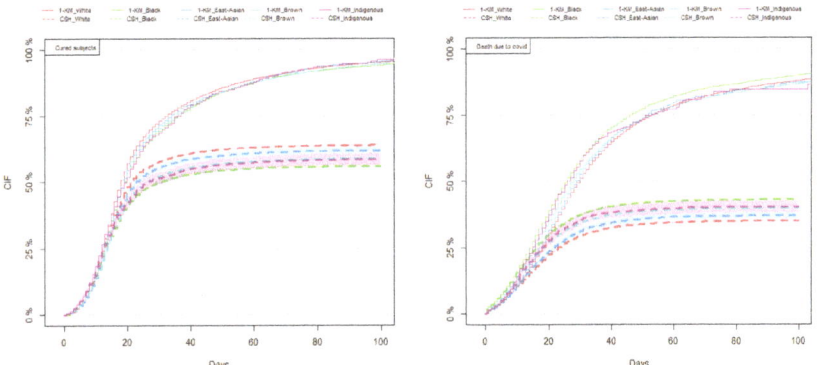

Figure 12. Cumulative Incidence Function (CIF) curves for exposure race.

Table 1. Results from the CSH model for COVID-19 mortality (COVID-19 death is the main event of interest): regression coefficient estimates ('Estimates'), hazard ratios (HR), standard errors (SE), 95% confidence intervals ('Lower CI', 'Upper CI'), p-values.

Exposures	Estimates	HR	SE	Lower CI	Upper CI	p-Value
Asthma (Yes)	−0.115	0.891	0.029	0.842	0.943	<0.001
Diabetes (Yes)	0.077	1.080	0.011	1.058	1.104	<0.001
Obesity (Yes)	0.034	1.034	0.018	0.998	1.072	0.060
Other.risk (Yes)	0.058	1.060	0.011	1.038	1.082	<0.001
Immuno (Yes)	0.252	1.286	0.024	1.227	1.348	<0.001
Kidney (Yes)	0.238	1.269	0.019	1.222	1.317	<0.001
Neuro (Yes)	0.270	1.310	0.019	1.262	1.360	<0.001
Flu.vaccine (Yes)	−0.063	0.939	0.011	0.918	0.960	<0.001
Hepatic.dis (Yes)	0.247	1.280	0.040	1.184	1.384	<0.001
Age: Old (60–70 Years)	0.276	1.318	0.018	1.272	1.366	<0.001
Age: Old-old (>70 Years)	0.712	2.038	0.017	1.973	2.104	<0.001
Age: Young (<40 Years)	−0.333	0.717	0.031	0.675	0.761	<0.001
Age: Young-Old (40–50 Years)	−0.120	0.887	0.026	0.844	0.933	<0.001
Sex (Male)	0.060	1.062	0.011	1.040	1.085	<0.001
ICU (Yes)	0.430	1.537	0.011	1.504	1.571	<0.001
Pneumo (Yes)	0.133	1.143	0.019	1.100	1.187	<0.001
Race: Black	0.198	1.219	0.023	1.165	1.275	<0.001
Race: East Asian	0.064	1.066	0.049	0.968	1.174	0.194
Race: Brown	0.149	1.160	0.011	1.135	1.187	<0.001
Race: Indigenous	0.315	1.370	0.096	1.136	1.653	<0.001

'Other.risk' = other risk factors; 'immuno' = immunosuppression, which is decreased from immunological system function; 'kidney' = chronic kidney disease; 'neuro' = neurological diseases; 'flu.vaccine' = flu vaccine last campaign; 'hepatic.dis' = chronic liver disease; 'pneumo' = lung chronic disease; 'ICU' = admitted to Intensive Care Unit.

3.4. Regression Analysis under the CSH and SDH Approaches

To analyze the effect of the exposures on the CIF, it was found that there were confounding effects among the symptoms and some of the patients' risk factors. Thus, we separated those confounding exposures and investigated the remaining risk factors on the cause of interest. In particular, the stepwise variable selection techniques were applied based on the AIC and likelihood ratio test under the Cox proportional hazard assumptions for the CSH and SDH approaches. The data were analyzed in R statistical software, version 4.1.1 [24]. The final regression model included the following variables: asthma, diabetes, obesity, other.risk, immuno, kidney, neuro, flu.vaccine, hepatic.dis, age, sex, ICU, pneumo, and race.

3.4.1. Regression Analysis for the CSH Approach

From Table 1, the worst outcome was observed for the age group Old-old (>70 years) with a hazard ratio around two-folds higher (HR: 2.038, CI: 1.973–2.104) as compared with the reference group of Medium-Old age (50–60 years). Furthermore, subjects who were admitted to the Intensive Cure Unit (ICU) had a significantly higher COVID-19 mortality than those not admitted in the ICU (HR: 1.537, CI: 1.504–1.571). Moreover, the exposures diabetes, other risks, and male sex had hazard ratios of 1.8, 1.06, and 1.06, respectively, indicating a mortality increase of 6–8% with respect to the their respective reference levels. In addition, it was found that the subjects who had been vaccinated for the flu were associated with a 6% decreased COVID-19 mortality than those who had not been vaccinated (HR: 0.939, CI: 0.918–0.960). Moreover, subjects with a state of decreased immunological system function (Immuno), chronic kidney disease, neurological disease, and chronic liver disease (hepatic.dis1) had an increased COVID-19 mortality of approximately 27–31% (HRs = 1.286, 1.269, 1.310, and 1.28, respectively) as compared with those who had no such disease status. Additionally, the rate of dying due to COVID-19 was significantly higher for all races as compared with White subjects, and, in particular, it was 37% and 22% higher, respectively, for Indigenous and Black subjects (HR: 1.370 and CI: 1.136–1.653, HR: 1.219 and CI: 1.165–1.275). Subsequently, the mortality rate for Black and Brown subjects was also found to be significantly higher than that for White subjects. Interestingly, subjects with asthma were found to have a lower COVID-19 mortality (HR: 0.891, CI: 0.842–0.943) with respect to subjects without this chronic disease. This may be justified by the fact that subjects with asthma were faster hospitalized and received extra care during hospitalization. Thus, the the risk of dying lessened.

3.4.2. Regression Analysis for the SDH Approach

This section explores the performance of the SDH approach, i.e., the Fine–Gray model. This approach makes it possible to obtain both the naive and the robust model-based standard errors. Here, only robust standard errors are reported. It is observed from Table 2 that the estimated coefficients for COVID-19 death deviate slightly from those obtained from the CSH regression model. The differences in the estimated parameters reflect the different underlying assumptions under competing risks survival data. Moreover, note that the CSH model describes the effect on the COVID-19 mortality rate, whereas the Fine–Gray model describes the effect on the cumulative risk of dying from COVID-19, transformed on the scale of the link function. The estimates derived from the Fine–Gray model have no simple interpretation, but they follow the same direction as the CSH model.

Table 2. Results from the Fine–Gray model (SDH approach) for the cumulative incidence of COVID-19 death (main event of interest): regression coefficient estimates ('Estimates'), hazard ratios (HR), robust standard errors (Robust SE), 95% confidence intervals ('Lower CI', 'Upper CI'), p-values.

Exposures	Estimates	HR	SE	Lower CI	Upper CI	p-Value
Asthma (Yes)	−0.115	0.891	0.029	0.842	0.943	<0.001
Diabetes (Yes)	0.078	1.081	0.011	1.058	1.105	<0.001
Obesity (Yes)	0.037	1.037	0.018	1.001	1.075	<0.050
Other.risk (Yes)	0.056	1.058	0.011	1.036	1.081	<0.001
Immuno (Yes)	0.241	1.272	0.025	1.211	1.338	<0.001
Kidney (Yes)	0.235	1.265	0.021	1.216	1.317	<0.001
Neuro (Yes)	0.267	1.306	0.021	1.253	1.361	<0.001
Flu.vaccine (Yes)	−0.062	0.940	0.011	0.919	0.961	<0.001
Hepatic.dis (Yes)	0.244	1.276	0.044	1.171	1.391	<0.001
Age: Old (60–70 Years)	0.278	1.321	0.017	1.277	1.367	<0.001
Age: Old-old (>70 Years)	0.710	2.035	0.016	1.971	2.101	<0.001
Age: Young (<40 Years)	−0.335	0.715	0.030	0.674	0.759	<0.001
Age: Young-Old (40–50 Years)	−0.120	0.887	0.025	0.845	0.931	<0.001
Sex (Male)	0.061	1.063	0.011	1.041	1.086	<0.001
ICU (Yes)	0.434	1.543	0.011	1.510	1.577	<0.001
Pneumo (Yes)	0.132	1.141	0.020	1.097	1.188	<0.001
Race: Black	0.198	1.219	0.024	1.164	1.277	<0.001
Race3: East Asian	0.053	1.054	0.047	0.961	1.157	0.264
Race4: Brown	0.144	1.155	0.012	1.129	1.181	<0.001
Race: Indigenous	0.323	1.381	0.106	1.121	1.701	<0.003

'Other.risk' = other risk factors; 'immuno' = immunosuppression, which is decreased from immunological system function; 'kidney' = chronic kidney disease; 'neuro' = neurological diseases; 'flu.vaccine' = flu vaccine last campaign; 'hepatic.dis' = chronic liver disease; 'pneumo' = lung chronic disease; 'ICU' = admitted to Intensive Care Unit.

3.5. Comparison of Model Predictions between the CSH and SDH Approaches

From the results obtained with the fitted regression models, the model-based predictions can be undertaken and compared by analyzing the subject's risk with some specific given values for the exposures. As an illustration, let us predict the cumulative risk (CIF) of dying from COVID-19 for subjects with a certain flu vaccination status and chronic liver disease status, under both the CSH and SDH approaches. Figure 13 shows two specific groups based on the two considered subjects' risk factors: group 1 is related to those who had been vaccinated for the flu and had no chronic liver disease; group 2 refers to those who had not been vaccinated for the flu and had chronic liver disease. From Figure 13 (left panel), it is observed that, at the beginning of the study, the CIF curves between the two groups appear to be similar until day 10. Then, the discrepancy of the CIF probabilities increases over time. In particular, in the CSH approach, the cumulative risk (CIF) of COVID-19 death reaches 50% in 25 days for group 1, and in 30 days for group 2. Moreover, the CIF probability gap is almost similar from 30 days to more than 100 days. On the contrary, in the SDH approach, after 10 days, the discrepancy in the CIF probabilities between the two groups increased similarly until 25 days, but then later on and up to 70 days, this gap is even more than that under the CSH approach, and finally, after that period, the CIF curves appear to be flat (see Figure 13, right panel).

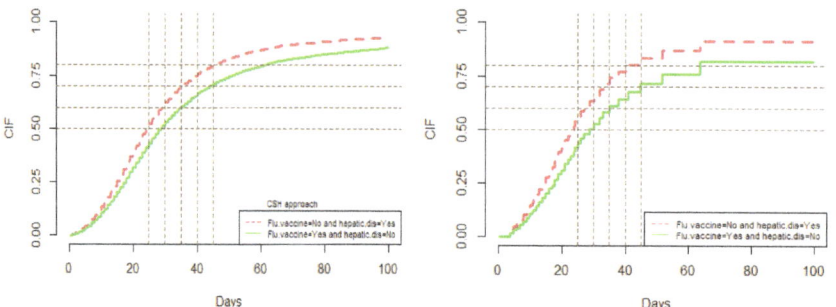

Figure 13. Predictions of cumulative risk (CIF) of dying from COVID-19 for subjects with flu vaccine and saturation status under the CSH approach (**left** panel) and the SDH approach (**right** panel).

4. Discussion

We used competing risk survival analyses to estimate the CIF of dying from COVID-19 and the CIF of dying from other causes in subjects with COVID-19 who had been monitored from the time they first showed symptoms to the time they died or left the study because they were cured. In the preliminary stage of this paper (Sections 3.2 and 3.3), the time to become cured is regarded as one of the events of interest and explored on its own. However, in the later part of this paper, while considering the competing risk settings, this event is considered as censored since the primary focus is on the causes of mortality, in particular on COVID-19 death. This setting goes against the presumption of non-informative censoring. In particular, the likelihood of death among cured patients is not indicative of the likelihood of death among patients remaining hospitalized. Regression models, on the other hand, are particularly pertinent when the techniques are examined using the IPCW technique because they can also take dependent censoring into consideration.

Since the cumulative risk curves did not cross over the studied time, Figures 3–12 generally satisfied the proportional hazards assumption. In comparison with the CSH approach, the K–M estimation strategy clearly overestimated the CIF functions over time. A significant overestimation gap was observed between the two approaches, particularly for COVID-19 deaths. The exposures of asthma, diabetes, obesity, other.risk, immuno, kidney, neuro, flu.vaccine, hepatic.dis, age, sex, ICU, pneumo, and race significantly increase the probability of death due to COVID-19. The highest hazard ratio, equal to 2.03, was observed for subjects with age greater than 70 years compared with the age group 50–60 years. The Fine–Gray model (SDH approach) yielded estimated coefficients for death due to COVID-19 that differed slightly from the CSH model's results. The disparities in the predicted parameters from the two approaches mirrored the differing underlying model assumptions for the competing risks setting.

Furthermore, from the fitted regression models, model-based predictions were undertaken and evaluated by assessing a certain subject's risk with the desired specified exposures. In the COVID-19 application, it was found that the SDH approach provides slightly higher estimated Cumulative Incidence Functions as compared with the CSH approach. Nowadays, competing risks data are found in many fields, ranging from medicine, where several types of oncology and therapeutic outcomes are studied simultaneously over time, to epidemiology, demography, and reliability, where, e.g., failure may be due to the breakdown of a mechanical device for several different causes. Therefore, the construction of accurate regression models for competing risks data based on the two discussed approaches (the CSH and SDH methods) is of potentially great interest in many other contexts and applied fields. Furthermore, based on the fitted models, all desired individual predictions of CIFs can be computed on the same data, i.e., the training data, but also on new data, i.e., the testing data, and prediction accuracy can be measured. The competing risks regression models described in this paper provide a very important foundation for ob taining accurate personalized predictions, on which, e.g., many current statistical learning and personalized medicine techniques are based.

In this study, the ICU is considered a time-constant exposure, but more appropriately, one may also consider the ICU as a time-dependent covariate. Moreover, it could also be of interest to study its time-varying effect on the CIFs. Due to the presence of acute severe respiratory failure in a significant proportion of COVID-19 cases, hospitalization, admission to the Intensive Care Unit, and intubation are frequently necessary to treat these cases [25]. Thus, alternative approaches such as extension to multi-state regression models [26] or direct regression models based on binomial regression [27] could help with predicting such an objective, allowing one to model time-dependent covariates and time-varying coefficients. Furthermore, although the estimation technique by [11] is efficient to estimate the proportional SDHs, alternative approaches such as pseudo-value and binomial regression approaches have more flexibility to model the CIF directly through different link functions. Nevertheless, the interpretation of the regression parameters in all these approaches is direct but not straightforward, depending on the chosen link function. However, computation and graphical representation of the CIF curves between different risk factor groups are straightforward and always possible to help one make personalized individual clinical decisions.

The data may not be comprehensive for all of the Brazilian COVID-19 population due to possible errors in the compilation and registration of the information by the diseased patients or the Personnel of the Ministry offices. The registration of cured patients was recorded only on some days and not continuously, providing some possible underestimation of the CIF for cured subjects. In addition, our analyses have the limitation that they do not further investigate the difference in the CIF between subjects who have one or more risk factors from subjects without risk factors, and they do not account for delayed entry into the ICU.

Author Contributions: Conceptualization, M.A.H.; methodology, G.C.; software and data validation, G.C. and M.A.H.; formal analysis, M.A.H.; investigation, G.C.; resources, G.C.; data curation, M.A.H.; original draft preparation, M.A.H.; draft review and editing, G.C.; data visualization, M.A.H.; supervision, G.C.; project administration, G.C.; funding acquisition, G.C. All authors have read and agreed to the published version of the manuscript.

Funding: This research was funded by the Italian Ministry of University and Research (MIUR) with the grant PRIN2017 (20178S4EK9).

Data Availability Statement: The COVID-19 data from 1 January 2020 to 30 April 2021 can be obtained from the Ministry of Health of Brazil.

Acknowledgments: The authors highly acknowledge the University of Padua for administrative and technical support.

Conflicts of Interest: The authors declare no conflict of interest. The funders had no role in the design of the study; in the collection, analyses, or interpretation of data; in the writing of the manuscript; or in the decision to publish the results.

Abbreviations

The following abbreviations are used in this manuscript:

COVID-19	Coronavirus Disease-2019
SARS-CoV2	Severe Acute Respiratory Syndrome Coronavirus 2
CRs	Competing Risks
CIF	Cumulative Incidence Function
K–M	Kaplan–Meier
CSH	Cause-Specific Hazard
SDH	Sub-Distribution Hazard
AIC	Akaike information criterion
ICU	Intensive Care Unit
HR	Hazard Ratio
IPCW	Inverse of the Probability of Censoring Weights

References

1. Ge, H.; Wang, X.; Yuan, X.; Xiao, G.; Wang, C.; Deng, T.; Yuan, Q.; Xiao, X. The epidemiology and clinical information about COVID-19. *Eur. J. Clin. Microbiol. Infect. Dis.* **2020**, *39*, 1011–1019. [CrossRef] [PubMed]
2. Ghosh, S.; Samanta, G.P.; Mubayi, A. Comparison of regression approaches for analyzing survival data in the presence of competing risks. *Lett. Biomath.* **2021**, *8*, 29–47.
3. Zuccaro, V.; Celsa, C.; Sambo, M.; Battaglia, S.; Sacchi, P.; Biscarini, S.; Valsecchi, P.; Pieri, T.C.; Gallazzi, I.; Colaneri, M.; et al. Competing-risk analysis of coronavirus disease 2019 in-hospital mortality in a Northern Italian centre from SMAtteo COVID-19 REgistry (SMACORE). *Sci. Rep.* **2021**, *11*, 1137. [CrossRef] [PubMed]
4. Salinas-Escudero, G.; Carrillo-Vega, M.F.; Granados-García, V.; Martínez-Valverde, S.; Toledano-Toledano, F.; Garduño-Espinosa, J. A survival analysis of COVID-19 in the Mexican population. *BMC Public Health* **2020**, *20*, 1616.
5. Nijman, G.; Wientjes, M.; Ramjith, J.; Janssen, N.; Hoogerwerf, J.; Abbink, E.; Blaauw, M.; Dofferhoff, T.; van Apeldoorn, M.; Veerman, K.; et al. Risk factors for in-hospital mortality in laboratory-confirmed COVID-19 patients in the Netherlands: A competing risk survival analysis. *PLoS ONE* **2021**, *16*, e0249231. [CrossRef]
6. Rathouz, P.J.; Valencia, V.; Chang, P.; Morton, D.; Yang, H.; Surer, O.; Fox, S.; Meyers, L.A.; Matsui, E.C.; Haynes, A.B. Survival analysis methods for analysis of hospitalization data: Application to COVID-19 patient hospitalization experience. *medRxiv* 2021, Preprint. [CrossRef]
7. Kim, H.T. Cumulative incidence in competing risks data and competing risks regression analysis. *Clin. Cancer Res.* **2007**, *13*, 559–565. [CrossRef]
8. Guo, C.; So, Y. Cause-specific analysis of competing risks using the PHREG procedure. *SAS Glob. Forum* **2018**, *2018*, 18.
9. Kaplan, E.L.; Meier, P. Nonparametric estimation from incomplete observations. *JASA* **1958**, *53*, 457–481. [CrossRef]
10. Andersen, P.K.; Borgan, Ø.; Gill, R.D.; Keiding, N. *Statistical Models Based on Counting Processes*; Springer Science & Business Media: Berlin/Heidelberg, Germany, 2012.
11. Fine, J.P.; Gray, R.J. A proportional hazards model for the subdistribution of a competing risk. *JASA* **1999**, *94*, 496–509. [CrossRef]
12. Satagopan, J.M.; Ben-Porat, L.; Berwick, M.; Robson, M.; Kutler, D.; Auerbach, A.D. A note on competing risks in survival data analysis. *Br. J. Cancer* **2004**, *9*, 1229–1235. [CrossRef] [PubMed]
13. Bender, R.; Augustin, T.; Blettner, M. Generating survival times to simulate Cox proportional hazards models. *Stat. Med.* **2005**, *24*, 1713–1723. [CrossRef] [PubMed]
14. Klein, J.P.; Van, H.; Hans, C.; Ibrahim, J.G.; Scheike, T.H. *Handbook of Survival Analysis*; CRC Press: Boca Raton, FL, USA, 2014.
15. Haller, B. The Analysis of Competing Risks Data with a Focus on Estimation of Cause-Specific and Subdistribution Hazard Ratios from a Mixture Model. Ph.D. Thesis, LMU School, Munchen, Germany, 2014.
16. Aalen, O. Nonparametric inference for a family of counting processes. *Ann. Stat.* **1978**, *6*, 701–726. [CrossRef]
17. Aalen, O.O.; Johansen, S. An empirical transition matrix for non-homogeneous Markov chains based on censored observations. *Scand. J. Statist.* **1978**, *5*, 141–150.
18. Prentice, R.L.; Kalbfleisch, J.D.; Peterson, A.V., Jr.; Flournoy, N.; Farewell, V.T.; Breslow, N.E. The analysis of failure times in the presence of competing risks. *Biometrics* **1978**, *34*, 541–554. [CrossRef]
19. Cheng, S.; Fine, J.P.; Wei, L. Prediction of cumulative incidence function under the proportional hazards model. *Biometrics* **1998**, *54*, 219–228. [CrossRef]
20. Bickel, P.J.; Doksum, T. *Mathematical Statistics*, 2nd ed.; Springer Prentice-Hall: Berlin/Heidelberg, Germany, 2001.
21. Demidenko, E. *Mixed Models: Theory and Applications with R*; John Wiley & Sons: Hoboken, NJ, USA, 2013.
22. Demidenko, E. Sample size determination for logistic regression revisited. *Stat. Med.* **2007**, *26*, 3385–3397. [CrossRef]
23. Gray, R.J. A class of K-sample tests for comparing the cumulative incidence of a competing risk. *Ann. Stat.* **1988**, *16*, 1141–1154. [CrossRef]
24. *R Core Team: A Language and Environment for Statistical Computing*; R Foundation for Statistical Computing: Vienna, Austria, 2019. Available online: Https://www.R-project.org/ (accessed on 19 August 2023).
25. Lai, C.C.; Shih, T.P.; Ko, W.C.; Tang, H.J.; Hsueh, P.R. Severe acute respiratory syndrome coronavirus 2 (SARS-CoV-2) and coronavirus disease-2019 (COVID-19): The epidemic and the challenges. *Int. J. Antimicrob. Agents* **2020**, *55*, 105924. [CrossRef]
26. Cortese, G.; Gerds, T.A.; Andersen, P.K. Comparing predictions among competing risks models with time–Dependent covariates. *Stat. Med.* **2013**, *32*, 3089–3101. [CrossRef]
27. Scheike, T.H.; Mei-Jie, Z.; Gerds, T.A. Cumulative Incidence Probability by Direct Binomial Regression. *Biometrika* **2008**, *49*, 205–220. [CrossRef]

Disclaimer/Publisher's Note: The statements, opinions and data contained in all publications are solely those of the individual author(s) and contributor(s) and not of MDPI and/or the editor(s). MDPI and/or the editor(s) disclaim responsibility for any injury to people or property resulting from any ideas, methods, instructions or products referred to in the content.

Article

Assessing the Impact of Time-Varying Optimal Vaccination and Non-Pharmaceutical Interventions on the Dynamics and Control of COVID-19: A Computational Epidemic Modeling Approach

Yan Li [1], Samreen [2], Laique Zada [3,*], Emad A. A. Ismail [4], Fuad A. Awwad [4] and Ahmed M. Hassan [5]

1. School of Mathematics and Data Sciences, Changji University, Changji 831100, China; shanghaijiaotong23@163.com
2. Department of Mathematics, Abdul Wali Khan University Mardan, Mardan 23200, Pakistan; samreen_fareed@awkum.edu.pk
3. Department of Mathematics, University of Peshawar, Peshawar 25120, Pakistan
4. Department of Quantitative Analysis, College of Business Administration, King Saud University, P.O. Box 71415, Riyadh 11587, Saudi Arabia; emadali@ksu.edu.sa (E.A.A.I.); fawwad@ksu.edu.sa (F.A.A.)
5. Faculty of Engineering, Future University in Egypt, New Cairo 11835, Egypt; ahmed.hassan.res@fue.edu.eg
* Correspondence: laique@uop.edu.pk

Citation: Li, Y.; Samreen; Zada, L.; Ismail, E.A.A.; Awwad, F.A.; Hassan, A.M. Assessing the Impact of Time-Varying Optimal Vaccination and Non-Pharmaceutical Interventions on the Dynamics and Control of COVID-19: A Computational Epidemic Modeling Approach. *Mathematics* **2023**, *11*, 4253. https://doi.org/10.3390/math11204253

Academic Editors: Cristiano Maria Verrelli and Fabio Della Rossa

Received: 30 August 2023
Revised: 3 October 2023
Accepted: 4 October 2023
Published: 11 October 2023

Copyright: © 2023 by the authors. Licensee MDPI, Basel, Switzerland. This article is an open access article distributed under the terms and conditions of the Creative Commons Attribution (CC BY) license (https://creativecommons.org/licenses/by/4.0/).

Abstract: Vaccination strategies remain one of the most effective and feasible preventive measures in combating infectious diseases, particularly during the COVID-19 pandemic. With the passage of time, continuous long-term lockdowns became impractical, and the effectiveness of contact-tracing procedures significantly declined as the number of cases increased. This paper presents a mathematical assessment of the dynamics and prevention of COVID-19, taking into account the constant and time-varying optimal COVID-19 vaccine with multiple doses. We attempt to develop a mathematical model by incorporating compartments with individuals receiving primary, secondary, and booster shots of the COVID-19 vaccine in a basic epidemic model. Initially, the model is rigorously studied in terms of qualitative analysis. The stability analysis and mathematical results are presented to demonstrate that the model is asymptotically stable both locally and globally at the COVID-19-free equilibrium state. We also investigate the impact of multiple vaccinations on the COVID-19 model's results, revealing that the infection risk can be reduced by administrating the booster vaccine dose to those individuals who already received their first vaccine doses. The existence of backward bifurcation phenomena is studied. A sensitivity analysis is carried out to determine the most sensitive parameter on the disease incidence. Furthermore, we developed a control model by introducing time-varying controls to suggest the optimal strategy for disease minimization. These controls are isolation, multiple vaccine efficacy, and reduction in the probability that different vaccine doses do not develop antibodies against the original virus. The existence and numerical solution to the COVID-19 control problem are presented. A detailed simulation is illustrated demonstrating the population-level impact of the constant and time-varying optimal controls on disease eradication. Using the novel concept of human awareness and several vaccination doses, the elimination of COVID-19 infections could be significantly enhanced.

Keywords: COVID-19 pandemic; multiple vaccine doses; sensitivity analysis; time-varying optimal controls; Pontryagin maximum principle

MSC: 93D05; 34A34; 49J15; 92B10

1. Introduction

The world's economy, communities, and public health have been profoundly impacted by the emergence of the novel coronavirus, SARS-CoV-2, which led to the COVID-19 pandemic. The first case of infection with this virus appeared in December 2019 in Wuhan, Hubei province of China [1]. This disease quickly spread outside of China only a few weeks after it first appeared. Initial reports of the disease outside China came from Japan and Thailand [2]. The COVID-19 pandemic has had a significant impact on society in different aspects. From a global health perspective, the virus has led to millions of worldwide cases of infection and loss of life. It has strained healthcare systems, particularly in regions with limited resources and capacity. This global disease also has far-reaching impacts on the economy, with disruptions in worldwide supply chains, business closures, and job losses. Socially, COVID-19 has led to school closures, travel restrictions, and changes in social behaviors, affecting individuals' mental health and well-being [3]. Extensive scientific investigation has been carried out to comprehend the dynamics and transmission of the SARS-CoV-2 virus [4]. There are many different indications and symptoms that COVID-19 might exhibit. On the other hand, some patients may remain symptom-free, and some may experience mild and then severe disease symptoms. Fever, coughing, sore throat, shortness of breath, feelings of exhaustion or low energy, muscle discomfort, and body pains are among the main signs and symptoms of COVID-19 [5,6]. The presence and intensity of symptoms can differ from person to person, and some people may not experience any symptoms at all or only experience minor ones. Different countries employ various techniques to stop the spread of infections, and the majority of countries adhere to similar rules such as social distancing, isolation, and self-quarantine [7].

Vaccination remains one of the effective interventions against severe infectious diseases [8–10]. The careful application of non-pharmaceutical therapies and vaccination programs work together to reduce the spread of COVID-19 globally. To prevent the disease, multiple dosages of vaccines are given. The majority of vaccinations are given to a person as primary (first time), secondary (second time), and booster shots (third time). The first course of vaccination doses administered to people who have never had any vaccination doses before is referred to as primary vaccination. The subsequent dosage given after the original immunization is referred to second-time vaccine and is also referred to as a second dose. The third-time vaccine is an additional dosage administered to people who have finished their first vaccine course to strengthen and prolong their immunological response [7].

In order to investigate the complex dynamics of infectious disease, different methodologies have been developed. The implementation of mathematical models is a valuable tool that has been utilized successfully to present different aspects of infectious diseases. Usually, these models include classical (ordinary and partial) derivatives [11–13], stochastic derivatives [14,15] and fractional derivatives [16,17]. In particular, to better explore the dynamic aspects of COVID-19, several epidemic models have been developed [18–20]. The impact of face mask use by the general public to curtail the COVID-19 pandemic was studied in [21]. Augusto et al. studied the changing behavior of the COVID-19 model [22]. The global impact of the first year of COVID-19 vaccination programs was studied in [23]. The impact of vaccination on two variants of COVID-19, alpha and delta, was studied in [24]. Ngonghala et al. [25] considered the omicron and delta variants of COVID-19 in the presence of multiple vaccinations. Their study revealed that treatment leads to a reduction in hospitalization rates, and the potential for COVID-19 elimination is increased when investments in control resources are directed toward promoting mask usage and vaccine intervention. In [26], the authors analyzed the mitigation of the pandemic via double-dose vaccination using an epidemic modeling approach. The outcomes of their study indicated that primary and secondary vaccination alone is not adequate for infection reduction, and thus it is essential to provide the booster shot (third time) of the vaccine for a better eradication of the infection. Recently, a similar fractional and fractal-fractional study with an exponential-type kernel analyzing the dynamics of COVID-19 under vaccination was presented in [27].

In this paper, we develop a mathematical model based on the SEVIHR type of compartmental model and incorporate three vaccine compartments for the first, second, and booster shots of COVID-19 vaccines. This study stands out from previous literature by considering multiple vaccination compartments in relation to the rates of antibody production against the original virus. We also used the assumption that vaccinated individuals might become infected with the virus if the antibodies have not been developed even after vaccinating to account for breakthrough infections. This enabled us to illustrate the reproductive number causing a significant widespread issue of the disease. The outcomes of these simulations aid those who are skeptical of vaccination in making thoughtful decisions. The section-wise description of the present work is as follows: The formulation of the COVID-19 model is presented in Section 2. Section 3 covers the basic qualitative properties and results of the proposed model. Section 4 presents the role of basic reproduction numbers, vaccination coverage, and bifurcation analysis. A simulation of the proposed model with constant control measures is given in Section 5. The sensitivity analysis is performed in Section 6. The formulation of the optimal control problem is given in Section 6. A simulation of the optimal control problem estimating the optimal solution is presented in Section 8. Finally, Section 9 presents the conclusion.

2. Modeling the Dynamics of COVID-19 with Multiple Vaccine Doses

We divide the total population into the following compartments: 1. susceptible individuals S; 2. exposed individuals E; 3. individuals with first-time vaccination V_1; 4. individuals with second-time vaccination V_2; 5. individuals with booster shots V_3; 6. infected individuals I; 7. hospitalized individuals I_H; and 8. R represents the recovered individuals. Therefore, the entire population can be expressed as

$$N(t) = S(t) + E(t) + V_1(t) + V_2(t) + V_3(t) + I(t) + I_H(t) + R(t).$$

The group of susceptible people is generated as a result of the birth rate θ, which reduces due to the transmission to the vaccinated class V_1 upon receiving the first vaccine dose at a rate ζ_1. Further, this class experiences a decrease after becoming infected at the contact rate α. All population groups experience natural mortality at a rate of μ. Thus, we obtain the following differential equation.

$$S'(t) = \theta - \frac{\alpha I}{N}S - (\zeta_1 + \mu)S,$$

Individuals in the exposed class are generated as a result of effective contacts between individuals in the infectious class I with those in classes S, V_1, V_2, and V_3 at the contact rates α. The symbols δ_1, δ_2, and δ_3 are the respective probabilities that vaccine recipients in the V_1, V_2 and V_3 groups do not develop antibodies to the original viruses after 28 days of inoculation. Therefore, the individuals susceptible to the original viruses considered in the model are S, $\delta_1 V_1$, $\delta_2 V_2$ and $\delta_3 V_3$. The transmission rate of the exposed individuals to the infectious class is denoted by σ and as a result, we derive the following equation for this class:

$$E'(t) = \left(\frac{\alpha IS}{N} + \frac{\alpha \delta_1 I V_1}{N} + \frac{\alpha \delta_2 I V_2}{N} + \frac{\alpha \delta_3 I V_3}{N}\right) - (\sigma + \mu)E,$$

Individuals in the susceptible class move to the class of first-time vaccinated individuals after receiving the initial vaccination at a rate of ζ_1. Subsequently, the individuals in this group reduce due to the contact rate α with infectious individuals and the administration of the second dose at the rate ζ_2. We obtain the following mathematical form for the dynamics of the first-time vaccinated class.

$$V_1'(t) = \zeta_1 S - \frac{\alpha \delta_1 I V_1}{N} - (\zeta_2 + \mu)V_1,$$

The population of second-time-vaccinated individuals is initially formed by administering a second dose to individuals in V_1 class at a rate ζ_2. This group experiences a decrease due to the contact rate α with infectious individuals and the administration of booster shots at the rate ζ_3. We obtain the following equation for the dynamics of the second-dose-vaccinated class.

$$V_2'(t) = \zeta_2 V_1 - \frac{\alpha \delta_2 I V_2}{N} - (\zeta_3 + \mu) V_2.$$

The class of third-time-vaccinated individuals is initiated through by administrating the booster shot to individuals in the V_2 class at a rate of ζ_3. This population experiences a decrease due to several factors, such as the contact rate α with infected people and the natural mortality rate μ. Thus, we obtain the following differential equation.

$$V_3'(t) = \zeta_3 V_2 - \frac{\alpha \delta_3 I V_3}{N} - \mu V_3.$$

The exposed individuals become infected and move to I class at the rate σ. The population in this class declined because of the hospitalization at a rate η, and the COVID-19 and natural deaths at rates denoted by γ and μ, respectively. Thus, we derive the following expression.

$$I'(t) = \sigma E - (\eta + \gamma + \mu) I.$$

The fraction $0 < b < 1$ of the ηI moves to R because of their natural immunity, while the remaining $(1 - b)$ are hospitalized. The individuals in the I_H class convalesce after proper treatment and join the recovered class at the recovery rate d. Thus, we obtain the following equations for the dynamics of hospitalized and recovered individuals.

$$I_H'(t) = \eta(1 - b) I - (d + \gamma + \mu) I_H,$$

$$R'(t) = \eta b I + d I_H - \mu R.$$

In the result of the above discussion, the compartmental model for COVID-19 transmission is summarized as follows:

$$\begin{aligned}
S'(t) &= \theta - \frac{\alpha I}{N} S - (\zeta_1 + \mu) S, \\
E'(t) &= \left(\frac{\alpha I S}{N} + \frac{\alpha \delta_1 I V_1}{N} + \frac{\alpha \delta_2 I V_2}{N} + \frac{\alpha \delta_3 I V_3}{N} \right) - (\sigma + \mu) E, \\
V_1'(t) &= \zeta_1 S - \frac{\alpha \delta_1 I V_1}{N} - (\zeta_2 + \mu) V_1, \\
V_2'(t) &= \zeta_2 V_1 - \frac{\alpha \delta_2 I V_2}{N} - (\zeta_3 + \mu) V_2, \\
V_3'(t) &= \zeta_3 V_2 - \frac{\alpha \delta_3 I V_3}{N} - \mu V_3, \\
I'(t) &= \sigma E - (\gamma + \eta + \mu) I, \\
I_H'(t) &= \eta(1 - b) I - (\mu + \gamma + d) I_H, \\
R'(t) &= \eta b I + d I_H - \mu R.
\end{aligned} \quad (1)$$

subject to non-negative initial conditions, $S(0) = S_0, E(0) = E_0, V_1(0) = V_1, V_2(0) = V_2, V_3(0) = V_3, I(0) = I_0, I_H(0) = I_{H_0}, R(0) = R_0$. Let

$$\lambda = \frac{(\alpha S + \alpha \delta_1 V_1 + \alpha \delta_2 V_2 + \alpha \delta_3 V_3)}{N},$$

and

$q_1 = (\zeta_1 + \mu)$, $q_2 = (\sigma + \mu)$, $q_3 = (\zeta_2 + \mu)$, $q_4 = (\zeta_3 + \mu)$, $q_5 = (\eta + \gamma + \mu)$, $q_6 = (d + \gamma + \mu)$.

Then, (1) becomes

$$\begin{aligned}
S'(t) &= \theta - \frac{\alpha I}{N}S - q_1 S, \\
E'(t) &= \lambda I - q_2 E, \\
V_1'(t) &= \zeta_1 S - \frac{\alpha \delta_1 I}{N} V_1 - q_3 V_1, \\
V_2'(t) &= \zeta_2 V_1 - \frac{\alpha \delta_2 I}{N} V_2 - q_4 V_2, \\
V_3'(t) &= \zeta_3 V_2 - \frac{\alpha \delta_3 I}{N} V_3 - \mu V_3, \\
I'(t) &= \sigma E - q_5 I, \\
I_H'(t) &= \eta(1-b)I - q_6 I_H, \\
R'(t) &= \eta b I + d I_H - \mu R.
\end{aligned} \qquad (2)$$

The transmission between various compartments are shown in the Figure 1.

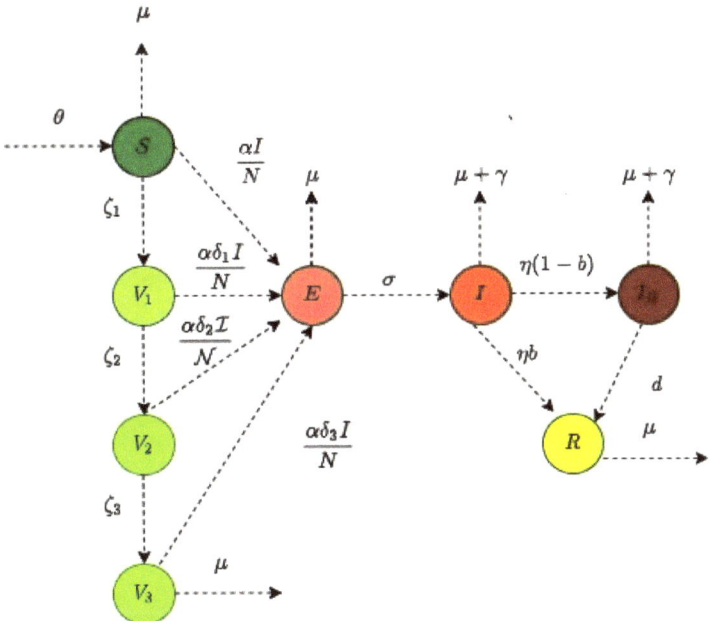

Figure 1. Flow-Chart of the COVID-19 model with vaccinations (2).

3. Qualitative Analysis of the Model

This section presents necessary mathematical aspects of the proposed COVID-19 transmission model having multiple vaccine compartments. We proceed as follows:

3.1. Positivity and Boundedness

3.1.1. Positivity

Theorem 1. *Given $S_0 \geq 0, E_0 \geq 0, V_{1_0} \geq 0, V_{2_0} \geq 0, V_{3_0} \geq 0, I_0 \geq 0, I_{H_0} \geq 0, R_0 \geq 0$. Then, the solution $(S(t), E(t), V_1(t), V_2(t), V_3(t), I(t), I_H(t), R(t))$ of model (2) are positive for all $t > 0$.*

Proof. First equation of the model (2) gives

$$S'(t) = \theta - \frac{\alpha I}{N} S - q_1 S,$$

$$S'(t) \geq -(\tilde{\alpha} - q_1)S, \quad \tilde{\alpha}(t) = \frac{\alpha I}{N}.$$
(3)

The integrating factor is given by $\exp\left(\int_0^t \tilde{\alpha}(s)ds + q_1 t\right)$, multiplying the inequality (3) by the integrating factor, we obtain,

$$S'(t) \exp\left(\int_0^t \tilde{\alpha}(s)ds + q_1 t\right) \geq 0.$$
(4)

The solution of (4) implies

$$S'(t) \geq S_0 \exp\left(-\left(\int_0^t \tilde{\alpha}(s)ds + q_1 t\right)\right) > 0.$$

In similar pattern, it can be shown that $E \geq 0, V_1 \geq 0, V_2 \geq 0, V_3 \geq 0, I \geq 0, I_H \geq 0, R \geq 0$ in the model (2). This implies that $(S(t), E(t), V_1(t), V_2(t), V_3(t), I(t), I_H(t), R(t))$ are all non-negative for non-negative initial conditions. □

3.1.2. Boundedness

Theorem 2. *The solution $(S(t), E(t), V_1(t), V_2(t), V_3(t), I(t), I_H(t), R(t))$ of the model (2) are bounded.*

Proof. Adding all the equations of the proposed model (2), we have

$$N'(t) = \theta - \mu N - \gamma(I + I_H),$$

$$N'(t) \leq \theta - \mu N.$$
(5)

Solving the inequality in (5), we have the following steps.

$$\int de^{\mu t} N(t) \leq \int \theta e^{\mu t} dt,$$

$$e^{\mu t} N(t) \leq \frac{\theta}{\mu} e^{\mu t} + C,$$

$$N(t) \leq e^{-\mu t}\left(\frac{\theta}{\mu} e^{\mu t} + C\right).$$

Using the initial condition $N(t) = N(0)$, at $t = 0$ implies that

$$N(t) \leq e^{-\mu t}\left(\frac{\theta}{\mu}e^{\mu t} + \left(N(0) - \frac{\theta}{\mu}\right)\right),$$

$$\leq e^{-\mu t}N(0) + \frac{\theta}{\mu}(1 - e^{-\mu t}),$$

$$\Rightarrow \lim_{t \to \infty} N(t) \leq \frac{\theta}{\mu}.$$

Hence, $N(t)$ is bounded by $\frac{\theta}{\mu}$, and by using the comparison theorem [28], we deduce that $N(t) \leq \frac{\theta}{\mu}$, if $N(0) \leq \frac{\theta}{\mu}$. Therefore, $\frac{\theta}{\mu}$ remains the upper bound of the region

$$\Omega = \begin{pmatrix} (S(t), E(t), V_1(t), V_2(t), V_3(t), I(t), I_H(t), R(t)) \\ : S(t) + E(t) + V_1(t) + V_2(t) + V_3(t) + I(t) + I_H(t) + R(t) \leq \frac{\theta}{\mu} \end{pmatrix}.$$ Thus, the region Ω is also positively invariant [29]. □

3.2. Equilibria and the Threshold Parameter of the Model

The COVID-19 compartmental model (2) exhibits two equilibria, namely, the COVID-free equilibrium (CFE), denoted as $\mathcal{P}^0$. This equilibrium can be given by the following expressions:

$$\mathcal{P}^0 = \left(S^0, E^0, V_1^0, V_2^0, V_3^0, I^0, I_H^0, R^0\right) = \left(\frac{\theta}{q_1}, 0, \frac{\theta \zeta_1}{q_1 q_3}, \frac{\theta \zeta_1 \zeta_2}{q_1 q_3 q_4}, \frac{\theta \zeta_1 \zeta_2 \zeta_3}{\mu q_1 q_2 q_3}, 0, 0, 0\right).$$

Taking the next generation approach into account [30], the threshold number $\mathcal{R}_0$ is formulated as follows:

$$\mathcal{R}_0 = \frac{\alpha \sigma}{q_2 q_5 N^0}\left(S^0 + \delta_1 V_1^0 + \delta_2 V_2^0 + \delta_3 V_3^0\right).$$

3.3. Local Stability at COVID-Free Equilibrium Point

Theorem 3. *The CFE point of the proposed model (2) is locally asymptotically stable (LAS) if $\mathcal{R}_0 < 1$ and unstable when $\mathcal{R}_0 > 1$.*

Proof. To establish the desired outcome, we need to demonstrate that the Jacobian of the linearized system at the CFE has negative eigenvalues. The matrix $J(\mathcal{P}^0)$ can be evaluated as

$$J(\mathcal{P}^0) = \begin{bmatrix} -q_1 & 0 & 0 & 0 & 0 & -\frac{\alpha \mu}{q_1} & 0 & 0 \\ 0 & -q_2 & 0 & 0 & 0 & \frac{\alpha \mu}{q_1} + \frac{\alpha \mu \delta_1 \zeta_1}{q_1 q_3} + \frac{\alpha \mu \delta_2 \zeta_1 \zeta_2}{q_1 q_3 q_4} + \frac{\alpha \delta_3 \zeta_1 \zeta_2 \zeta_3}{q_1 q_3 q_4} & 0 & 0 \\ \zeta_1 & 0 & -q_3 & 0 & 0 & -\frac{\alpha \mu \delta_1}{q_1 q_3} & 0 & 0 \\ 0 & 0 & \zeta_2 & -q_4 & 0 & -\frac{\alpha \mu \delta_2 \zeta_1 \zeta_2}{q_1 q_3 q_4} & 0 & 0 \\ 0 & 0 & 0 & \zeta_3 & -\mu & -\frac{\alpha \delta_3 \zeta_1 \zeta_2 \zeta_3}{q_1 q_3 q_4} & 0 & 0 \\ 0 & \sigma & 0 & 0 & 0 & -q_5 & 0 & 0 \\ 0 & 0 & 0 & 0 & 0 & (1-b)\eta & -q_6 & 0 \\ 0 & 0 & 0 & 0 & 0 & b\eta & d & -\mu \end{bmatrix}.$$

The eigenvalues of $J(\mathcal{P}^0)$ are $-\mu, -\mu, q_1, q_3, q_4, q_6$, and the roots of the 2nd degree equation $c_0 \lambda^2 + c_1 \lambda + c_2 = 0$. Here, the coefficients are defined in terms of the basic reproduction number $\mathcal{R}_0$ as

$$c_0 = 1,$$
$$c_1 = q_2 + q_5,$$
$$c_2 = q_2 q_5 (1 - \mathcal{R}_0).$$

Since, $c_0, c_1 > 0$ and $c_2 > 0$ if $\mathcal{R}_0 < 1$. Therefore, the well-known Ruth–Hurwitz criteria demonstrate that the system (2) is LAS. □

3.4. Global Asymptotic Stability of (GAS) for Special Case

The GAS of the model at CFE point is provided for a special case. Consider the special case of the model (2) with $\gamma = \delta_1 = \delta_2 = \delta_3 = 0$. The assumption regarding the subsequent parameters is made to facilitate mathematical analysis. The feasible region for the special case of the model (2) is constructed as

$$\Omega_* = \left\{ (S, E, V_1, V_2, V_3, I, I_H, R) \in \mathbb{R}_+^* : S \leq S^0, V_1 \leq V_1^0, V_2 \leq V_2^0, V_3 \leq V_3^0 \right\}.$$

It can be shown that Ω_* is a positive invariant set and attracts the solution of system (2) with respect to the special case. Moreover, the basic reproductive number of the reduced model is as follows:

$$\hat{\mathcal{R}}_0 = \mathcal{R}_0|_{\gamma=\delta_1=\delta_2=\delta_3=0} = \frac{\alpha\sigma S^0}{q_1 q_2 q_5^* N^0}, \quad \text{where } q_5^* = \eta + \mu, q_6^* = \mu + d.$$

Theorem 4. *The CFE of the special case of the model (2) with $\gamma = \delta_1 = \delta_2 = \delta_3 = 0$ is GAS in Ω_* whenever, $\hat{\mathcal{R}}_0 \leq 1$.*

Proof.

$$L(t) = A_1 \frac{(S-S^0)^2}{2S^0} + A_2 E + A_3 I,$$

$$L' = A_1 \frac{(S-S^0)}{S^0} S' + A_2 E' + A_3 I',$$

$$L' = A_1 \frac{(S-S^0)}{S^0} \left(-\left(\frac{\alpha I}{N}\right)(S-S^0) - q_1(S-S^0) - \frac{\alpha I}{N} S^0 \right) + A_2 \left(\frac{\alpha IS}{N} - q_2 E \right) + A_3 (\sigma E - q_5^* I)$$

$$\leq -A_1 \frac{\alpha(S-S^0)^2}{S^0} I + A_2 \left(\alpha I \frac{S^0}{N^0} - q_2 E \right) + A_3 (\sigma E - q_5^* I),$$

$$= -A_1 \frac{\alpha(S-S^0)^2}{S^0} I + (\sigma A_3 - q_2 A_2) E + \left(\alpha \frac{S^0}{N^0} A_2 - q_5^* A_3 \right) I,$$

$$= -A_1 \frac{\alpha(S-S^0)^2}{S^0} I + (\sigma A_3 - q_2 A_2) E + q_5^* A_3 \left(\frac{\alpha S^0}{q_5^* A_3 N^0} A_2 - 1 \right) I,$$

$$L' \leq -(S-S^0)^2 I + q_2(\hat{\mathcal{R}}_0 - 1) E,$$

where,

$$A_1 = \frac{S^0}{\alpha}, \quad A_2 = \frac{\sigma A_3}{q_2}, \quad A_3 = 1.$$

Thus, $L'(t) \leq 0$ if $\hat{\mathcal{R}}_0 \leq 1$ further $L'(t) = 0$ if $E = I = I_H = 0$. So, the number of infected individuals becomes zero as $t \to \infty$. Using $E = I = I_H = 0$ in the above model we have $S \to \frac{\theta}{q_1}, V_1 \to \frac{\theta \zeta_1}{q_1 q_3}, V_2 \to \frac{\theta \zeta_1 \zeta_2}{q_1 q_3 q_4}, V_3 \to \frac{\theta \zeta_1 \zeta_2 \zeta_3}{\mu q_1 q_3 q_4}$ and $R \to 0$ as $t \to \infty$. Thus, using Lyapunov stability theorem, every solution of the given model with non-negative initial condition approaches to $\mathcal{P}^0$ as $t \to \infty$ in Ω. Thus, it follows that the system (2) is GAS. □

4. Interpretation of Vaccination Coverages Based on $\mathcal{R}_0$

We provide the critical rate of vaccine coverage that could lead to eradication of the infection, denoted as $\mathcal{R}_0(\zeta_1, \zeta_2, \zeta_3) = \mathcal{R}_{0_e}$. In a scenario with no vaccine, i.e., when $\zeta_1 = 0$, $\zeta_2 = 0$, and $\zeta_3 = 0$, then $\mathcal{R}_0$ is reduced to

$$\mathcal{R}_{0_e} = \mathcal{R}_0(0,0,0) = \frac{\alpha\sigma}{q_2 q_5}. \tag{6}$$

After some rearrangement $\mathcal{R}_0(\zeta_1)$ and with $\zeta_2 = \zeta_3 = 0$ can be written as

$$\mathcal{R}_0(\zeta_1) = \frac{\alpha\sigma(\zeta_1 \delta_1 + \mu)}{q_2 q_5 (\zeta_1 + \mu)}.$$

Thus,

$$\mathcal{R}_0(\infty) = \lim_{\zeta_1 \to \infty} \mathcal{R}_0(\zeta_1) = \frac{\alpha\sigma}{q_2 q_5}\delta_1 = \mathcal{R}_{0_e}\delta_1.$$

The partial derivative with respect to ζ_1 leads to the following form:

$$\frac{\partial \mathcal{R}_0}{\partial \zeta_1} = -\frac{\mathcal{R}_{0_e}\mu(1-\delta_1)}{q_1^2} < 0.$$

Therefore, $\mathcal{R}_{0_e}\delta_1 \leq \mathcal{R}_0(\zeta_1) \leq \mathcal{R}_{0_e}$, and hence, $\mathcal{R}_{0_e} < 1$ implies $\mathcal{R}_0(\zeta_1) < 1$. Further, if $\mathcal{R}_{0_e} > 1$, it is important to note that

$$\mathcal{R}_0(\infty) < 1 \Leftrightarrow \mathcal{R}_0 \delta_1 < 1 \Leftrightarrow \delta_1 > \delta_1^* = \frac{1}{\mathcal{R}_{0_e}}.$$

This interpretation suggests that when the value of δ_1 is low, and if the effective reproductive number $\mathcal{R}_{0_e} > 1$, the disease might not spread extensively provided that the first-dose vaccination coverage is substantial. Similarly, if $\zeta_1, \zeta_2 \neq 0$,

$$\mathcal{R}_0(\zeta_1, \zeta_2) = \frac{\mathcal{R}_{0_e}(\mu^2 + \mu\delta_1\zeta_1 + (\mu + \delta_2\zeta_1)\zeta_2)}{(\zeta_1 + \mu)(\zeta_2 + \mu)}.$$

After some rearrangement, we have

$$\mathcal{R}_0(\zeta_1, \zeta_2) = \frac{1}{\mu q_3}\mathcal{R}_0(\zeta_1, 0) + \frac{\mathcal{R}_{0_e}(\mu + \delta_2\zeta_1)\zeta_2}{q_1 q_3}.$$

Thus,

$$\mathcal{R}_0(\zeta_1, \infty) = \lim_{\zeta_2 \to \infty} \mathcal{R}_0(\zeta_1, \zeta_2) = \frac{\mathcal{R}_{0_e}(\mu + \delta_2\zeta_1)}{q_1}.$$

The partial differentiation with respect to ζ_2 yields

$$\frac{\partial \mathcal{R}_0}{\partial \zeta_2} = -\frac{\mathcal{R}_{0_e}(\delta_1 - \delta_2)\zeta_1\mu}{q_1 q_2 q_3^2 q_5} < 0.$$

Therefore, $\frac{\mathcal{R}_{0_e}(\mu+\delta_2\zeta_1)}{q_1} \leq \mathcal{R}_0(\zeta_1, \zeta_2) \leq \mathcal{R}_{0_e}$, and hence $\mathcal{R}_{0_e} < 1$ implies $R_0(\zeta_1, \zeta_2) < 1$. Further if $\mathcal{R}_{0_e} > 1$, it is important to note that

$$\mathcal{R}_0(\zeta_1, \infty) < 1 \Leftrightarrow \frac{\mathcal{R}_{0_e}(\mu + \delta_2\zeta_1)}{q_1} < 1 \Leftrightarrow \delta_2 > \delta_2^* = \frac{q_1}{\mathcal{R}_{0_e}} - \frac{\mu}{\zeta_1}.$$

The above interpretation implies that when the value of δ_2 is minimal and if the effective reproductive number $\mathcal{R}_{0_e} > 1$, there is a possibility of infection elimination if the second vaccination coverage is substantial.

Finally, in a scenario if not only first and second vaccine doses are administrated, but additionally, a booster shot is provided, i.e., when $\zeta_1, \zeta_2, \zeta_3 \neq 0$ then,

$$\mathcal{R}_0(\zeta_1, \zeta_2, \zeta_3) = \frac{\mu \mathcal{R}_0(\zeta_1, \zeta_2, 0)}{q_4} + \frac{\mathcal{R}_{0_e}(\mu^2 + \mu\delta_1\zeta_1 + (\mu + \delta_3\zeta_1)\zeta_2)\zeta_3}{q_1 q_3 q_4}.$$

Thus,

$$\mathcal{R}_0(\zeta_1, \zeta_2, \infty) = \lim_{\zeta_3 \to \infty} \mathcal{R}_0(\zeta_1, \zeta_2, \zeta_3) = \frac{\mathcal{R}_{0_e}(\mu^2 + \mu\delta_1\zeta_1 + (\mu + \delta_3\zeta_1)\zeta_2)}{q_1 q_3}.$$

Taking partial derivative with respect to ζ_3 yields

$$\frac{\partial \mathcal{R}_0}{\partial \zeta_3} = -\frac{\alpha \sigma \mu (\delta_2 - \delta_3)\zeta_1\zeta_2}{q_1 q_2 q_3 q_4^2 q_5} < 0.$$

Therefore, $\frac{\mathcal{R}_{0_e}(\mu^2 + \mu\delta_1\zeta_1 + (\mu + \delta_3\zeta_1)\zeta_2)}{q_1 q_3} \leq \mathcal{R}_0(\zeta_1, \zeta_2, \zeta_3) \leq \mathcal{R}_{0_e}$, and hence $\mathcal{R}_{0_e} < 1$ implies $\mathcal{R}_0(\zeta_1, \zeta_2, \zeta_3) < 1$. Further, if $\mathcal{R}_{0_e} > 1$ it is important to note that

$$\mathcal{R}_0(\zeta_1, \zeta_2, \infty) < 1 \Leftrightarrow \frac{\mathcal{R}_{0_e}(\mu^2 + \mu\delta_1\zeta_1 + (\mu + \delta_3\zeta_1)\zeta_2)}{q_1 q_3} < 1 \Leftrightarrow \delta_3 > \delta_3^* = \frac{q_1 q_3}{\mathcal{R}_{0_e}\zeta_1} - \frac{\mu(\mu + \delta_1\zeta_1 + \zeta_2)}{\zeta_1}.$$

This implies that when δ_3 is minimal and the effective reproductive number $\mathcal{R}_{0_e} > 1$, the disease has the potential to be eliminated through the administration of a booster shot to a person who had already received the initial dose.

Backward Bifurcation Analysis

In this subsection, we discuss the existence of backward bifurcation of the system (2). To analyze this, we follow the center manifold theory. If we take α as bifurcation parameter then at $\mathcal{R}_0 = 1$ we have,

$$\alpha^* = \frac{q_1 q_2 q_3 q_4 q_5}{\sigma(\mu q_3 q_4 + \zeta_1(\mu q_4 \delta_1 + \zeta_2(\mu \delta_2 + \delta_3 \zeta_3)))}.$$

The variables in the system (2) are subjected to the following variations so that $S = x_1$, $E = x_2, V_1 = x_3, V_2 = x_4, V_3 = x_5, I = x_6, I_H = x_7$, and $R = x_8$. Further, using the vector notation $x = (x_1, x_2, x_3, x_4, x_5, x_6, x_7, x_8)^{tr}$. COVID-19 (2) can be written equivalently as $\frac{dx}{dt} = f$, where $f = (f_1, f_2, f_3, f_4, f_5, f_6, f_7, f_8)^{tr}$ as shown below:

$$x_1'(t) = \theta - \frac{\alpha x_6}{N}x_1 - q_1 x_1,$$

$$x_2'(t) = \left(\frac{\alpha x_6}{N}x_1 + \frac{\alpha \delta_1 x_6}{N}x_3 + \frac{\alpha \delta_2 x_6}{N}x_4 + \frac{\alpha \delta_3 x_6}{N}x_5\right) - q_2 x_2,$$

$$x_3'(t) = \zeta_1 x_1 - \frac{\alpha \delta_1 x_6}{N}x_3 - q_3 x_3,$$

$$x_4'(t) = \zeta_2 x_3 - \frac{\alpha \delta_2 x_6}{N}x_4 - q_4 x_4,$$

$$x_5'(t) = \zeta_3 x_4 - \frac{\alpha \delta_3 x_6}{N}x_5 - \mu x_5,$$

$$x_6'(t) = \sigma x_2 - q_5 x_6,$$

$$x_7'(t) = \eta(1-b)x_6 - q_6 x_7,$$

$$x_8'(t) = \eta b x_6 + d x_7 - \mu x_8,$$

where $N = \sum_{i=1}^{8} x_i$.

The Jacobian matrix evaluated at CFE point $\mathcal{P}^0$ with α^* is

$$J(\mathcal{P}^0) = \begin{bmatrix} -q_1 & 0 & 0 & 0 & 0 & -\frac{\alpha^* \mu}{q_1} & 0 & 0 \\ 0 & -q_2 & 0 & 0 & 0 & \frac{\alpha^* \mu}{q_1} + \frac{\alpha^* \mu \delta_1 \zeta_1}{q_1 q_3} + \frac{\alpha^* \mu \delta_2 \zeta_1 \zeta_2}{q_1 q_3 q_4} + \frac{\alpha^* \delta_3 \zeta_1 \zeta_2 \zeta_3}{q_1 q_3 q_4} & 0 & 0 \\ \zeta_1 & 0 & -q_3 & 0 & 0 & -\frac{\alpha^* \mu \delta_1 \zeta_1}{q_1 q_3} & 0 & 0 \\ 0 & 0 & \zeta_2 & -q_4 & 0 & -\frac{\alpha^* \mu \delta_2 \zeta_1 \zeta_2}{q_1 q_3 q_4} & 0 & 0 \\ 0 & 0 & 0 & \zeta_3 & -\mu & -\frac{\alpha^* \delta_3 \zeta_1 \zeta_2 \zeta_3}{q_1 q_3 q_4} & 0 & 0 \\ 0 & \sigma & 0 & 0 & 0 & -q_5 & 0 & 0 \\ 0 & 0 & 0 & 0 & 0 & \eta(1-b) & -q_6 & 0 \\ 0 & 0 & 0 & 0 & 0 & b\eta & d & -\mu \end{bmatrix},$$

The Jacobian matrix has a simple eigenvalue calculated at α^*. The right and left eigenvectors are denoted by $W = (w_1, w_2, w_3, w_4, w_5, w_6, w_7, w_8)$ and $V = (v_1, v_2, v_3, v_4, v_5, v_6, v_7, v_8)$, respectively, where

$$w_1 = -\frac{\mu q_2 q_3 q_4 q_5 w_6}{\sigma q_1(\mu q_3 q_4 + \zeta_1(\mu \delta_1 q_4 + \zeta_2(\mu \delta_2 + \delta_3 \zeta_3)))}, \quad w_2 = \frac{q_5 w_6}{\sigma},$$

$$w_3 = -w_6\left(\frac{\mu q_2 q_5 \delta_1 \zeta_1 q_4}{\sigma q_3(\mu q_3 q_4 + \zeta_1(\mu \delta_1 q_4 + \zeta_2(\mu \delta_2 + \delta_3 \zeta_3)))} + \frac{\mu q_2 q_5 \zeta_1 q_3 q_4}{\sigma q_1 q_3(\mu q_3 q_4 + \zeta_1(\mu \delta_1 q_4 + \zeta_2(\mu \delta_2 + \delta_3 \zeta_3)))}\right),$$

$$w_4 = -\frac{\mu q_2 q_5 w_6 \zeta_1 \zeta_2(\delta_1 q_1 q_4 + q_2(\mu + \delta_2 q_1 + \zeta_3))}{\sigma q_1 q_3 q_4(\mu^2 q_4 + \mu \delta_1 \zeta_1 q_4 + \zeta_2(\mu(\mu + \delta_2 \zeta_1) + (\mu + \delta_3 \zeta_1)\zeta_3))},$$

$$w_5 = -\frac{q_2 q_5 w_6 \zeta_1 \zeta_2 \zeta_3(\mu \delta_1 q_1 q_4 + q_3(\mu \delta_2 q_1 + (\mu + \delta_3 q_1)q_4))}{\mu \sigma q_1 q_3 q_4(\mu^2 q_4 + \mu \delta_1 \zeta_1 q_4 + \zeta_2(\mu(\mu + \delta_2 \zeta_1) + (\mu + \delta_3 \zeta_1)\zeta_3))},$$

$$w_6 > 0, \quad w_7 = \frac{\eta(1-b)w_6}{q_6}, \quad w_8 = \frac{(d\eta + b\gamma\eta + b\eta\mu)}{\mu q_6}w_6,$$

and,

$$v_1 = 0, v_2 > 0, v_3 = v_4 = v_5 = 0, v_6 = \frac{q_2 v_2}{\sigma}, v_7 = v_8 = 0.$$

Furthermore, bifurcation coefficients $\bar{a}$, and $\bar{b}$ of the proposed model (2) evaluated at $(\mathcal{P}^0, \alpha^*)$ are calculated as

$$\bar{a} = \sum_{k,i,j=1}^{8} v_k w_i w_j \frac{\partial^2 f_k(\mathcal{P}^0, \alpha^*)}{\partial x_i \partial x_j}, \quad \bar{b} = \sum_{k,i=1}^{8} v_k w_i \frac{\partial^2 f_k(\mathcal{P}^0, \alpha^*)}{\partial x_i \partial \beta}.$$

So we have

$$\bar{a} = \frac{-2\alpha^* \mu v_2 w_6}{\theta q_1 q_3 q_4} \left(q_3 q_4 \left(\begin{array}{c} \zeta_1 w_1 + \mu(w_2 + w_3 + w_4 + w_5 + w_6 + w_7 + w_8) - q_1(w_3 \delta_1 + w_4 \delta_2 + w_5 \delta_3) \\ + (w_1 + w_2 + w_3 + w_4 + w_5 + w_6 + w_7 + w_8) \zeta_1 (\mu q_4 \delta_1 + \zeta_2(\mu \delta_2 + \zeta_3 \delta_3)) \end{array} \right) \right) < 0,$$

$$\bar{b} = \frac{v_2 w_6 (\mu q_3 q_4 + \zeta_1 (\mu q_4 \delta_1 + \zeta_2(\mu \delta_2 + \delta_3 \zeta_3)))}{q_1 q_3 q_4} > 0.$$

The presence of the backward bifurcation in the system depends on the sign of $\bar{b}$, as indicated in [31]. Specifically, if $\bar{a} > 0$ and $\bar{b} > 0$, the system will experience a backward bifurcation. In the context of backward bifurcation, the endemic equilibrium and the CFE coexist with one being stable and the other being unstable when $\mathcal{R}_0 < 1$. For the COVID-19 model (2), the biological significance of backward bifurcation lies in the fact that the condition $\mathcal{R}_0 < 1$ is essential, but not solely adequate for mitigating the infection from the community. The outcome depends on the initial population size.

5. Estimating the Time Series Solution

This section focuses on the numerical simulation of the epidemic model (2) to analyze the dynamics of various state variables for $\mathcal{R}_0 > 1$ and $\mathcal{R}_0 < 1$. The model without optimal controls is numerically solved using the well-known Runge–Kutta iterative scheme of fourth order. Simulation is conducted in Matlab version R2022b for the 0–700 days. Initially, the model is simulated with the parameter values given in Table 1 such that $\mathcal{R}_0 < 1$, and the resulting plots are depicted in Figures 2 and 3. The figures show that the population curves converge to the COVID-19-free equilibrium state. Similarly, in Figures 4 and 5, a few values of the model's parameters provided in Table 1 are changed in a way that $\mathcal{R}_0 > 1$ These graphical interpretations show that the solution trajectories converge to an endemic state. Moreover, we analyzed the dynamics of the model with different initial values of state variables for $\mathcal{R}_0 > 1$ and $\mathcal{R}_0 < 1$ as shown in the Figures 6 and 7, respectively.

Table 1. Description with numerical values of the model's parameters.

Parameter	Meaning	Value	Reference
θ	humans' recruitment rate	7,828,143	[32]
μ	natural death rate	0.011380	[32]
ζ_1	rate of first COVID vaccine dose	0.710	[32]
ζ_2	rate of second COVID vaccine dose	0.650	Assumed
ζ_3	rate of booster shot	0.290	[32]
δ_1	possibility that after 28 days of vaccinations, those who received the first vaccine dose has not produced immunity to the original virus	4.7%	[33]
δ_2	possibility that after 28 days of vaccinations, those who received the 2nd vaccine dose has not produced immunity to the original virus	0.2%	[33]
δ_3	possibility that after 28 days of vaccinations, those who received the booster shot has not produced immunity to the original virus	0.1%	[33]
σ	transmission rate of exposed to infectious class	1/5.2	[26]
b	recovery rate of infectious people	0.76210	[26]
η	flow rate of infectious individuals	0.0720	[32]
α	effective contacts rate	0.750	Assumed
γ	mortality rate due to infection	0.000010	Assumed
d	recovery rate of hospitalization individuals	0.01070	Assumed

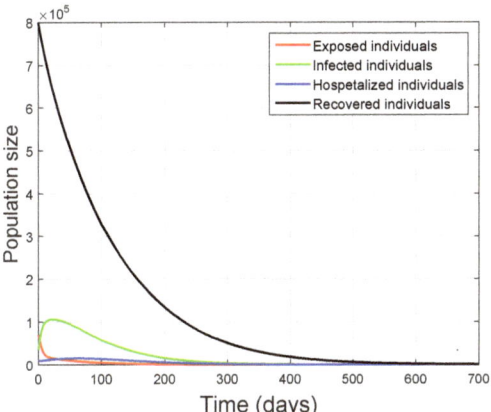

Figure 2. Graphical results of the proposed model for exposed, infected, hospitalized and recovered compartments when $\mathcal{R}_0 < 1$.

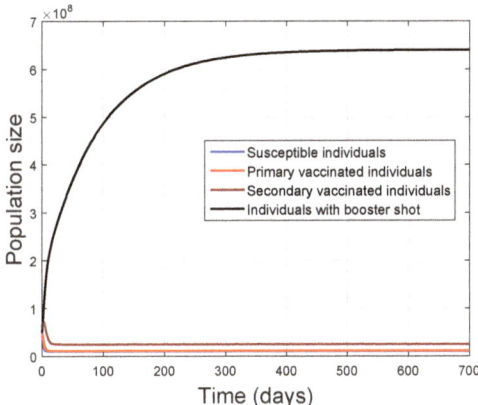

Figure 3. Simulation of the proposed model for susceptible and vaccinated compartments using $\mathcal{R}_0 < 1$.

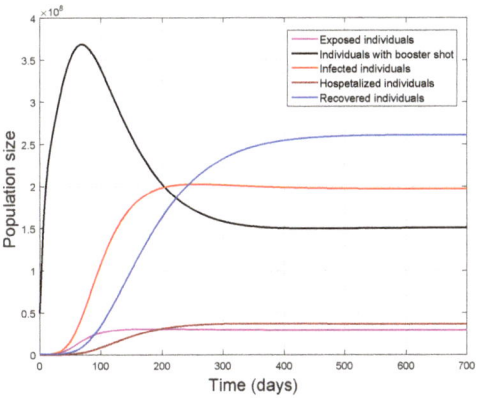

Figure 4. Simulation of the proposed model for individuals with booster short, exposed, infected, hospitalized and recovered compartments when $\mathcal{R}_0 > 1$.

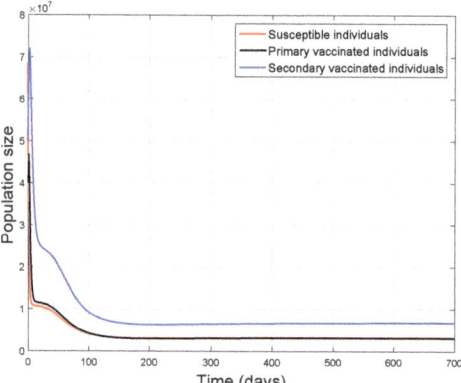

Figure 5. Simulation of the proposed model for susceptible and primary and secondary vaccinated compartments when $\mathcal{R}_0 > 1$.

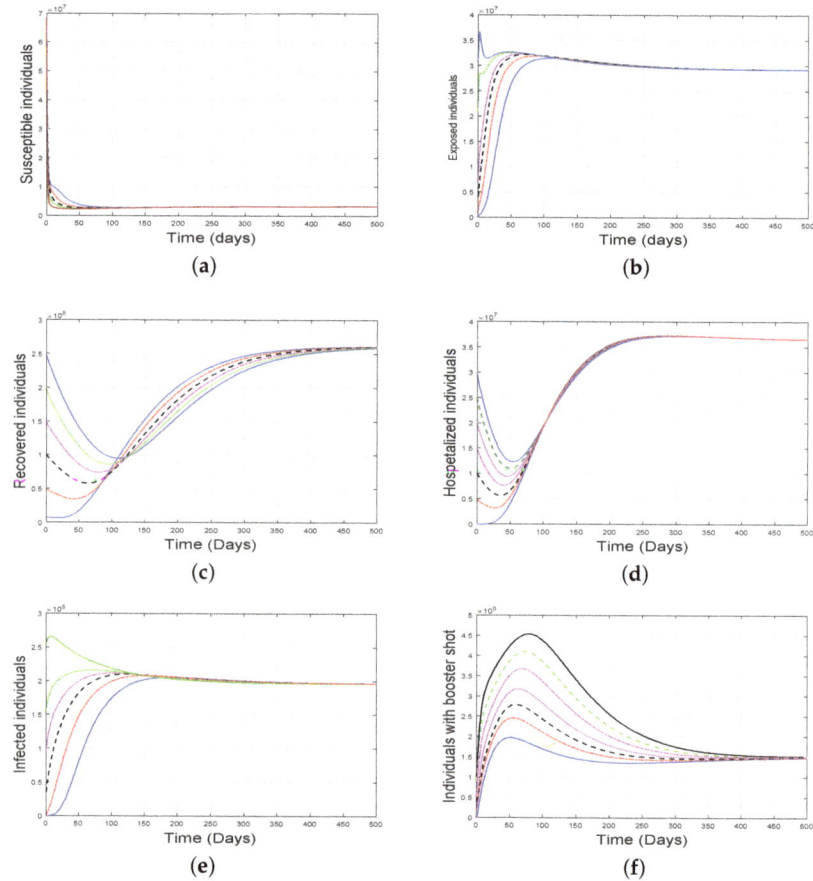

Figure 6. Simulation of the proposed COVID-19 model with different initial values of the state variables and $\mathcal{R}_0 > 1$, where (**a**) susceptible, (**b**) exposed, (**c**) recovered, (**d**) hospitalized, (**e**) infected, (**f**) shows the individuals with booster short. The curves with different colors show the dynamics by choosing different values of the corresponding state variables.

In the next section, the application of optimal control theory for the mitigation of pandemic is carried out. Before establishing effective optimal controls, we perform the sensitivity analysis to investigate the most influential factors on the transmission of disease.

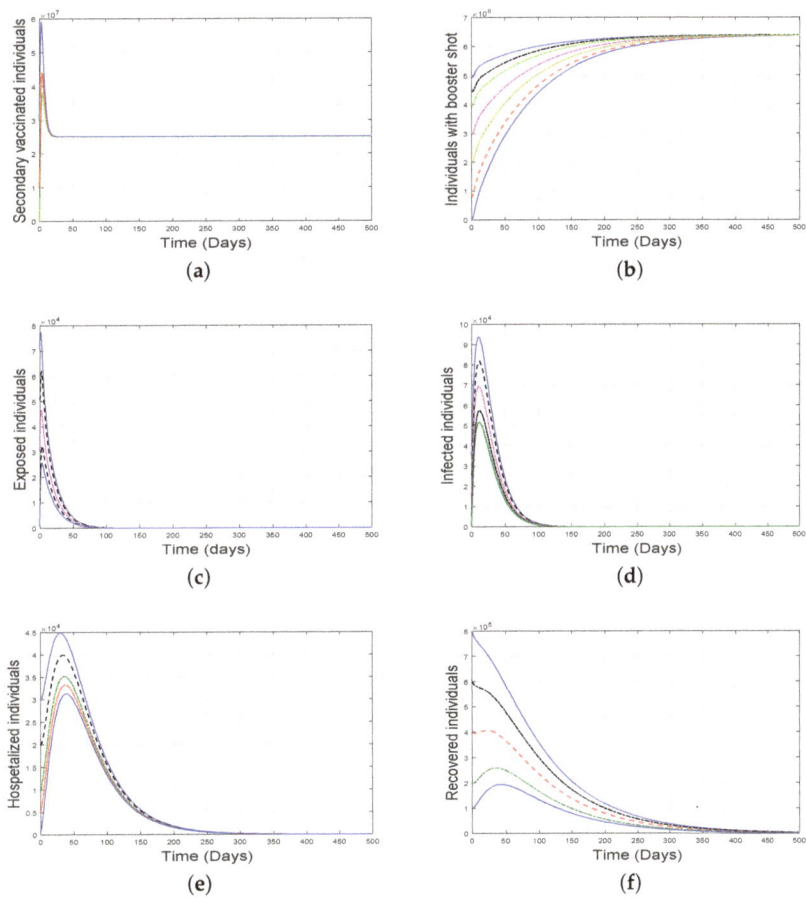

Figure 7. Dynamics of the COVID-19 model with different initial values of the state variables and $\mathcal{R}_0 < 1$, where (**a**) secondary vaccinated, (**b**) individuals with booster short, (**c**) exposed, (**d**) infected, (**e**) hospitalized, (**f**) R]recovered individuals. The curves with different colors show the dynamics by choosing different values of the corresponding state variables.

6. Sensitivity Analysis

Sensitivity theory is a powerful tool that provides insights into the model's behavior and helps in better understanding the influence of the variations in the input parameters on the respective output of the model. This technique is used in various fields, including economics, engineering, and biological sciences, to assess the potential factors of the problem under consideration. The analysis examines how small changes in the input values of the model affect the corresponding output. In the epidemiological modeling approach, sensitivity analysis provides valuable insights into identifying the parameters that play a substantial role in influencing disease transmission and control. In this study, we specifically performed sensitivity analysis of some crucial parameters. We employ the well-known normalized parametric scheme based on the forward sensitivity index of the model's parameters, as described by Chitnis et al. [34]. A positive (or negative) index illustrates that the parameter has a direct (or inverse) effect on $\mathcal{R}_0$.

Definition 1. *The normalized sensitivity index, which is used to quantify the relative change in $\mathcal{R}_0$ concerning changes in model's various parameters, is defined as*

$$Y_x = \frac{x}{|\mathcal{R}_0|} \times \frac{\partial \mathcal{R}_0}{\partial x}. \tag{7}$$

Applying Equation (7), the Table 2 shows the normalized sensitivity index assessing the proportional change of $\mathcal{R}_0$ with respect to the model parameters. The parameters $\alpha, \sigma, \mu, \delta_1, \delta_2,$ and δ_3 directly effect $\mathcal{R}_0$. This shows that the value of $\mathcal{R}_0$ will grow with an increase in the parameters above (or decrease respectively). The parameters $\gamma, \eta, \zeta_1, \zeta_2, \zeta_3$ have an inverse relationship with $\mathcal{R}_0$. The graphical representation of sensitivity indices are depicted in the plot (Figure 8).

Table 2. Sensitivity index of the selected model's parameters.

Parameter	Index
Y_α	1
Y_σ	0.0558698
Y_μ	0.73501
Y_{δ_1}	0.0452883
Y_{δ_2}	0.0041564
Y_{δ_3}	0.0529594
Y_γ	-0.000119918
Y_η	-0.863413
Y_{ζ_1}	-0.881821
Y_{ζ_2}	-0.0435263
Y_{ζ_3}	-0.00199973

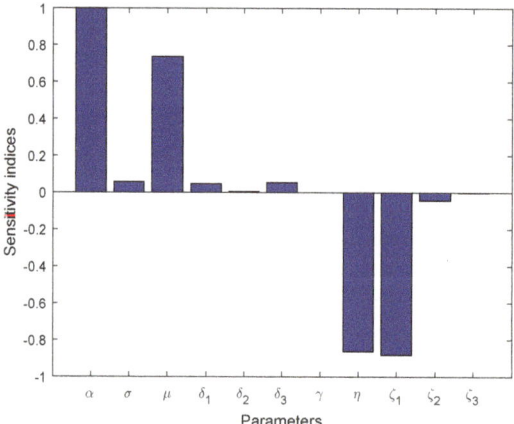

Figure 8. Bar graph of sensitivity indices of the COVID-19 model.

The significant insights gained from the sensitivity analysis can be helpful in various applications such as

- Identifying potential parameters: This helps identify the input parameter(s) that play a substantial role in influencing disease dynamics.
- This helps understand how uncertainties in input parameters can propagate to uncertainties in the model's predictions.
- Optimization: In optimization problems, the sensitivity indices of model parameters can guide the setting of appropriate optimal interventions.

- Decision-making: Understanding the sensitivity of the model to various inputs can assist decision-makers in making informed choices.

7. Optimal Control Analysis of COVID-19 Model

In this section, we introduce an optimal control strategy for the model (2). The aim of optimal control is to minimize the COVID-19 disease by providing the following control strategies.

- $u_1(t)$: The first control is used to reduce the number of effective contacts between infected and susceptible individuals.
- $u_2(t)$: The second control is used to enhance the first vaccine dose efficacy.
- $u_3(t)$: The third control is used to reduce the possibility that after 28 days of vaccination, those who received the first dose do not develop immunity to the original virus.
- $u_4(t)$: This control is used to reduce the possibility that after 28 days of vaccination, those who received the second dose do not develop immunity to the original virus.
- $u_5(t)$: The fifth control is used to reduce the possibility that after 28 days of vaccination, those who received the third dose do not develop immunity to the original virus.
- $u_6(t)$: The sixth control is used to enhance the second-time vaccination rate.
- $u_7(t)$: This control is used to enhance the third-time vaccination rate.

By employing the controls mentioned above, this section formulates an optimal control problem that elucidates how time-dependent control strategies contribute to disease eradication. The control model is established in (8). Based on the sensitivity index, the desired controls are selected. Hence, the resulting control model is structured as follows:

$$
\begin{aligned}
S'(t) &= \theta - \tfrac{\alpha I}{N} S(1 - u_1(t)) - (u_2(t) + \mu)S, \\
E'(t) &= \left(\tfrac{\alpha I S}{N}(1 - u_1(t)) + \tfrac{\alpha \delta_1 I V_1}{N}(1 - u_3(t)) + \tfrac{\alpha \delta_2 I V_2}{N}(1 - u_4(t)) + \tfrac{\alpha \delta_3 I V_3}{N}(1 - u_5(t)) \right) - (\sigma + \mu)E, \\
V_1'(t) &= u_2(t)S - \tfrac{\alpha \delta_1 I V_1}{N}(1 - u_3(t)) - (u_6(t) + \mu)V_1, \\
V_2'(t) &= u_6(t)V_1 - \tfrac{\alpha \delta_2 I V_2}{N}(1 - u_4(t)) - (u_7(t) + \mu)V_2, \\
V_3'(t) &= u_7(t)V_2 - \tfrac{\alpha \delta_3 I V_3}{N}(1 - u_5(t)) - \mu V_3, \\
I'(t) &= \sigma E - (\eta + \mu + \gamma)I, \\
H'(t) &= \eta(1 - b)I - (\gamma + \mu + d)I_H, \\
R'(t) &= \eta b I + d I_H - \mu R.
\end{aligned} \quad (8)
$$

with the same initial conditions given in (2). The respective objective functional is described as

$$
J(u_1, u_2, u_3, u_4, u_5, u_6, u_7) = \int_0^T \left(\begin{array}{c} A_1 E + A_2 V_1 + A_3 V_2 + A_4 V_3 + \tfrac{A_5 u_1^2}{2} + \tfrac{A_6 u_2^2}{2} + \tfrac{A_7 u_3^2}{2} \\ + \tfrac{A_8 u_4^2}{2} + \tfrac{A_9 u_5^2}{2} + \tfrac{A_{10} u_6^2}{2} + \tfrac{A_{11} u_7^2}{2} \end{array} \right) dt, \quad (9)
$$

where, A_1, A_2, A_3, A_4 are the balancing constants associated with the suggested variables of the objective function, A_5, A_6, A_7, A_8 are the the cost factors while T represents the final time. We used the quadratic objective functional due to the non-linearity of the intervention considered for the mitigation of the pandemic. For a more comprehensive understanding, please refer to the work and associated references [35–37]. Our primary goal is to identify the optimal controls

$$\bar{u}_i(t) \; for \; i = 1, 2, 3, 4, 5, 6, 7,$$

so that,

$$J(\bar{u}_1, \bar{u}_2, \bar{u}_3, \bar{u}_4, \bar{u}_5, \bar{u}_6, \bar{u}_7) = \min_{\Xi}\{J(u_1, u_2, u_3, u_4, u_5, u_6, u_7)\}.$$

The corresponding control set is given by

$$\Xi = \{(u_1, u_2, u_3, u_4, u_5, u_6, u_7) : [0, T] \to [0, 1] \ (u_1, u_2, u_3, u_4, u_5, u_6, u_7) \text{ is a Lebesgue measurable}\}. \qquad (10)$$

The Lagrangian and Hamiltonian for the provided control system (8) are shown as $\mathcal{L}$ and $\mathcal{H}$, respectively, and are given as follows:

$$\mathcal{L} = A_1 E + A_2 V_1 + A_3 V_2 + A_4 V_3 + \frac{1}{2}\left(A_5 u_1^2 + A_6 u_2^2 + A_7 u_3^2 + A_8 u_4^2 + A_9 u_5^2 + A_{10} u_6^2 + A_{11} u_7^2\right), \qquad (11)$$

and

$$\begin{aligned}
\mathcal{H} = & A_1 E + A_2 V_1 + A_3 V_2 + A_4 V_3 + \frac{1}{2}\left(A_5 u_1^2 + A_6 u_2^2 + A_7 u_3^2 + A_8 u_4^2 + A_9 u_5^2 + A_{10} u_6^2 + A_{11} u_7^2\right) + \\
& \lambda_1\left(\theta - \frac{\alpha IS}{N}(1-u_1) - (u_2+\mu)S\right) + \lambda_2\left(\begin{array}{l}\frac{\alpha I}{N}\left(S(1-u_1) + V_1\delta_1(1-u_3) + V_2\delta_2(1-u_4) + \right. \\ V_3\delta_3(1-u_5)\right) \\ -(\sigma+\mu)E\end{array}\right) \\
& +\lambda_3\left(u_2 S - \frac{\alpha I V_1 \delta_1}{N}(1-u_3) - (u_6+\mu)V_1\right) + \lambda_4\left(V_1 u_6 - \frac{IV_2\delta_2\alpha}{N}(1-u_4) - (u_7+\mu)V_2\right) + \\
& \lambda_5\left(u_7 V_3 - \frac{\alpha I V_3 \delta_3}{N}(1-u_5) - \mu V_3\right) + \lambda_6(\sigma E - (\mu+\eta+\gamma)I) + \lambda_7(\eta(1-b)I - (\mu+d+\gamma)I_H) \\
& +\lambda_8(\eta b I + d I_H - \mu R),
\end{aligned} \qquad (12)$$

where, λ_m for $m = 1, 2, 3, \ldots, 8$ represents the adjoint variables.

Solution of Optimal Control Problem

In this section, the solution to the optimal control COVID-19 as outlined in (8) is established. To achieve this, famous Pontryagin's principle [38,39] is employed. The desired optimal solution is denoted by $(\bar{u}_1, \bar{u}_2, \bar{u}_3, \bar{u}_4, \bar{u}_5, \bar{u}_6, \bar{u}_7)$. Furthermore, the following are the corresponding necessary optimality conditions used in the solution procedure stated as

$$\begin{cases} \frac{dz}{dt} = \frac{\partial}{\partial \lambda_m}\mathcal{H}(t, \bar{u}_i, \lambda_m), \\ \frac{\partial}{\partial u}\mathcal{H}(t, \bar{u}_i, \lambda_m) = 0, \\ \frac{d\lambda_m(t)}{dt} = -\frac{\partial}{\partial z}\lambda_m(t, \bar{u}_i, \lambda_m). \end{cases} \qquad (13)$$

The criteria's mentioned in (13) and the following theorem has been utilized to derive the solution of the optimal system.

Theorem 5. *The controls* $(\bar{u}_1, \bar{u}_2, \bar{u}_3, \bar{u}_4, \bar{u}_5, \bar{u}_6, \bar{u}_7)$ *and the solution* $(\bar{S}, \bar{E}, \bar{V}_1, \bar{V}_2, \bar{V}_3, \bar{I}, \bar{I}_H, \bar{R})$ *of the control system (8) minimizing the objective functional in the problem, then there exist adjoint variables (co-state variables)* $\lambda_m, m = 1, 2, \ldots, 8$. *Further, the transversality conditions* $\lambda_m(T) = 0, \ m = 1, 2, 3 \ldots 8,$ *such that*

$$\lambda_1'(t) = \mu\lambda_1 + u_2(\lambda_1 - \lambda_3) + (\lambda_1 - \lambda_2)\left(\frac{\alpha I(\tilde{N}-\tilde{S})}{\tilde{N}^2}\right)(1-u_1) + (\lambda_2 - \lambda_3)(1-u_3)$$

$$\left(\frac{\alpha \tilde{V}_1 \delta_1 I}{\tilde{N}^2}\right) + (\lambda_2 - \lambda_4)(1-u_4)\left(\frac{\alpha \tilde{V}_2 \delta_2 I}{\tilde{N}^2}\right) + (\lambda_2 - \lambda_5)(1-u_5)\left(\frac{\alpha \tilde{V}_3 \delta_3 I}{\tilde{N}^2}\right),$$

$$\lambda_2'(t) = -A_1 + \mu\lambda_2 + \sigma(\lambda_2 - \lambda_6) + (\lambda_2 - \lambda_1)(1-u_1)\frac{\alpha I \tilde{S}}{\tilde{N}^2} + (\lambda_2 - \lambda_3)(1-u_3)$$

$$\left(\frac{\alpha \tilde{V}_1 \delta_1 I}{\tilde{N}^2}\right) + (\lambda_2 - \lambda_4)(1-u_4)\left(\frac{\alpha \tilde{V}_2 \delta_2 I}{\tilde{N}^2}\right) + (\lambda_2 - \lambda_5)(1-u_5)\left(\frac{\alpha \tilde{V}_3 \delta_3 I}{\tilde{N}^2}\right),$$

$$\lambda_3'(t) = -A_2 + \mu\lambda_3 + (\lambda_3 - \lambda_4)u_6 + (\lambda_2 - \lambda_1)(1-u_1)\frac{\alpha I \tilde{S}}{\tilde{N}^2} + (\lambda_3 - \lambda_2)(1-u_3)$$

$$\left(\frac{\alpha \delta_1 I(\tilde{N}-\tilde{V}_1)}{\tilde{N}^2}\right) + (\lambda_2 - \lambda_4)(1-u_4)\left(\frac{\alpha \tilde{V}_2 \delta_2 I}{\tilde{N}^2}\right) + (\lambda_2 - \lambda_5)(1-u_5)\left(\frac{\alpha \tilde{V}_3 \delta_3 I}{\tilde{N}^2}\right),$$

$$\lambda_4'(t) = -A_3 + \mu\lambda_4 + u_7(\lambda_4 - \lambda_5) + (\lambda_2 - \lambda_1)(1-u_1)\frac{\alpha I \tilde{S}}{\tilde{N}^2} + (\lambda_2 - \lambda_3)(1-u_3)$$

$$\left(\frac{\alpha \tilde{V}_1 \delta_1 I}{\tilde{N}^2}\right) + (\lambda_4 - \lambda_2)(1-u_4)\left(\frac{\alpha \delta_2 I(\tilde{N}-\tilde{V}_2)}{\tilde{N}^2}\right) + (\lambda_2 - \lambda_5)(1-u_5)\left(\frac{\alpha \tilde{V}_3 \delta_3 I}{\tilde{N}^2}\right),$$

$$\lambda_5'(t) = -A_4 + \mu\lambda_5 + (\lambda_2 - \lambda_1)(1-u_1)\frac{\alpha I \tilde{S}}{\tilde{N}^2} + (\lambda_2 - \lambda_3)(1-u_3)\left(\frac{\alpha \tilde{V}_1 \delta_1 I}{\tilde{N}^2}\right) + (\lambda_2 - \lambda_4)(1-u_4) \quad (14)$$

$$\left(\frac{\alpha \tilde{V}_2 \delta_2 I}{\tilde{N}^2}\right) + (\lambda_5 - \lambda_2)(1-u_5)\left(\frac{\alpha \delta_3 I(\tilde{N}-\tilde{V}_3)}{\tilde{N}^2}\right),$$

$$\lambda_6'(t) = (\lambda_1 - \lambda_2)(1-u_1)\left(\frac{\alpha \tilde{S}(\tilde{N}-I)}{\tilde{N}^2}\right) + (\lambda_3 - \lambda_2)(1-u_3)\left(\frac{\alpha \delta_1 \tilde{V}_1(\tilde{N}-I)}{\tilde{N}^2}\right) + (\lambda_4 - \lambda_2)(1-u_4)$$

$$\left(\frac{\alpha \delta_2 \tilde{V}_2(\tilde{N}-I)}{\tilde{N}^2}\right) + (\lambda_5 - \lambda_2)(1-u_5)\left(\frac{\alpha \delta_3 \tilde{V}_3(\tilde{N}-I)}{\tilde{N}^2}\right) + (\mu+\gamma)\lambda_6 + \eta(\lambda_6 - \lambda_7) +$$

$$b\eta(\lambda_7 - \lambda_8),$$

$$\lambda_7'(t) = d(\lambda_7 - \lambda_8) + (\mu+\gamma)\lambda_7 + (\lambda_2 - \lambda_1)(1-u_1)\frac{\alpha I \tilde{S}}{\tilde{N}^2} + (\lambda_2 - \lambda_3)(1-u_3)\left(\frac{\alpha \tilde{V}_1 \delta_1 I}{\tilde{N}^2}\right)$$

$$+(\lambda_2 - \lambda_4)(1-u_4)\left(\frac{\alpha \tilde{V}_2 \delta_2 I}{\tilde{N}^2}\right) + (\lambda_2 - \lambda_5)(1-u_5)\left(\frac{\alpha \tilde{V}_3 \delta_3 I}{\tilde{N}^2}\right),$$

$$\lambda_8'(t) = \mu\lambda_8 + (\lambda_2 - \lambda_1)(1-u_1)\frac{\alpha I \tilde{S}}{\tilde{N}^2} + (\lambda_2 - \lambda_3)(1-u_3)\left(\frac{\alpha \tilde{V}_1 \delta_1 I}{\tilde{N}^2}\right)$$

$$+(\lambda_2 - \lambda_4)(1-u_4)\left(\frac{\alpha \tilde{V}_2 \delta_2 I}{\tilde{N}^2}\right) + (\lambda_2 - \lambda_5)(1-u_5)\left(\frac{\alpha \tilde{V}_3 \delta_3 I}{\tilde{N}^2}\right).$$

Furthermore, the associated optimal controls $\bar{u}_1, \bar{u}_2, \bar{u}_3, \bar{u}_4, \bar{u}_5, \bar{u}_6,$ and $\bar{u}_7$ are given by

$$\bar{u}_1 = \frac{I\tilde{S}\alpha(\lambda_2 - \lambda_1)}{A_5 N}, \quad \bar{u}_2 = \frac{\tilde{S}(\lambda_1 - \lambda_3)}{A_6}, \quad \bar{u}_3 = \frac{I\tilde{V}_1 \alpha \delta_1(\lambda_2 - \lambda_3)}{A_7 N}, \quad \bar{u}_4 = \frac{I\tilde{V}_2 \alpha \delta_2(\lambda_2 - \lambda_4)}{A_8 N},$$

$$\bar{u}_5 = \frac{I\tilde{V}_3 \alpha \delta_3(\lambda_2 - \lambda_5)}{A_9 N}, \quad \bar{u}_6 = \frac{\tilde{V}_1(\lambda_3 - \lambda_4)}{A_{10}}, \quad \bar{u}_7 = \frac{\tilde{V}_2(\lambda_4 - \lambda_5)}{A_{11}}.$$

(15)

Proof. By using the condition stated in (13), the transversality conditions and results given in (14) are obtained for the Hamiltonian function given in (12) using $S = \tilde{S}, E = \tilde{E}, V_1 = \tilde{V}_1, V_2 = \tilde{V}_2, V_3 = \tilde{V}_3, I = \tilde{I}, I_H = \tilde{I}_H$, and $R = \tilde{R}$. Moreover, using the condition $\frac{\partial \mathcal{H}(t, \tilde{u}_j, \lambda_m)}{\partial u_j} = 0$ given in (13), the optimal controls $\tilde{u}_1, \tilde{u}_2, \tilde{u}_3, \tilde{u}_4, \tilde{u}_5, \tilde{u}_6$, and $\tilde{u}_7$ shown in (15) are derived. □

8. Estimating the Optimal Solution

This section aims to analyze the significant impact of the proposed time-varying controls on disease incidence and potential mitigation. For this purpose, the COVID-19 control model is simulated with the aforementioned time-varying control variables to display their impact on the disease dynamics. An effective iterative approach known as RK4 is used to carry out the simulation process. The weights and balancing constants are chosen as $A_1 = 10$, $A_2 = 0.1$, $A_3 = 0.1$, $A_4 = 0.01$, $A_5 = 50$, $A_6 = 10$, $A_7 = 30$, $A_8 = 100$, $A_9 = 50$, $A_{10} = 20$, and $A_{11} = 100$, while the simulation's parameters are taken from Table 1. It is important to note that the numerical values of weighed and balanced constants are taken for the sake of simulation. The red dashed curves illustrate the dynamical behavior of various populations under varied control measures (implementing all the control measures at the time), while the black solid curves exhibit the changing behavior with constant controls. The dynamics of susceptible, primary vaccinated, exposed, and secondary vaccinated individuals with and without time-varying controls are analyzed in Figure 9, while a similar analysis of individuals with a booster shots, hospitalized, infected, and recovered individuals are presented in Figure 10. Figures 11 and 12 demonstrate the corresponding control profiles. When the suggested optimal controls are actively utilized, the population in the susceptible class reduces while the population in all vaccinated classes increases significantly and reaches its maximum level with time. However, the population in the exposed and infected classes dramatically decreased and vanished after 60 and 90 days, respectively, with the implementation of time-varying controls. Because of averting infection, the populations in the hospitalized and recovered classes also reduced significantly, as shown in Figure 10c,d. The time-varying personal protection control u_1 is utilized at the maximum level from the initial time to approximately 220 days and decreases gradually until the end of the considered time level, as shown in Figure 11a. The time-varying control for first-time vaccine efficacy enhancement u_2 is initially at the maximum level until the first 100 days and then immediately reduces until the end, as can be seen in Figure 11b. Figures 11c,d and 12a illustrate the intensity of controls u_3, u_4, and u_5 evaluating the reduction in the possibility that first, second, and third vaccine doses do not develop immunity to the original viruses, respectively. It can be observed that all of these controls are initially maintained at the maximum level and then gradually reduced until the end of the considered time interval. The implementation level of controls used for the enhancement of second- and third-time vaccinations u_6 and u_7 are shown in Figure 12b,c, respectively. These controls are implemented at the maximum level from the start to the end of the time level under consideration.

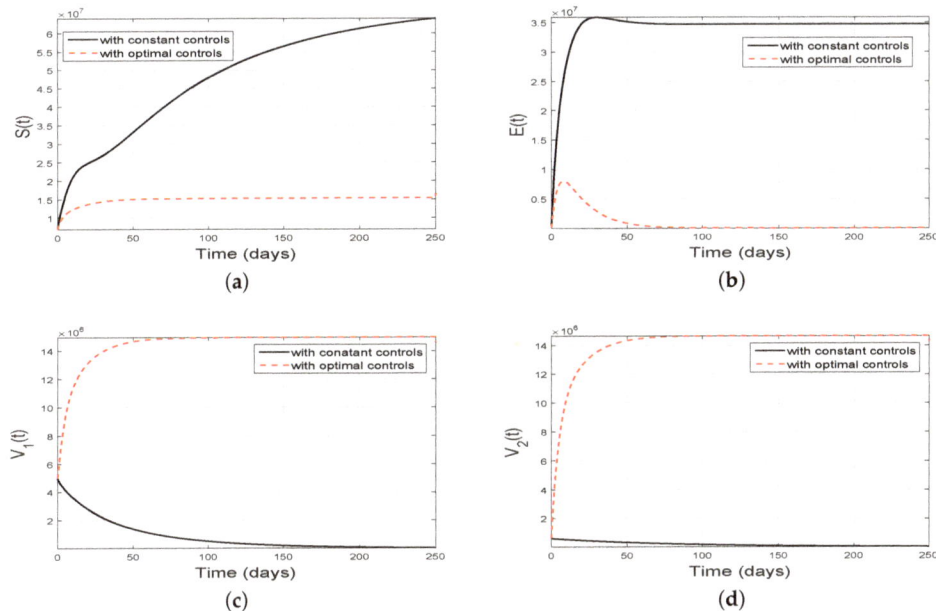

Figure 9. Simulation of (**a**) susceptible, (**b**) exposed, (**c**) primary and (**d**) secondary vaccinated individuals in the model (8) with optimal and with constant controls. The constant values of first, second and booster COVID-19 vaccine controls are considered as 0.7, 0.6, and 0.29, respectively.

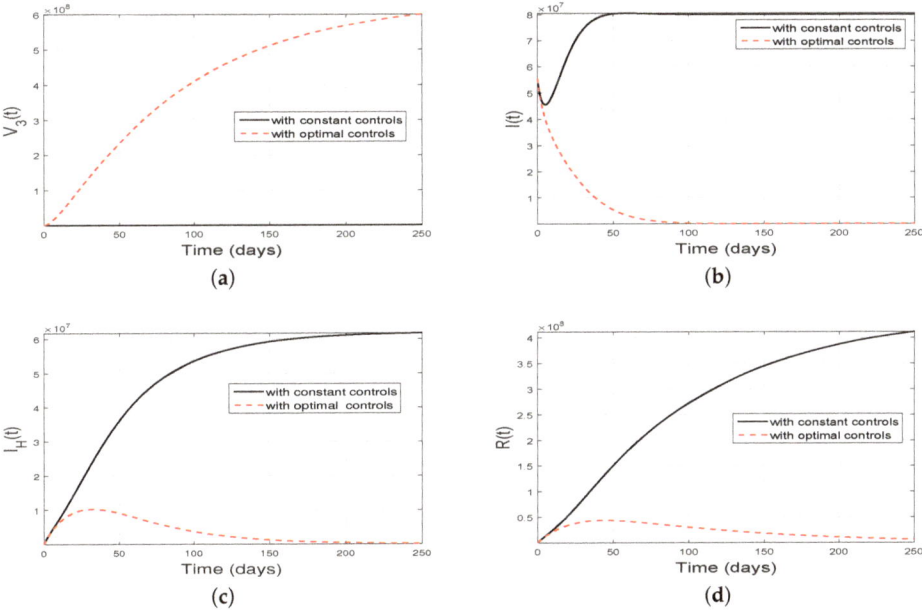

Figure 10. Simulation of people with (**a**) booster short, (**b**) infected, (**c**) hospitalized and (**d**) recovered compartment in the model (8) with optimal and with constant controls. The constant values of first, second and booster COVID-19 vaccine controls are considered as 0.7, 0.6, and 0.29, respectively.

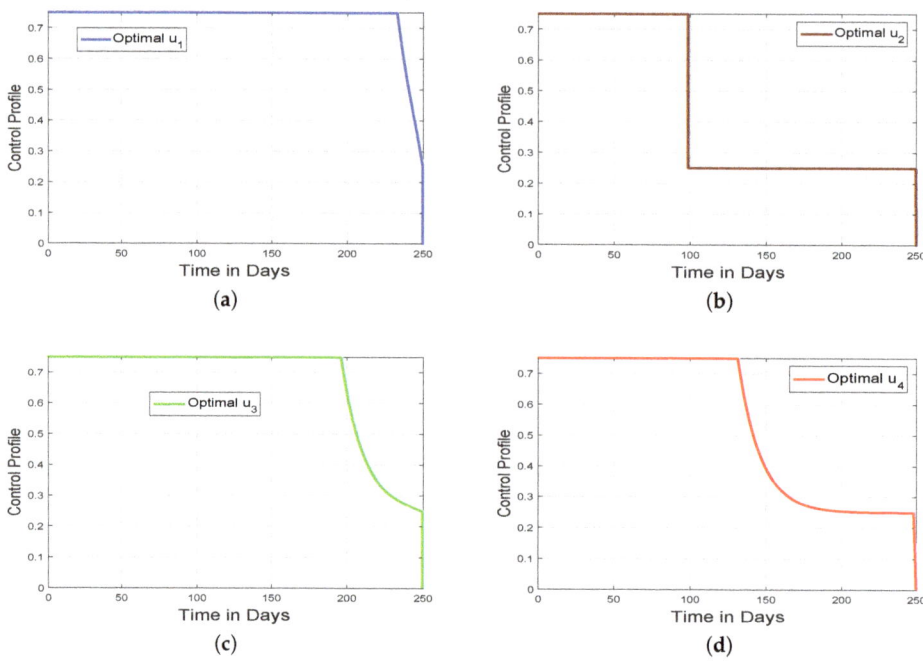

Figure 11. The corresponding optimal control profiles with controls (**a**) $u_1(t)$, (**b**) $u_2(t)$, (**c**) $u_3(t)$ and (**d**) $u_4(t)$.

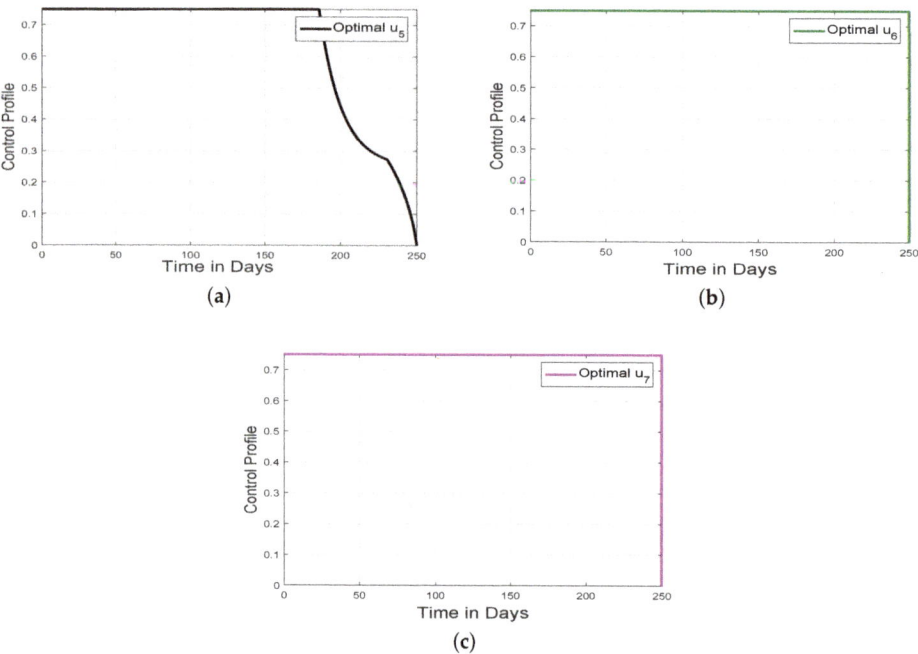

Figure 12. The corresponding optimal control profiles with controls (**a**) $u_5(t)$, (**b**) $u_6(t)$ and (**c**) $u_7(t)$

9. Conclusions

Vaccination remains one of the most effective preventive interventions and is a critical component in mitigating disease outbreaks. In this study, we developed a novel mathematical model to assess the impact of administering multiple constant and time-varying vaccines, including the first and second doses and booster shots, on infection incidence and persistence. Initially, the necessary mathematical assessment of the model is presented and basic reproduction is calculated. To curb infection, we determined the critical vaccination convergence rate, which is dependent on the reproduction number. Moreover, using backward bifurcation analysis, we concluded that bifurcation depends on the sign of $\bar{a}$. To measure each parameter's relative influence on disease spread, the sensitivity indices of the proposed model parameters for the basic reproduction number are tabulated using a normalized approach. Furthermore, we reconstructed the model using optimal control theory to identify the best control strategy for minimizing infection. We introduced seven time-varying controls, where $u_1(t)$ is the personal protection control used for the reduction in effective contacts of infected individuals with susceptible, $u_2(t)$, $u_6(t)$, $u_7(t)$ controls are used for the 1st, 2nd, and 3rd vaccine dose efficacy enhancement. The controls $u_3(t)$, $u_4(t)$, $u_5(t)$ are used to reduce the possibility that after 28 days of vaccination, those who received the first, second, and third doses do not develop immunity to the original virus. Using the well-known Pontryagin's maximum principle, the necessary optimal conditions are determined. The implementation of the suggested time-varying optimal controls has been found to play a crucial role in effectively minimizing the risk of disease transmission. Moreover, by effectively adjusting vaccination strategies, such as the timing and frequency of doses, the model shows that the spread of the disease can be significantly controlled. These time-varying controls allow for a more adaptive and responsive approach, enabling healthcare authorities to tailor vaccination campaigns on the basis of evolving epidemiological conditions, the emergence of new variants, and the overall vaccination coverage in the population. This flexibility empowers the public health system to optimize vaccination efforts and better control the spread of the infection, ultimately leading to a reduced incidence of the disease and improved public health outcomes.

Author Contributions: Y.L.: Conceptualization, formal analysis, writing—review and editing, S. and L.Z.: Software, methodology, visualization, validation, writing—review & editing, E.A.A.I. and F.A.A.: Conceptualization, supervision, project administration, formal analysis, writing—review and editing, funding acquisition, A.M.H.: Supervision, investigation, funding acquisition, resources. All authors have read and agreed to the published version of the manuscript.

Funding: This project is funded by King Saud University, Riyadh, Saudi Arabia.

Data Availability Statement: Not applicable.

Acknowledgments: Researchers Supporting Project number (RSPD2023R1060), King Saud University, Riyadh, Saudi Arabia.

Conflicts of Interest: The authors declare no conflict of interest.

References

Andersen, K.G.; Rambaut, A.; Lipkin, W.I.; Holmes, E.C.; Garry, R.F. The proximal origin of SARS-CoV-2. *Nat. Med.* **2020**, *26*, 450–452. [CrossRef] [PubMed]

World Health Organization. *Novel Coronavirus (2019-nCoV): Situation Report*; World Health Organization: Geneva, Switzerland, 2020; p. 11.

Pfefferbaum, B.; North, C.S. Mental health and the COVID-19 pandemic. *N. Engl. J. Med.* **2020**, *383*, 510–512. [CrossRef] [PubMed]

Tu, Y.F.; Chien, C.S.; Yarmishyn, A.A.; Lin, Y.Y.; Luo, Y.H.; Lin, Y.T.; Lai, W.Y.; Yang, D.M.; Chou, S.J.; Yang, Y.P.; et al. A review of SARS-CoV-2 and the ongoing clinical trials. *Int. J. Mol. Sci.* **2020**, *21*, 2657. [CrossRef]

Guan, W.J.; Ni, Z.Y.; Hu, Y.; Liang, W.H.; Ou, C.Q.; He, J.X.; Liu, L.; Shan, H.; Lei, C.L.; Hui, D.S.; et al. Clinical characteristics of coronavirus disease 2019 in China. *N. Engl. J. Med.* **2020**, *382*, 1708–1720. [CrossRef]

Lauer, S.A.; Grantz, K.H.; Bi, Q.; Jones, F.K.; Zheng, Q.; Meredith, H.R.; Azman, A.S.; Reich, N.G.; Lessler, J. The incubation period of coronavirus disease 2019 (COVID-19) from publicly reported confirmed cases: Estimation and application. *Ann. Intern. Med.* **2020**, *172*, 577–582. [CrossRef]

7. World Health Organization. *Non-Pharmaceutical Public Health Measures for Mitigating the Risk and Impact of Epidemic and Pandemic Influenza: Annex: Report of Systematic Literature Reviews (No. WHO/WHE/IHM/GIP/2019.1)*; World Health Organization: Geneva, Switzerland, 2019.
8. Polack, F.P. Safety and Efficacy of the BNT162b2 mRNA COVID-19 Vaccine. *N. Engl. J. Med.* **2020**, *383*, 2603–2615. [CrossRef] [PubMed]
9. Al-arydah, M. Mathematical modeling and optimal control for COVID-19 with population behavior. *Math. Meth. Appl. Sci.* **2023**, 1–15. [CrossRef]
10. Al-arydah, M.; Berhe, H.; Dib, K.; Madhu, K. Mathematical modeling of the spread of the coronavirus under strict social restrictions. *Math. Meth. Appl. Sci.* **2021**, 1–11. [CrossRef]
11. Aatif, A.; Ullah, S.; Khan, M.A. The impact of vaccination on the modeling of COVID-19 dynamics: A fractional order model. *Nonlinear Dyn.* **2022**, *110*, 3921–3940.
12. Zarin, R. Numerical study of a nonlinear COVID-19 pandemic model by finite difference and meshless methods. *Partial. Differ. Equations Appl. Math.* **2022**, *6*, 100460. [CrossRef]
13. Alshehri, A.; Ullah, S. A numerical study of COVID-19 epidemic model with vaccination and diffusion. *Math. Biosci. Eng.* **2023**, *20*, 4643–4672. [CrossRef]
14. Ali, I.; Khan, S.U. Dynamics and simulations of stochastic COVID-19 epidemic model using Legendre spectral collocation method. *AIMS Math.* **2023**, *8*, 4220–4236. [CrossRef]
15. Din, A.; Amine, S.; Allali, A. A stochastically perturbed co-infection epidemic model for COVID-19 and hepatitis B virus. *Nonlinear Dyn.* **2023**, *111*, 1921–1945. [CrossRef]
16. Rahat, Z.; Khan, A.; Yusuf, A.; Sayed, A.-K.; Inc, M. Analysis of fractional COVID-19 epidemic model under Caputo operator. *Math. Methods Appl. Sci.* **2023**, *46*, 7944–7964.
17. Ravichandran, C.; Logeswari, K.; Khan, A.; Abdeljawad, T.; Goamez-Aguilar, J.F. An epidemiological model for computer virus with Atangana-Baleanu fractional derivative. *Results Phys.* **2023**, *51*, 106601. [CrossRef]
18. Lou, J.; Zheng, H.; Zhao, S.; Cao, L.; Wong, E.L.; Chen, Z.; Chan, R.W.; Chong, M.K.; Zee, B.C.; Chan, P.K.; et al. Quantifying the effect of government interventions and virus mutations on transmission advantage during COVID-19 pandemic. *J. Infect. Public Health* **2022**, *15*, 338–342. [CrossRef]
19. Wang, Y.; Wang, P.; Zhang, S.; Pan, H. Uncertainty modeling of a modified SEIR epidemic model for COVID-19. *Biology* **2022**, *11*, 1157. [CrossRef]
20. Liu, P.; Huang, X.; Zarin, R.; Cui, T.; Din, A. Modeling and numerical analysis of a fractional order model for dual variants of SARS-CoV-2. *Alex. Eng. J.* **2023**, *65*, 427–442. [CrossRef]
21. Eikenberry, S.E.; Mancuso, M.; Iboi, E.; Phan, T.; Eikenberry, K.; Kuang, Y.; Kostelich, E.; Gumel, A.B. To mask or not to mask: Modeling the potential for face mask use by the general public to curtail the COVID-19 pandemic. *Infect. Dis. Model.* **2020**, *5*, 293–308. [CrossRef] [PubMed]
22. Agusto, F.B.; Erovenko, I.V.; Fulk, A.; Abu-Saymeh, Q.; Romero-Alvarez, D.; Ponce, J.; Sindi, S.; Ortega, O.; Saint Onge, J.M.; Peterson, A.T. To isolate or not to isolate: The impact of changing behavior on COVID-19 transmission. *BMC Public Health* **2022**, *22*, 138.
23. Watson, O.J.; Barnsley, G.; Toor, J.; Hogan, A.B.; Winskill, P.; Ghani, A.C. Global impact of the first year of COVID-19 vaccination: A mathematical modelling study. *Lancet Infect. Dis.* **2022**, *22*, 1293–1302. [CrossRef] [PubMed]
24. Eyre, D.W.; Taylor, D.; Purver, M.; Chapman, D.; Fowler, T.; Pouwels, K.B.; Walker, A.S.; Peto, T.E. Effect of COVID-19 vaccination on transmission of alpha and delta variants. *N. Engl. J. Med.* **2022**, *386*, 744–756. [CrossRef] [PubMed]
25. Ngonghala, C.N.; Taboe, H.B.; Safdar, S.; Gumel, A.B. Unraveling the dynamics of the Omicron and Delta variants of the 2019 coronavirus in the presence of vaccination, mask usage, and antiviral treatment. *Appl. Math. Model.* **2023**, *114*, 447–465. [CrossRef] [PubMed]
26. Peter, O.J.; Panigoro, H.S.; Abidemi, A.; Ojo, M.M.; Oguntolu, F.A. Mathematical model of COVID-19 pandemic with double dose vaccination. *Acta Biotheor.* **2023**, *71*, 9. [CrossRef] [PubMed]
27. Wang, Y.; Ullah, S.S.; Khan, I.U.; AlQahtani, S.A.; Hassan, A.M. Numerical assessment of multiple vaccinations to mitigate the transmission of COVID-19 via a new epidemiological modeling approach. *Results Phys.* **2023**, *52*, 106889. [CrossRef]
28. Lakshmikantham, V.; Leela, S.; Martynyuk, A.A. *Stability Analysis of Nonlinear Systems*; M. Dekker: New York, NY, USA, 1989; pp. 249–275.
29. LaSalle, J.P.; Lefschetz, S. *The Stability of Dynamical Systems (SIAM, Philadelphia, 1976)*; Zhonghuai Wu Yueyang Vocational Technical College Yueyang: Hunan, China, 1976.
30. Van den Driessche, P.; Watmough, J. Reproduction numbers and sub-threshold endemic equilibria for compartmental models of disease transmission. *Math. Biosci.* **2002**, *180*, 29–48. [CrossRef] [PubMed]
31. Castillo-Chavez, C.; Song, B. Dynamical models of tuberculosis and their applications. *Math. Biosci. Eng.* **2004**, *1*, 361–404. [CrossRef] [PubMed]
32. Akinwande, N.I.; Ashezua, T.T.; Gweryina, R.I.; Somma, S.A.; Oguntolu, F.A.; Usman, A.; Abdurrahman, O.N.; Kaduna, F.S.; Adajime, T.P.; Kuta, F.A.; et al. Mathematical model of COVID-19 transmission dynamics incorporating booster vaccine program and environmental contamination. *Heliyon* **2022**, *8*, e11513. [CrossRef]

33. Kim, Y.R.; Choi, Y.J.; Min, Y. A model of COVID-19 pandemic with vaccines and mutant viruses. *PLoS ONE* **2022**, *17*, e0275851. [CrossRef]
34. Chitnis, N.; Hyman, J.M.; Cushing, J.M. Determining important parameters in the spread of malaria through the sensitivity analysis of a mathematical model. *Bull. Math. Biol.* **2008**, *70*, 1272–1296. [CrossRef]
35. Saif, U.; Khan, M.A. Modeling the impact of non-pharmaceutical interventions on the dynamics of novel coronavirus with optimal control analysis with a case study. *Chaos Solitons Fractals* **2020**, *139*, 110075.
36. Agusto, F.B.; Khan, M.A. Optimal control strategies for dengue transmission in Pakistan. *Math. Biosci.* **2018**, *305*, 102–121. [CrossRef]
37. Saif, U.; Khan, M.A.; Gmez-Aguilar, J.F. Mathematical formulation of hepatitis B virus with optimal control analysis. *Optim. Control. Appl. Methods* **2019**, *40*, 529–544.
38. Pontryagin, L.S. *Mathematical Theory of Optimal Processes*; CRC Press: Boca Raton, FL, USA, 1987.
39. Fleming, W.H.; Rishel, R.W. *Deterministic and Stochastic Optimal Control*; Springer Science and Business Media: Berlin/Heidelberg, Germany, 2012; Volume 1.

Disclaimer/Publisher's Note: The statements, opinions and data contained in all publications are solely those of the individual author(s) and contributor(s) and not of MDPI and/or the editor(s). MDPI and/or the editor(s) disclaim responsibility for any injury to people or property resulting from any ideas, methods, instructions or products referred to in the content.

Article

Two-Age-Structured COVID-19 Epidemic Model: Estimation of Virulence Parameters through New Data Incorporation

Cristiano Maria Verrelli [1,*] and Fabio Della Rossa [2]

1 Electronic Engineering Department, University of Rome Tor Vergata, Via del Politecnico 1, 00133 Rome, Italy
2 Department of Electronic, Information and Biomedical Engineering, Politecnico di Milano, 20133 Milan, Italy; fabio.dellarossa@polimi.it
* Correspondence: verrelli@ing.uniroma2.it; Tel.: +39-(0)6-72597410

Abstract: The COVID-19 epidemic has required countries to implement different containment strategies to limit its spread, like strict or weakened national lockdown rules and the application of age-stratified vaccine prioritization strategies. These interventions have in turn modified the age-dependent patterns of social contacts. In our recent paper, starting from the available age-structured real data at the national level, we identified, for the Italian case, specific virulence parameters for a two-age-structured COVID-19 epidemic compartmental model (under 60, and 60 years and over) in six different diseases transmission scenarios under concurrently adopted feedback interventions. An interpretation of how each external scenario modifies the age-dependent patterns of social contacts and the spread of COVID-19 disease has been accordingly provided. In this paper, which can be viewed as a sequel to the previous one, we mainly apply the same general methodology therein (involving the same dynamic model) to new data covering the three subsequent additional scenarios: (i) a mitigated coordinated intermittent regional action in conjunction with the II vaccination phase; (ii) a super-attenuated coordinated intermittent regional action in conjunction with the II vaccination phase; and (iii) a last step towards normality in conjunction with the start of the III vaccination phase. As a new contribution, we show how meaningful updated information can be drawn out, once the identification of virulence parameters, characterizing the two age groups within the latest three different phases, is successfully carried out. Nevertheless, differently from our previous paper, the global optimization procedure is carried out here with the number of susceptible individuals in each scenario being left free to change, to account for reinfection and immunity due to vaccination. Not only do the slightly different estimates we obtain for the previous scenarios not impact any of the previous considerations (and thus illustrate the robustness of the procedure), but also, and mainly, the new results provide a meaningful picture of the evolution of social behaviors, along with the goodness of strategic interventions.

Keywords: COVID-19 epidemic; model identification; parameter estimation; compartmental model; vaccine effects; global optimization

MSC: 37M10; 34H05; 37M05; 62P25; 93B30; 91C05; 00A06

Citation: Verrelli, C.M.; Della Rossa, F. Two-Age-Structured COVID-19 Epidemic Model: Estimation of Virulence Parameters through New Data Incorporation. *Mathematics* **2024**, *12*, 825. https://doi.org/10.3390/math12060825

Academic Editor: Hongyu LIU

Received: 26 January 2024
Revised: 1 March 2024
Accepted: 7 March 2024
Published: 12 March 2024

Copyright: © 2024 by the authors. Licensee MDPI, Basel, Switzerland. This article is an open access article distributed under the terms and conditions of the Creative Commons Attribution (CC BY) license (https://creativecommons.org/licenses/by/4.0/).

1. Introduction

The worldwide reaction to the unprecedented challenges posed by the COVID-19 pandemic has been marked by a diverse array of strategic initiatives and interventions implemented with the collective goal of not only containing the transmission of the virus but also mitigating the far-reaching impacts it has had on public health, socio-economic structures, and the overall fabric of global societies. Scientific research, particularly that grounded in mathematical models, has played a pivotal role in steering these interventions, providing valuable insights and innovative ideas that have informed decision-making processes and enhanced the effectiveness of public health strategies on a global scale. These proposed strategies have encompassed a spectrum of measures, including widespread lockdown and social distancing protocols [1,2], the promotion and enforcement of mask-wearing, extensive testing, and contact tracing efforts [3–5], the rapid development and deployment of vaccines [6,7], and public awareness campaigns [8–10].

Moreover, scientific research has been instrumental in comprehending the multifaceted impact of each proposed measure, offering a nuanced understanding from both health and socio-economic perspectives [11–13]. This dual assessment has been crucial in shaping a holistic approach, ensuring that interventions not only address the immediate health crisis but also consider the broader implications on societies and economies [14–16]. By integrating scientific findings into policy discussions, nations have been better equipped to navigate the intricate balance between safeguarding public health and minimizing socio-economic disruptions, exemplifying a commitment to evidence-based decision-making in the face of this complex global health crisis.

The aspect of utmost significance that has garnered scientific and societal attention for COVID-19 is its mortality rate. Although all age groups have been susceptible to COVID-19 infection, older age groups have faced an elevated risk of severe symptoms and mortality [17–19]: the case fatality rate among adults over 60 years has been estimated to be four times higher than that of young adults [20]. This stark reality has prompted a series of studies aimed at understanding how the virus spreads within a population, considering the variations in age distribution. For instance, ref. [21] introduced a compartmental model to predict the number of infected, hospitalized, and deceased individuals in a population divided into 17 age classes, parameterizing it using COVID-19 infection data from Switzerland. Taking a step further in analyzing the virus spread across different age groups, ref. [22] proposed a partial differential equation model, forecasting the disease progression in three countries spanning different continents: the United States of America, the United Arab Emirates, and Algeria. These studies also provided recommendations for interventions based on age including prioritizing vaccination for older individuals and implementing stricter age-dependent social distancing measures.

Italy, being at the forefront of the pandemic's impact, underwent a series of rigorous measures and interventions to combat the unprecedented challenges posed by the novel coronavirus [2]. In our study [23], we ventured into uncharted territory by pioneering the development of a two-age-structured COVID-19 epidemic model. This model utilizes real data at the national level to discern how the various phases of the pandemic in Italy has changed our typical social behaviors between the elder and the young population. Leveraging available age-structured real data, we uncovered valuable insights into the nuanced relationships between age demographics, social contacts, and the transmission of COVID-19. The model not only offered a comprehensive understanding of the disease dynamics but also facilitated an interpretation of how external scenarios, such as public health interventions and societal behaviors, modified age-dependent patterns of social contacts.

As a logical progression of our earlier work [23], this paper serves as a consequential sequel, in which we not only incorporate new and updated data but also perform the global optimization procedure by leaving the number of susceptible individuals in each scenario free to change, to account for reinfection and vaccination-owing immunity. The present work does not thus present a contribution to the theory of epidemic models, namely, no new model is proposed. Instead, it presents how to use the model of [23] to further interpret reality wisely. Changing the parameter identification procedure generates slightly different estimates for the previous scenarios, which, notably, do not impact any of the previously drawn considerations, thus illustrating the robustness of the procedure and the goodness of our previous results. Through the lens of our two-age-structured model, we aim to provide a nuanced analysis of the dynamic interplay between strategic actions, societal behaviors, and the evolving patterns of COVID-19 transmission. In essence, this paper serves as a bridge between the theoretical constructs of our initial model and the evolving realities of the pandemic. By applying our established framework to fresh sets of data, we seek to unravel the intricate ways in which the implemented interventions have shaped the trajectory of the virus. This effort is not merely an academic exercise; it is a timely exploration of the real-world implications of strategic actions and behavioral adaptations in the ongoing battle against COVID-19.

2. Incorporation of New Data into the Methodology of [23]

2.1. Retrospect

The real data that were analyzed in [23] had been taken from the Italian context, where the following subsequent-in-time different strategies were implemented:

(a) A strict national lockdown rule (scenario a, from $t_0^a = 9$ March 2020 to $t_e^a = 28$ April 2020) that removes social contacts in workplaces, schools, markets, and other public areas;

(b–d) A weakened feedback social distancing and contact reduction intervention, which is composed of a weakened lockdown phase (scenario b, from $t_0^b = 7$ May 2020 to $t_e^b = 3$ June 2020), a low-distancing phase (scenario c, from $t_0^c = 9$ June 2020 to $t_e^c = 8$ September 2020), and a low-distancing + workplace/school-contacts re-activation phase (scenario d, from $t_0^d = 15$ September 2020 to $t_e^d = 27$ October 2020), with a progressive release of the population back to their daily routine;

(e) A coordinated intermittent regional action (scenario e, from $t_0^e = 7$ November 2020 to $t_e^e = 29$ December 2020), where social distancing measures are put in place or relaxed independently by each region based on the ratio between hospitalized individuals and the specific regional health system capacity;

(f) Direct mRNA-vaccination of subjects—especially the elderly—(scenario f, from $t_0^f = 5$ January 2021 to $t_e^f = 12$ May 2021) at highest risk for severe outcomes, along with Vaxzevria vaccination of young subjects within specific occupational categories (to protect subjects at highest risk for severe outcomes indirectly).

2.2. The New Three Subsequent Scenarios

We now identify another three scenarios, characterized by a successive weakening of the rules adopted during lockdown (until the abolition of the mandatory use of masks in public places and hospitals) during the vaccination campaign. The three such scenarios are characterized by:

(G) A mitigated coordinated intermittent regional action in conjunction with the II vaccination phase (scenario G, from $t_0^G = 19$ May 2021 to $t_e^G = 14$ July 2021): during this phase, only people with a COVID-19 vaccination certificate (*Green Pass*) can leave their city of residence, and schools that were teaching online until scenario (f) resume normal operation (in presence);

(H) A super-attenuated coordinated intermittent regional action, without mobility restrictions, where even large events (such as sports competitions, congresses, fairs, private parties) can be held in the regions with a sufficiently low number of cases in conjunction with the II vaccination phase (scenario H, from $t_0^H = 21$ July 2021 to $t_e^H = 22$ September 2021);

(I) The last step towards normality (normal reopening of schools, with the vaccination certificate only mandatory for teachers, complete lifting of the obligation to use face masks, reopening of entertainment activities such as discos and ballrooms) and start of the III vaccination phase (scenario I, from $t_0^I = 29$ September 2021 to $t_e^I = 10$ November 2021).

All the data are taken from the official Ministerial website https://www.epicentro iss.it/coronavirus/aggiornamenti (accessed on 26 September 2023), which report: (i) the cumulative detected cases on a weekly scale, $C(t)$, divided per age (to compute $C_y(t)$ and $C_o(t)$); and (ii) the number of recovered people (not divided per age), $R(t)$.

3. Model and Simplifying Assumptions

In this section, for the sake of exhaustiveness, we recall the deterministic compartmental model proposed in [23]. The model was proposed to investigate how different epidemic phases, characterized by different political strategies used to contain the epidemics, affected the age-dependent patterns of social contacts and the spread of COVID-19. It is a natural extension of the classical SIR model in which the fluxes between the susceptible and the infected compartments are assumed to be proportional to the encounter rate. Each compartment is subdivided into different age groups. In order to avoid issues related to the lack of parameter identifiability, modeling and analysis were limited to two age groups. The two age groups correspond to individuals below the age of 60 and those aged 60 and older. As already highlighted in [23], at least four reasons guide the choice of such two groups. 1. Such a division highlights a division of active and retired populations, with different patterns of social interactions leading to different transmission dynamics. 2. Empirical estimates based on population-level data recognize a sharp difference in fatality rates between young and old people. 3. Priority for vaccination in Italy has been given to people older than 60 years. 4. A closer look at the Italian data reveals that this choice leads to a uniform division of the number of COVID-19 cases most uniformly.

We recall that the model aims to identify the parameters in the aforementioned time windows. More specifically, the length of the periods for our model equates, on average, to a couple of months, and always less than 4 months. At the beginning of each time window, moreover, the initial conditions are estimated from the data. In light of this, the following simplifying assumptions are here further specified:

- Aging, reproduction, and natural death have negligible effects so that the variation in the number of susceptible in the time window only depends on infection.
- Infected people cannot be infected another time in the time window.
- Effect of vaccines against infection is negligible in the time window.

The above assumptions limit the maximum length of the time windows; in other words, the model cannot be used to make any forecast on long time windows, since it neglects fundamental characteristics such as aging, reinfection, and vaccination. These assumptions, however, hold in the short time, and aging, becoming susceptible after an infection, or gaining immunity with vaccination are captured by the model at the beginning of the time window at which the estimation of the current susceptible subjects is carried out. The resulting model is accordingly given by:

$$
\begin{aligned}
S_y(t+1) &= S_y(t) - S_y(t)(v_{11}I_y(t) + v_{12}I_o(t))/N(t) \\
S_o(t+1) &= S_o(t) - S_o(t)(v_{21}I_y(t) + v_{22}I_o(t))/N(t) \\
I_y(t+1) &= (1 - \tau_1 - \gamma)I_y(t) + S_y(t)(v_{11}I_y(t) + v_{12}I_o(t))/N(t) \\
I_o(t+1) &= (1 - \tau_2 - \gamma)I_o(t) + S_o(t)(v_{21}I_y(t) + v_{22}I_o(t))/N(t) \\
C_y(t+1) &= C_y(t) + \tau_1 I_y(t) \\
C_o(t+1) &= C_o(t) + \tau_2 I_o(t)
\end{aligned}
\quad (1)
$$

in which:

- t is the time, measured in days;
- S_i, I_i, $i = y, o$ are the numbers of susceptible and infected for the two age classes, respectively;
- C_y and C_o are the numbers of reported cases for the two age classes;
- $N(t)$ is the number of persons who are not quarantined, hospitalized, or dead at time t.

A schematic of the model is proposed in Figure 1, in which each state variable is represented by a node, and the arrows represent the fluxes between the state variables. The parameters v_{ij}, $i, j = 1, 2$ represent the virulence of the virus among the different age classes, while $1/\tau_i$, $i = 1, 2$ is the average time for disease identification, and $\gamma = 0.07$ is the rate of asymptomatic infected who recover without being reported. As for the SIR model, an explicit solution to Equation (1) is not available. On the other hand, being a dynamical model, it can provide us with the forward evolution of the time series of the state variables, once an initial condition is given. However, note that the presented model needs the time series of active people, $N(t)$, to be simulated. This series cannot be reconstructed from the state variables, since the number of quarantined, hospitalized, recovered, or dead are not taken into account. As in [23], the scope of this model is not to make a prediction, but just to estimate its parameters to overview people's reactions, subject to the new three different scenarios.

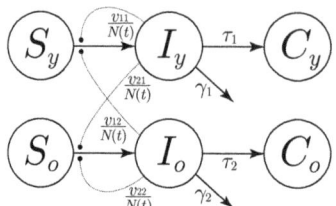

Figure 1. Schematic of model (1). Each node represents a state variable of the model ($S_i, I_i, C_i, i = y, o$ are the numbers of susceptible, infected, and reported cases for the two age classes, respectively), with each arrow representing a flux—proportional to the reported parameter and the departing node value—toward the arriving node. The dotted lines represent nonlinear proportional interactions that also modulate the flux.

4. Estimation of Model Parameters

The methodology proposed in [23] is aimed at identifying the changes in behavior and social interactions between older and young people based on the number of COVID-19 reported cases. The key parameters used to achieve this goal are:

- v_{11}^i, characterizing the intra-juvenile virulence;
- v_{12}^i, characterizing the juvenile–elder virulence;
- v_{21}^i, characterizing the elder–juvenile virulence;
- v_{22}^i, characterizing the intra-elder virulence;
- $1/\tau_1^i$, denoting the average time for disease identification in young subjects;
- $1/\tau_2^i$, denoting the average time for disease identification in old subjects;
- I_{t0y}^i, representing the young subjects infected at the beginning of the scenario time window;
- I_{t0o}^i, representing the old subjects infected at the beginning of the scenario time window.

The parameters are identified in the new aforementioned scenarios by adapting the procedure proposed in [23], which uses the relationship

$$N(t) = N(0) - (C_y(t) + C_o(t)) + R(t)$$

representing the number of active people each day, while fitting the model (namely, parameters and initial conditions) to the real data by minimizing a cost function. More specifically, starting from the time window that identifies scenario a, i.e., $t \in [t_0^i, t_e^i], i = a$, we compute the trajectory of the model

$$S_y^i(t+1) = S_y^i(t) - S_y^i(t)(v_{11}^i I_y^i(t) + v_{12}^i I_o^i(t))/N(t)$$
$$S_o^i(t+1) = S_o^i(t) - S_o^i(t)(v_{21}^i I_y^i(t) + v_{22}^i I_o^i(t))/N(t)$$
$$I_y^i(t+1) = (1 - \tau_1^i - \gamma)I_y^i(t) + S_y^i(t)(v_{11}^i I_y^i(t) + v_{12}^i I_o^i(t))/N(t)$$
$$I_o^i(t+1) = (1 - \tau_2^i - \gamma)I_o^i(t) + S_o^i(t)(v_{21}^i I_y^i(t) + v_{22}^i I_o^i(t))/N(t)$$

that starts from the initial conditions

$$S_y^i(t_0^i) = S_{t0y}^i, \quad I_y^i(t_0^i) = I_{t0y}^i,$$
$$S_o^i(t_0^i) = S_{t0o}^i, \quad I_o^i(t_0^i) = I_{t0o}^i.$$

Notice that this is different from what was performed in [23], where, for each time window i, the number of susceptible at the beginning of the time window was set to $S_c^i(t_0^i) = N_c(0) - C_c(t_0^i) - I^i(t_0^i)$, $c \in \{y, o\}$. As mentioned before, this is done to account for both the reinfection and the immunity against infection that is possibly provided by the vaccination. At last, we compute—as a cost—the relative error between the predicted and the real new daily cases. Finally, to guarantee the continuity of the identified solution, we impose the initial condition of the current scenario (parameters S_{t0y}^i, S_{t0o}^i, I_{t0y}^i, I_{t0o}^i) to be different from the final condition of the previous one by at most 30%.

The details of such a procedure are provided hereafter. Let $\bar{C}_y(t)$, $\bar{C}_o(t)$ denote the reported cases for the two age classes (the bar-notation specifies the real nature of the data); the new cases reported at time $t + 1$ are thus given by $\bar{C}_k(t + 1) - \bar{C}_k(t), k = y, o$. On the other hand, the new cases predicted—by the model—within the i-th time window come from the last equations of (1), so as to obtain $C_k(t+1) - C_k(t) = \tau_j I_k^i(t)$, $(k, j) = (y, 1), (o, 2)$. Accordingly, the cost associated with the i-th time window is given by

$$J_c^i = \sum_{t=t_0^i}^{t_e^i} \left(\frac{(\bar{C}_y(t+1) - \bar{C}_y(t)) - \tau_1^i I_y^i(t)}{\bar{C}_y(t)} \right)^2 + \left(\frac{(\bar{C}_o(t+1) - \bar{C}_o(t)) - \tau_2^i I_o^i(t)}{\bar{C}_o(t)} \right)^2.$$

Now, on the basis of the time series of the reported cases for the two age classes, $\bar{C}_y(t)$, $\bar{C}_o(t)$, and the time series of the recovered people, $\bar{R}(t)$, the number of active people is computed as

$$N(t) = N(t_0^a) - (\bar{C}_y(t) + \bar{C}_o(t)) + \bar{R}(t)$$

and the model parameters and initial conditions are then determined as solutions—through the fmincon routine in Matlab©—to the optimization problem

$$\min_{v_{jl}^i, \tau_j^i, S_{t0k}^i, I_{t0k}^i} \quad J_c = \sum_i J_{c'}^i \quad j, l = \{1, 2\}, k = \{y, o\}, i \in \{a, \ldots, I\}$$

such that
$v_{jl}^i \geq 0 \quad j, l = \{1, 2\}, i \in \{a, \ldots, I\}$
$\tau_j^i \geq 0 \quad j = \{1, 2\}, i \in \{a, \ldots, I\}$
$\frac{|S_{t0k}^i - S_k^h(t_e^h+1)|}{S_k^h(t_e^h+1)} < 0.3 \quad k = \{y, o\}, i \in \{b, \ldots, I\}, h \text{ just preceding } i$
$\frac{|I_{t0k}^i - I_k^h(t_e^h+1)|}{I_k^h(t_e^h+1)} < 0.3 \quad k = \{y, o\}, i \in \{b, \ldots, I\}, h \text{ just preceding } i$
$S_k^i(t_0^i) = S_{t0k}^i, \ I_k^i(t_0^i) = I_{t0k}^i \quad k = \{y, o\}, i \in \{a, \ldots, I\}$
$S_y^i(t+1) = S_y^i(t) - S_y^i(t)\frac{v_{11}^i I_y^i(t) + v_{12}^i I_o^i(t)}{N(t)} \quad t \in [t_0^i, t_e^i], i \in \{a, \ldots, I\}$
$S_o^i(t+1) = S_o^i(t) - S_o^i(t)\frac{v_{21}^i I_y^i(t) + v_{22}^i I_o^i(t)}{N(t)} \quad t \in [t_0^i, t_e^i], i \in \{a, \ldots, I\}$
$I_y^i(t+1) = (0.93 - \tau_1^i) I_y^i(t) + \frac{S_y^i(t)(v_{11}^i I_y^i(t) + v_{12}^i I_o^i(t))}{N(t)} \quad t \in [t_0^i, t_e^i], i \in \{a, \ldots, I\}$
$I_o^i(t+1) = (0.93 - \tau_2^i) I_o^i(t) + \frac{S_o^i(t)(v_{21}^i I_y^i(t) + v_{22}^i I_o^i(t))}{N(t)} \quad t \in [t_0^i, t_e^i], i \in \{a, \ldots, I\}$

that exhibits 90 free parameters (the others being fixed by the equality constraints).

Note that data are collected weekly so that the cost function is computed only at the time for which the data are present and not each day, t. The total number of data we use for our fitting procedure is 198, while the problem of estimation needs at least 90 data points (one for each of the parameters we are estimating). As in [23], a practical identifiability analysis of the parameters around the estimation point confirms that the values we obtained with this procedure can be locally determined from the data we used (the local minimum we have found has no directions on which the cost function does not significantly increase with respect to the parameter variations). Moreover, the obtained estimates turn out to be robust with respect to the hyper-parameters characterizing the optimization method, including the 30% barrier imposed to guarantee the continuity of the identified solution. The resulting picture of the age-dependent patterns of social contacts and of the spread of COVID-19 disease in the Italian context is reported in Figure 2. It highlights the following practical evidence: (i) the abrupt increase of detected cases is counteracted, from 9 March 2020 to 28 April 2020, by the strict national lockdown rule; (ii) the low-increasing profile of detected cases, from 7 May 2020 to 8 September 2020, corresponds to a weakened feedback social distancing and contact reduction intervention; (iii) the workplace/school-contacts re-activation phase, from 15 September 2020 to 27 October 2020, as well as the coordinated intermittent regional action, from 7 November 2020 to 29 December 2020, correspond to a new increase of the detected cases, in which the number of young cases is larger than old cases; (iv) the further increase in cases corresponds to the subsequent scenario, from 5 January 2021 to 12 May 2021, in which a direct mRNA vaccination of subjects—especially the elderly—at highest risk for severe outcomes, along with Vaxzevria vaccination of young subjects belonging to crucial occupational categories, is performed; a low-speed increase in detected cases corresponds to the final aggregate window proceeding, in order, from 19 May 2021 to 14 July 2021 (a mitigated coordinated intermittent regional action in conjunction with the II vaccination phase), from 21 July 2021 to 22 September 2021 (a super-attenuated coordinated intermittent regional action in conjunction with the II vaccination phase), and from 29 September 2021 to 10 November 2021 (normal reopening of schools, complete lifting of the obligation to use face masks, reopening of entertainment activities, and start of the III vaccination phase).

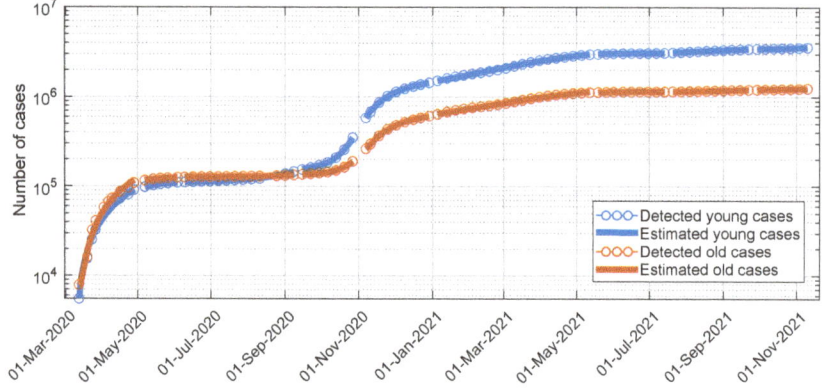

Figure 2. Data fitting for the compartmental model: actual and estimated cumulative profiles for young subjects infected and old subjects infected (logarithmic scale).

The parameters estimated in the different scenarios ($i \in \{a, \ldots, f\}$) v_{kl}^i, τ_l^k $k, l \in \{1, 2\}$ are reported in Table 1, along with the new ones estimated in the new three scenarios G, H and I. The same happens for the estimated initial conditions I_{t0y}^i, I_{t0o}^i, S_{t0y}^i, S_{t0o}^i that appear in Table 2. Note that, since the optimization procedure is global and since we have left the susceptible in each scenario free to change in order to consider reinfection and immunity due to vaccination, the obtained parameters differ slightly from the one reported in [23]. Even though such differences do not impact any of the considerations in [23], according to the data reported in Table 1 (values in red):

- The juvenile–elder virulence—when compared with [23]—will appear to be smaller in scenarios c (namely, low-feedback social distancing and contact reduction intervention and e (namely, decreased social contacts in schools at a national level and social distancing measures put in place or relaxed independently by each region) to identify successfully, within c and e, a sort of decoupling (that is lost in all the other scenarios) between the two age classes and recognize the benefits, within e, of a reduction in the social contacts in schools at a national level.
- The reasonable punctual reduction—when compared with [23]—of the intra-elder virulence during the summer of scenario c (now of the same magnitude as the intra-

juvenile virulence) will preserve the already established trend of such a parameter over the scenarios.
- The increase—when compared with [23]—of the average time for disease identification in old subjects within scenario c better shapes the behavior of such a parameter by consistently moving to scenario d (re-activation of social contacts in workplaces and schools) the moment in which the elderly paid a higher level of attention to symptoms while preserving the remaining, already established trend over the scenarios.

Table 1. Estimated parameters v_{kl}^i, τ_j^i, $k, l \in \{1, 2\}$ in the different scenarios $i \in \{a, \ldots, f, G, H, I\}$ [v_{11}^i for intra-juvenile virulence; v_{12}^i for juvenile–elder virulence; v_{21}^i for elder–juvenile virulence; v_{22}^i for intra-elder virulence; $1/\tau_1^i$ as average time for disease identification in young subjects; and $1/\tau_2^i$ as average time for disease identification in old subjects]. Parameter values that significantly differ from the one reported in [23] are reported in red. The new scenarios G–I are highlighted with a gray background.

		Scenario a			
v_{11}^a	v_{12}^a	v_{21}^a	v_{22}^a	τ_1^a	τ_2^a
0.7532	0.0000	1.3718	0.0001	0.349	0.2372
		Scenario b			
v_{11}^b	v_{12}^b	v_{21}^b	v_{22}^b	τ_1^b	τ_2^b
0.0016	0.5709	0.0211	0.4174	0.13	0.1928
		Scenario c			
v_{11}^c	v_{12}^c	v_{21}^c	v_{22}^c	τ_1^c	τ_2^c
0.5319	0.0018	0.0355	0.4953	0.1906	0.2327
		Scenario d			
v_{11}^d	v_{12}^d	v_{21}^d	v_{22}^d	τ_1^d	τ_2^d
0.1108	2.0435	0.0252	0.7988	0.24	0.32
		Scenario e			
v_{11}^e	v_{12}^e	v_{21}^e	v_{22}^e	τ_1^e	τ_2^e
0.6418	0.0256	0.2003	0.4538	0.3023	0.5258
		Scenario f			
v_{11}^f	v_{12}^f	v_{21}^f	v_{22}^f	τ_1^f	τ_2^f
0.6403	1.3348	0.2658	0.5990	0.4465	0.5754
		Scenario G			
v_{11}^G	v_{12}^G	v_{21}^G	v_{22}^G	τ_1^G	τ_2^G
0.2903	0.0014	0.0421	0.0024	0.1203	0.1134
		Scenario H			
v_{11}^H	v_{12}^H	v_{21}^H	v_{22}^H	τ_1^H	τ_2^H
2.1727	0.0013	0.4060	0.1909	0.8976	0.5847
		Scenario I			
v_{11}^I	v_{12}^I	v_{21}^I	v_{22}^I	τ_1^I	τ_2^I
2.2546	0.0019	0.7842	0.0042	0.9913	0.9991

Table 2. Estimated initial conditions I_{t0y}^i, I_{t0o}^i, S_{t0y}^i, S_{t0o}^i in the different scenarios $i \in \{a, \ldots, f, G, H, I\}$ [I_{t0y}^i for the initial young subjects infected; I_{t0o}^i for the initial old subjects infected; S_{t0y}^i for the initial young subjects susceptible; and S_{t0o}^i for the initial old subjects susceptible]. The new scenarios G–I are highlighted with a gray background.

Scenario a			
I_{t0y}^a	I_{t0o}^a	S_{t0y}^a	S_{t0o}^a
5.5299×10^3	0.546×10^3	3.3372×10^7	2.6981×10^7
Scenario b			
I_{t0y}^b	I_{t0o}^b	S_{t0y}^b	S_{t0o}^b
4.9100×10^3	3.2500×10^3	3.3372×10^7	2.6978×10^7
Scenario c			
I_{t0y}^c	I_{t0o}^c	S_{t0y}^c	S_{t0o}^c
3.2991×10^2	6.499×10^2	3.3377×10^7	2.6981×10^7
Scenario d			
I_{t0y}^d	I_{t0o}^d	S_{t0y}^d	S_{t0o}^d
8.6983×10^3	0.2895×10^3	3.3368×10^7	2.6981×10^7
Scenario e			
I_{t0y}^e	I_{t0o}^e	S_{t0y}^e	S_{t0o}^e
9.5000×10^4	1.3000×10^4	3.328×10^7	2.6969×10^7
Scenario f			
I_{t0y}^f	I_{t0o}^f	S_{t0y}^f	S_{t0o}^f
2.2000×10^4	0.5000×10^4	3.3355×10^7	2.6977×10^7
Scenario G			
I_{t0y}^G	I_{t0o}^G	S_{t0y}^G	S_{t0o}^G
3.6619×10^4	1.5647×10^4	2.9177×10^7	2.5467×10^7
Scenario H			
I_{t0y}^H	I_{t0o}^H	S_{t0y}^H	S_{t0o}^H
2.7712×10^3	0.3423×10^3	2.9078×10^7	2.5454×10^7
Scenario I			
I_{t0y}^I	I_{t0o}^I	S_{t0y}^I	S_{t0o}^I
9.3518×10^3	0.1367×10^3	2.8975×10^7	2.5437×10^7

Once the estimates of the model parameters have been obtained (Tables 1 and 2), the values of the *reproduction number*, $R_t^i[m]$, associated with model (1) in each scenario i [m stands for model-based computation], as the average number of new infections caused by an infected person, can be computed through the formula

$$R_t^i[m] = \frac{1}{N(t)}\sigma_1\left(\begin{bmatrix} S_y^i(t) & 0 \\ 0 & S_o^i(t) \end{bmatrix}\left(\frac{1}{\tau_1^i + \gamma}\begin{bmatrix} v_{11}^i & v_{12}^i \\ v_{11}^i & v_{12}^i \end{bmatrix} + \frac{1}{\tau_2^i + \gamma}\begin{bmatrix} v_{21}^i & v_{22}^i \\ v_{21}^i & v_{22}^i \end{bmatrix}\right)\right)$$

where $\sigma_1(\cdot)$ denotes the biggest among the moduli of the eigenvalues of the matrix argument. The resulting mean reproduction numbers, $R_t^i[m]$, over $i \in \{a, \ldots, f, G, H, I\}$ read 1.2, 0.7, 1.1, 1.3, 1, 1, 0.8, 1, 1. They are compatible with the maximum likelihood values of the national reproduction number in Figure 3, computed from raw data through the `EpiEstim` toolbox. This shows that model (1) is able to catch the main epidemic features along the considered scenarios.

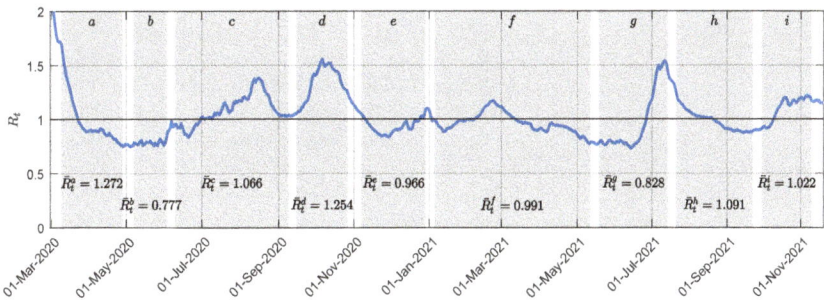

Figure 3. National reproduction number, R_t, within the considered time windows. Each shaded portion of the plane corresponds to a specific scenario. The mean value, $\bar{R}_t^i$ (among the values corresponding to our sampling) within each time window, $i \in \{a, \ldots, f, G, H, I\}$, is reported at the bottom of each shaded region.

5. Discussion

The following comments are reported. They provide a deep interpretation of the estimation results in Tables 1–3.

- All the estimates corresponding to the different scenarios, including the estimated I_{t0y}^i, I_{t0o}^i (initial young subjects infected; initial old subjects infected), allow the estimated profile to reproduce the actual one along the different scenarios satisfactorily, as shown in Figure 2.
- The number of susceptible individuals in both age groups continuously decreases in all the scenarios starting from scenario G (including additional scenario J), as an effect of the vaccination campaign starting within scenario f and continuing within G. However, a non-drastic reduction in such a number seems to confirm that the vaccination action principally protects against severe symptomatology rather than giving total immunity [24].
- Starting from scenario G, again as an effect of the vaccination campaign starting within scenario f and continuing within G, the juvenile–elder and the intra-elder virulences exhibited a large reduction. The major strength of the vaccination action for the elderly, however, allows for relatively large elder–juvenile virulence values.
- The average time for disease identification in young subjects in all the scenarios, a–f and G–I, ranges from 1 to 9 days across the scenarios. Notably, scenarios b–c have an average time of approximately 5–7 days, whereas new scenario G has a slightly longer time of about 8 days. This variation can be attributed to the fact that, after the lockdown period and related concerns, young individuals tended to pay less attention to their symptoms, particularly in scenarios b and c (covering the summer period, from 7 May 2020 to 8 September 2020). The same phenomenon was observed in new scenario G, which occurred from $t_0^G = 19$ May 2021 to $t_e^G = 14$ July 2021. This period coincided with weakened, intermittent regional actions and the second vaccination stage (booster), along with no festivities, like Christmas and Easter. In contrast, scenarios H–I exhibited a substantial reduction in the average time for disease identification among young subjects, possibly due to more immediate reliance on testing in the presence of: (i) typical entertainment habits during the summer that saw a large increase in travel (compared with previous analogous scenario c); (ii) normal reopening of schools and entertainment activities such as discos and ballrooms. These scenarios also witnessed an increase in intra-juvenile virulence (2.1727, 2.2546) compared with G (0.2903).
- Even the average time for disease identification in old subjects in all the scenarios, a–f and G–I, varies from 1 to 9 days, with less than 4 days occurring in scenarios

d–f (in which the elderly paid a higher level of attention to symptoms) and H–I (corresponding to an increase of the intra-juvenile and elder–juvenile virulences).

- The elder–juvenile virulences exhibit a specific increasing trend starting from scenario e (namely, coordinated intermittent regional action), except for scenario G (namely, weakened intermittent regional actions, second vaccination stage (booster), and, mainly, no festivities, like Christmas and Easter), in which all the virulences show a relatively large reduction. The subsequent increase of intra-juvenile and elder–juvenile virulences seems to suggest a sort of greater decoupling of social habits between young subjects and old ones, and allows us to pose a question while recognizing scenario G as the most favourable one in terms of virulence values: what would have happened if, after the first vaccination campaign, the super-attenuated coordinated intermittent regional actions and the last step towards normality had been delayed until after the summer vacations, and actions like the ones in a–b, e had been performed to reduce the intra-juvenile virulence?

6. One and Two Years Later

Furthermore, new real data about the pandemic on the following time window:

(J) from $t_0^J = 1$ March 2023 to $t_e^J = 3$ May 2023

Are also taken in order to allow for a direct comparison with analogous scenario a two years before and with f two years before. The estimated parameters for additional scenario J (estimated by minimizing J_c^J) appear in Table 3.

Table 3. Estimated parameters for time window J (to be compared with a and f).

Time Window J from $t_0^J = 1$ March 2023 to $t_e^J = 3$ May 2023					
v_{11}^J	v_{12}^J	v_{21}^J	v_{22}^J	τ_1^J	τ_2^J
0.4451	0.1670	0.2065	0.1740	0.1453	0.2067
I_{t0y}^J		I_{t0o}^J		S_{t0y}^J	S_{t0o}^J
1.7655×10^3		0.5398×10^4		2.8915×10^7	2.5418×10^7

From $t_0^J = 1$ March 2023 to $t_e^J = 3$ May 2023: two years later than a, one year later than f. The average time for disease identification in young subjects is larger than a and f (about 7 days compared with the previous 2–4 days). The disease has become endemic with no more strong restrictions in social habits: (i) young subjects somehow pay less attention than old ones, for whom the average time changed from 2–3 days to just 5; (ii) by forming a completely novel picture, a balance between (not small) juvenile–elder and elder–juvenile virulences appears, as never happened before, with the intra-juvenile and intra-elder virulences concurrently decreasing, when compared with f, but again being non-small, again confirming that the vaccination action has definitely attenuated severe symptomatology rather than providing total immunity. Finally, the estimate of the average reproduction number in this scenario is 1.04, again confirming the endemic nature of COVID-19 in Italy.

7. Conclusions

The epidemiological model for COVID-19 developed in [23], which considers the epidemic within the younger age group and older age group separately, has been used to provide an updated insight into the different evolution of the epidemic in the considered two age groups, while simultaneously evaluating, through the estimation of crucial model parameters, the impact of changes in social distancing measures and vaccination action. The exact structure of the contact patterns in the general population is, in fact, still unknown to a large extent and deserves specific research efforts, to characterize better the effects of political choices that, over time, change the rules governing social distancing and behaviors.

The methodological contribution of this paper, compared with what was proposed in [23], is incremental: given the high probability of reinfection and to take into account the low probability of immunity to infection acquired by vaccination, we introduced the number of susceptible at the beginning of each time window within the set of parameters to be estimated. Such a number was fixed to the number of non-infected individuals in [23]. Interestingly and retrospectively, we have shown that the introduction of this degree of freedom did not significantly affect the results of the already analyzed scenarios *a-f* (in line with the results presented in [23]), showing the correctness of the hypotheses put forward in [23], as well as the robustness of the procedure. Furthermore, the new results provide a meaningful picture of the evolution of the social behaviors and the goodness of the performed strategic interventions.

The significance of this work thus lies in the application and validation of methodological frameworks proposed in [23] to newly updated data. By adopting the approach outlined in previous work, we showcase the enduring relevance of established methodologies in the ever-evolving landscape of the current pandemic. This underscores a critical message: the wealth of knowledge amassed through extensive scientific endeavors in recent years remains indispensable. As we navigate the dynamic challenges posed by COVID-19, it is imperative for policymakers and researchers alike to continuously leverage and update the methodological foundations already established. The findings of this study reinforce the notion that the synergy between established methodologies and real-time data analysis is paramount for informed decision-making. Through this, we advocate for an ongoing commitment to evidence-based practices, ensuring that our scientific insights persistently inform strategies and policies, ultimately contributing to the effective management of the *res publica*.

This paper underscores the significance of the extensive scientific endeavors conducted in recent years, emphasizing that the wealth of knowledge acquired must be consistently employed to interpret the ongoing situation. Serving as the conclusion to the Special Issue "New Challenges in Mathematical Modelling and Control of COVID-19 Epidemics: Analysis of Non-pharmaceutical Actions and Vaccination Strategies" we curated, it highlights the significance of the Special Issue itself, showing the importance of utilizing methodological achievements to navigate the current landscape effectively. By demonstrating how the overarching principles guide the interpretation of data, in our view, this paper further underscores the critical role that accumulated scientific insights play in addressing the complexities of the COVID-19 pandemic (as well as other pandemics) nowadays.

Future research efforts shall be devoted to analyze different model structures comparatively, even within the stochastic framework.

Author Contributions: Conceptualization: C.M.V. and F.D.R.; methodology: C.M.V. and F.D.R.; software and resources: F.D.R.; formal analysis: C.M.V. and F.D.R.; validation, investigation: C.M.V. and F.D.R.; writing—original draft: C.M.V.; writing—review and editing: C.M.V. and F.D.R. All authors have read and agreed to the published version of the manuscript.

Funding: This research received no external funding.

Data Availability Statement: The publicly archived datasets that were analyzed during the study are explicitly quoted in the paper.

Acknowledgments: The authors are grateful to E. Cottafava, in her quality of General Secretary of Fondazione GIMBE, Via Amendola, 2, 40121 Bologna, for her willingness to provide data concerning the profile of the Italian reproduction numbers over time.

Conflicts of Interest: The authors declare no conflict of interest.

References

1. Qian, M.; Jiang, J. COVID-19 and social distancing. *J. Public Health* **2020**, *30*, 259–261. [CrossRef]
2. Della Rossa, F.; Salzano, D.; Di Meglio, A.; De Lellis, F.; Coraggio, M.; Calabrese, C.; Guarino, A.; Cardona-Rivera, R.; De Lellis, P.; Liuzza, D.; et al. A network model of Italy shows that intermittent regional strategies can alleviate the COVID-19 epidemic. *Nat. Commun.* **2020**, *11*, 5106. [CrossRef]
3. Kretzschmar, M.E.; Rozhnova, G.; Bootsma, M.C.; van Boven, M.; van de Wijgert, J.H.; Bonten, M.J. Impact of delays on effectiveness of contact tracing strategies for COVID-19: A modelling study. *Lancet Public Health* **2020**, *5*, e452–e459. [CrossRef] [PubMed]
4. Keeling, M.J.; Hollingsworth, T.D.; Read, J.M. Efficacy of contact tracing for the containment of the 2019 novel coronavirus (COVID-19). *J. Epidemiol. Community Health* **2020**, *74*, 861–866. [CrossRef]
5. Tupper, P.; Otto, S.P.; Colijn, C. Fundamental limitations of contact tracing for COVID-19. *Facets* **2021**, *6*, 1993–2001. [CrossRef]
6. Italia, M.; Della Rossa, F.; Dercole, F. Model-informed health and socio-economic benefits of enhancing global equity and access to COVID-19 vaccines. *Sci. Rep.* **2023**, *13*, 21707. [CrossRef] [PubMed]
7. Kucharski, A.J.; Klepac, P.; Conlan, A.J.; Kissler, S.M.; Tang, M.L.; Fry, H.; Gog, J.R.; Edmunds, W.J.; Emery, J.C.; Medley, G.; et al. Effectiveness of isolation, testing, contact tracing, and physical distancing on reducing transmission of SARS-CoV-2 in different settings: A mathematical modelling study. *Lancet Infect. Dis.* **2020**, *20*, 1151–1160. [CrossRef]
8. Musa, S.S.; Qureshi, S.; Zhao, S.; Yusuf, A.; Mustapha, U.T.; He, D. Mathematical modeling of COVID-19 epidemic with effect of awareness programs. *Infect. Dis. Model.* **2021**, *6*, 448–460. [CrossRef]
9. Ancona, C.; Lo Iudice, F.; Garofalo, F.; De Lellis, P. A model-based opinion dynamics approach to tackle vaccine hesitancy. *Sci. Rep.* **2022**, *12*, 11835. [CrossRef] [PubMed]
10. Maji, C.; Al Basir, F.; Mukherjee, D.; Ravichandran, C.; Nisar, K. COVID-19 propagation and the usefulness of awareness-based control measures: A mathematical model with delay. *AIMs Math* **2022**, *7*, 12091–12105. [CrossRef]
11. Thunström, L.; Newbold, S.C.; Finnoff, D.; Ashworth, M.; Shogren, J.F. The benefits and costs of using social distancing to flatten the curve for COVID-19. *J. Benefit-Cost Anal.* **2020**, *11*, 179–195. [CrossRef]
12. Moosa, I.A. The effectiveness of social distancing in containing COVID-19. *Appl. Econ.* **2020**, *52*, 6292–6305. [CrossRef]
13. Silva, P.C.; Batista, P.V.; Lima, H.S.; Alves, M.A.; Guimarães, F.G.; Silva, R.C. COVID-ABS: An agent-based model of COVID-19 epidemic to simulate health and economic effects of social distancing interventions. *Chaos Solitons Fractals* **2020**, *139*, 110088. [CrossRef]
14. Childs, M.L.; Kain, M.P.; Kirk, D.; Harris, M.; Couper, L.; Nova, N.; Delwel, I.; Ritchie, J.; Mordecai, E.A. The impact of long-term non-pharmaceutical interventions on COVID-19 epidemic dynamics and control. *MedRxiv* **2020**. [CrossRef]
15. Goldsztejn, U.; Schwartzman, D.; Nehorai, A. Public policy and economic dynamics of COVID-19 spread: A mathematical modeling study. *PLoS ONE* **2020**, *15*, e0244174. [CrossRef] [PubMed]
16. Childs, M.L.; Kain, M.P.; Harris, M.J.; Kirk, D.; Couper, L.; Nova, N.; Delwel, I.; Ritchie, J.; Becker, A.D.; Mordecai, E.A. The impact of long-term non-pharmaceutical interventions on COVID-19 epidemic dynamics and control: The value and limitations of early models. *Proc. R. Soc. B* **2021**, *288*, 20210811. [CrossRef] [PubMed]
17. Onder, G.; Rezza, G.; Brusaferro, S. Case-fatality rate and characteristics of patients dying in relation to COVID-19 in Italy. *Jama* **2020**, *323*, 1775–1776. [CrossRef] [PubMed]
18. Ruan, Q.; Yang, K.; Wang, W.; Jiang, L.; Song, J. Clinical predictors of mortality due to COVID-19 based on an analysis of data of 150 patients from Wuhan, China. *Intensive Care Med.* **2020**, *46*, 846–848. [CrossRef] [PubMed]
19. Zou, Y.; Yang, W.; Lai, J.; Hou, J.; Lin, W. Vaccination and quarantine effect on COVID-19 transmission dynamics incorporating Chinese-spring-festival travel rush: Modeling and simulations. *Bull. Math. Biol.* **2022**, *84*, 30. [CrossRef] [PubMed]
20. Guo, Y.; Liu, X.; Deng, M.; Liu, P.; Li, F.; Xie, N.; Pang, Y.; Zhang, X.; Luo, W.; Peng, Y.; et al. Epidemiology of COVID-19 in older persons, Wuhan, China. *Age Ageing* **2020**, *49*, 706–712. [CrossRef]
21. Balabdaoui, F.; Mohr, D. Age-stratified discrete compartment model of the COVID-19 epidemic with application to Switzerland. *Sci. Rep.* **2020**, *10*, 21306. [CrossRef] [PubMed]
22. Bentout, S.; Tridane, A.; Djilali, S.; Touaoula, T.M. Age-structured modeling of COVID-19 epidemic in the USA, UAE and Algeria. *Alex. Eng. J.* **2021**, *60*, 401–411. [CrossRef]
23. Verrelli, C.M.; Della Rossa, F. Two-age-structured COVID-19 epidemic model: Estimation of virulence parameters to interpret effects of national and regional feedback interventions and vaccination. *Mathematics* **2021**, *9*, 2414. [CrossRef]
24. Mohammed, I.; Nauman, A.; Paul, P.; Ganesan, S.; Chen, K.H.; Jalil, S.M.S.; Jaouni, S.H.; Kawas, H.; Khan, W.A.; Vattoth, A.L.; et al. The efficacy and effectiveness of the COVID-19 vaccines in reducing infection, severity, hospitalization, and mortality: A systematic review. *Hum. Vaccines Immunother.* **2022**, *18*, 2027160. [CrossRef]

Disclaimer/Publisher's Note: The statements, opinions and data contained in all publications are solely those of the individual author(s) and contributor(s) and not of MDPI and/or the editor(s). MDPI and/or the editor(s) disclaim responsibility for any injury to people or property resulting from any ideas, methods, instructions or products referred to in the content.

MDPI AG
Grosspeteranlage 5
4052 Basel
Switzerland
Tel.: +41 61 683 77 34

Mathematics Editorial Office
E-mail: mathematics@mdpi.com
www.mdpi.com/journal/mathematics

Disclaimer/Publisher's Note: The statements, opinions and data contained in all publications are solely those of the individual author(s) and contributor(s) and not of MDPI and/or the editor(s). MDPI and/or the editor(s) disclaim responsibility for any injury to people or property resulting from any ideas, methods, instructions or products referred to in the content.

www.ingramcontent.com/pod-product-compliance
Lightning Source LLC
LaVergne TN
LVHW070425100526
838202LV00014B/1525